Economics of Industrial Ecology

Economics of Industrial Ecology

Materials, Structural Change, and Spatial Scales

edited by Jeroen C. J. M. van den Bergh and Marco A. Janssen

The MIT Press
Cambridge, Massachusetts
London, England

MC

© 2004 Massachusetts Institute of Technology

All Rights Reserved. No part of this book may be reproduced in any form by any electronic or mechanical means (including photocopying, recording, or information storage and retrieval) without permission in writing from the publisher.

MIT Press books may be purchased at special quantity discounts for business or sales promotional use. For information, please email special_sales@mitpress.mit.edu or write to Special Sales Department, The MIT Press, 5 Cambridge Center, Cambridge, MA 02142.

This book was set in Times New Roman on 3B2 by Asco Typesetters, Hong Kong.
Printed and bound in the United States of America.

Library of Congress Cataloging-in-Publication Data

Economics of industrial ecology : materials, structural change, and spatial scales / edited by Jeroen C. J. M. van den Bergh and Marco A. Janssen.
   p.   cm.
 Includes bibliographical references.
 ISBN 0-262-22071-7 (alk. paper)
  1. Industrial ecology. 2. Industrial ecology—Economic aspects. I. Bergh, Jeroen C. J. M. van den, 1965–. II. Janssen, Marco, 1969–.
TS161.E28   2005
658.4'08—dc22   2004053089

Printed on Recycled Paper.

10 9 8 7 6 5 4 3 2 1

Contents

Preface

In 1998 a project called MUSSIM (Material Use and Spatial Scales in Industrial Metabolism) was started with funding from the Netherlands Organization for Scientific Research (NWO). The project aimed to study the driving forces behind material flows at different spatial scales. The partners involved in the project were the Free University in Amsterdam (the Faculty of Economics and Business Administration and the Institute for Environmental Studies), Wageningen University (the Department of Environmental Economics), the Netherlands Bureau for Economic Policy Analysis (CPB), and the Dutch National Institute for Public Health and the Environment (RIVM). The results of this comprehensive study form the backbone of this book.

As part of the MUSSIM project, we organized two special sessions on economics and industrial ecology during the First International Conference on Industrial Ecology, held in November 2001 in Leiden, the Netherlands. In addition to presenting the results of our own work, we invited leading scholars in the field from other countries to present their work too. The two types of contributions were combined to produce the present book. All contributions have been reviewed by at least two referees.

The result is a volume with a unique collection of studies that combine elements of economics and industrial ecology in a methodologically rigorous way. A range of modeling techniques is employed, and applications involve a wide array of materials, sectors, and countries.

We are grateful to three anonymous reviewers for The MIT Press, whose suggestions helped us to improve the manuscript in a final stage, and to Patricia Ellman for saving us from grave errors in use of the (American) English language.

I Background

1 Introduction and Overview

Jeroen C. J. M. van den Bergh and Marco A. Janssen

1.1 Introduction

The up-and-coming field known as "industrial ecology" is currently dominated by descriptive and design studies of physical processes and technical solutions that leave out relevant economic conditions and mechanisms. The main motivation for this book is that such an approach is insufficient to provide policymakers and business managers with economically feasible, cost-effective, and socially supported instruments and solutions. The approach presented here therefore aims to integrate the natural science and technological dimensions of industrial ecology with an economic angle. Various authors in the field of industrial ecology have strongly recommended such a synthesis (e.g., Koenig and Cantlon 2000; Fischhoff and Small 2000; Brunner 2002). The main value of adding economics to industrial ecology can best be summarized as increasing policy realism. Combining economics with industrial ecology will entail three elements that are largely lacking from the current literature on industrial ecology:

1. Adding an economic context to industrial ecology, in the form of costs, benefits, economic efficiency considerations, (re)allocations, investments, market processes and distortions, economic growth, multisectoral interactions, international trade, and so forth.
2. Showing the usefulness of a range of concepts, theories, and methods for the integration of economics and industrial ecology in empirical applications. Examples are structural decomposition analysis, (general) equilibrium analysis, complex systems modeling, econometric-statistical analysis, dynamic input-output modeling, urban and regional economics, theories of the firm, and institutional and evolutionary economics.
3. Deriving general policy lessons based on the integration of economic and industrial ecology considerations at a theoretical level, as well as on

the insights of empirical case studies. In essence, this means combining lessons about physical, technical, and economic opportunities as well as about their limitations.

   To motivate the idea that industrial ecology lacks a coherent and thorough treatment of relevant economic dimensions, let us consider authoritative surveys of and outlets for insights and research in the field of industrial ecology. Socolow et al. (1994), Graedel and Allenby (2003), and Ayres and Ayres (2002) are very representative of the insights that the field has delivered so far, and the *Journal of Industrial Ecology* allows us to trace recent research themes and trends. Socolow et al. (thirty-six chapters in 530 pages) is so broad that it really holds the middle ground between a text on industrial ecology and one on environmental science, covering issues from the firm to the global level. It contains two chapters that deal with economics, one analyzing principal-agent problems at the level of firms, and another studying raw materials extraction and trade. Although useful, these chapters can certainly not be regarded as anything close to a complete treatment of economic dimensions as previously outlined. Graedel and Allenby (twenty-six chapters in 363 pages) more clearly steps away from traditional environmental science and is strong on firm-level and manufacturing issues but lacks any treatment of economic considerations, be it at the firm or the economy-wide level. Ayres and Ayres (forty-six chapters in 680 pages) includes an entire section on the theme "Economics and Industrial Ecology," which consists of seven very short chapters. But it turns out that several of these chapters do not really address economic issues at all, but instead focus on physical indicators (Total Material Requirement, or TMR), exergy, transmaterialization, and technology policy. The chapters that do deal with economic issues survey the inclusion of material flows in economic models, the empirical relationship between dematerialization and economic growth, and the literature on optimal resource extraction. Again, this is useful but does not represent a sufficiently broad coverage of economic issues.

   In effect, from reading both these books and the *Journal of Industrial Ecology*, one can easily obtain the impression that industrial ecology completely lacks economic considerations and instead is mainly about technological planning and design. At the aggregate level, this is perhaps most clearly reflected in the well-known Factor X debate ($X = 10$ in the case of Factor Ten Club 1994, and $X = 4$ in the case of von Weizsäcker et al. 1997). A scan of all the issues of the *Journal of Industrial Ecology* (twenty-three in seven volumes) delivered a disappointingly small number

of articles that address (and then often only tangentially) economic reasoning or methods (these articles are mentioned in the literature survey
presented in chapter 2). The planning-and-design perspective strongly
contrasts with the economics perspective, which emphasizes firm behavior
(strategies, routines, and input-mix decisions that affect material use),
market processes (liberalization and decentralization), and economy-wide
feedback involving prices, incomes, foreign trade, change in sector structure and intermediate deliveries, shifts in consumer expenditures, and
market-based policy instruments. One can conclude that, with regard to
the three elements listed earlier, (1) and (2) have received very little to
no attention in the literature on industrial ecology. Instead, all method-
related developments and empirical work have focused on noneconomic
methods like material and substance flow analysis, life cycle analysis, and
process and product design. Given the neglect of these first two elements,
element (3) has evidently not been accomplished.

Economists have long been interested in the impact that economies
have on natural resources and the environment and the negative feedback
this may provide to welfare and economic growth (van den Bergh 1999).
This interest has, inter alia, resulted in studies that examine the relationship between physical flows through the economy, on the one hand,
and market functioning, economic growth, international trade, and environmental regulation, on the other. Early work that combined economics and material flow analysis started with the methodological studies
by Ayres and Kneese (1969), Kneese, Ayres, and D'Arge (1970), and
Georgescu-Roegen (1971). Much of the subsequent work focused on
input-output and other types of economic models, with extensions to account for polluting residuals (James 1985). Boulding (1966) and Daly
(1968) can be seen as conceptual predecessors of this line of thought. In
the context of his "steady state," Daly (1977) proposes the idea of reducing or minimizing the physical resource "throughput" that runs through
the economy, which closely meets the approach of industrial ecology.
Daly and Townsend (1993) presents a collection of reprinted classic
articles on the boundary of philosophy, economics, and environmental
science, many of which are imbued with the spirit of industrial ecology.

The emergence of the field of industrial ecology since the 1990s has
stimulated a great deal of methodologically well-founded research that
is aimed at measuring, describing, predicting, redirecting, and reducing
physical flows in economies. The current book intends to offer a balanced
combination of environmental-physical, technological, and economic considerations. This is believed to provide the best basis for identifying

opportunities to reduce pressures on the environment that are linked to material flows, as well as to design public policies to foster such opportunities. A brief overview of the relevant literature on the interface between economics and industrial ecology is provided in chapter 2. In addition, the book presents new and original studies that try to accomplish a close link between economics, in particular environmental and resource economics, and industrial ecology. These have not yet influenced the dominant themes in industrial ecology—witness the critical evaluation from an economic perspective of the representative books and the field journal earlier in the chapter.

Research on industrial metabolism has emphasized the description of material flows in economic systems (Daniels and Moore 2002 and Daniels 2002 offer a good overview of all the methods and their applications). Studies along these lines provide interesting and useful information about the size of material flows and the identification of stocks in which certain undesirable materials accumulate. This leads to concepts like "chemical time bombs" (e.g., the accumulation of chemicals in river mud) and "waste mining" (i.e., beyond a certain point, material waste that has accumulated in certain locations can be turned into a resource suitable for profitable mining) (e.g., Bartone 1990; Allen and Behmanesh 1994). Nevertheless, several relevant issues cannot be addressed properly with a descriptive industrial ecology approach. One reason is that the "metabolism" of economic systems changes over time, which cannot be understood simply by the measurement of material flows. An understanding of the relationship between economic activities and material flows can help to unravel the socioeconomic causes of both the physical flows and the changes therein. An analytical approach to this problem requires that attention be given to, inter alia, the behavior of economic agents like producers and consumers, interactions among stakeholders in production chains, the spectrum of technological choices, and the role of trade and interregional issues. At a more abstract level, such an approach involves substitution and allocation mechanisms at the level of production technologies, firms, sectors, and the composition of demand. In other words, besides a description of material flows, the field of industrial ecology requires a comprehension of the processes behind the material flows. A deeper understanding of the processes leading to changes in material flows can also provide insight into how to develop both effective and efficient policies that lead to a reduction in harmful material flows. In addition, the notion of rebound effects can be seriously examined. This notion comprises the whole range of indirect technological and economic consequences of certain technological scenarios. Such indirect effects can be induced by, or run through, alterations in prices, changes in market demand and supply, substitution in production or consumption, growth in incomes, and changes in consumer expenditures. A better grip on the sign and magnitude of rebound effects is needed to counter simplistic technology-based arguments in favor of certain policy or management strategies and to temper naïve expectations about what can be achieved over certain periods of time with technology. This is relevant, for example, when estimating the impacts of rapid changes in information and communication technology on material flows.

This book provides a unique overview of different economic approaches to address problems associated with the use of materials in economic systems at different levels. Some of these approaches have economics at the core, whereas others add economic aspects to technological or natural-science-dominated approaches or applications. The major part of the book offers a collection of new studies that cover a wide range of approaches and methods for integrating physics and technological analysis and knowledge with economics. The scale varies from industrial parks in Denmark and the Netherlands to the international trade of waste. The variety of approaches can be explained by the fact that the diversity and complexity of topics on the boundary of economics and industrial ecology—indicated by the dimensions: materials, technology, physics, economics, and varying (spatial and aggregation) scales—cannot be completely covered with a single approach. As a result, the book reflects a pluralistic approach.

The contributions have been organized in four themes (parts II to V). The first theme (chapters 3 and 4) is concerned with the historical analysis of structural change. The second theme (chapters 5 to 8) covers a range of models that try to predict future structural change under different policy scenarios. The third theme (chapters 9 and 10) addresses two models that can be used to examine waste management and recycling opportunities. Finally, the fourth theme (chapters 11 and 12) adopts a local-scale perspective by focusing on the dynamics of eco-industrial parks. Chapter 13 closes the book with a summary and synthesis of policy implications.

1.2   Historical Analysis of Structural Change

Chapters 3 and 4 offer different (statistical and decomposition) approaches to the historical analysis of the impact of structural change on material flows through the economy. In chapter 3, Ayres, Ayres, and

Warr challenge the widespread idea of dematerialization by presenting data on the major commodity flows in the U.S. economy since 1900, in both mass and exergy terms. Based on these data, the U.S. economy turns out not to be "dematerializing," to any degree that has environmental significance. Since 1900 it has exhibited a slow and modest long-term increase in materials consumption per capita, except during the Depression and World War II. The trends with regard to resource productivity (gross domestic product [GDP] per unit of materials consumption) are moderately increasing overall. Ayres et al. argue that policy should focus not on reducing the total mass of materials consumed, but on reducing the need for consumables, especially intermediates.

In chapter 4, Hoekstra and van den Bergh present a quantitative historical analysis based on the method of input-output (I/O) structural decomposition analysis (SDA). They open with an overview of the literature on SDA. This method uses historical I/O data to identify the relative importance of a wide range of drivers. SDA has been applied in studies of energy use and energy-related emissions, but only once (in unpublished work) of material flows. The authors carry out a decomposition of the use of iron and steel and plastics products in the Netherlands for the period 1990–1997. This is based on a new data set, unique in the world, which incorporates hybrid-unit input-output tables. The data were constructed especially for the present purpose in cooperation with the national statistical office of the Netherlands, through a construction process that is briefly discussed. Finally, an illustration is given of how the results of SDA can serve as an input to a forecasting scenario analysis.

1.3   Projective Analysis of Structural Change

Chapters 5 to 8 present case studies on regional and national scales that employ a range of modeling approaches. In chapter 5, Ruth, Davidsdottir, and Amato describe a model that combines engineering and econometric techniques for the analysis of the dynamics of large industrial systems. A transparent dynamic computer modeling approach is chosen to integrate information from these analyses in ways that foster the participation of stakeholders from industry and government agencies in all stages of the modeling process—from problem definition and determination of system boundaries to the generation of scenarios and the interpretation of results. Three case studies of industrial energy use in the United States are presented: one each for the iron and steel, pulp and paper, and ethylene industries.

In chapter 6, Foran and Poldy describe two analytical frameworks that have been applied to Australia: namely, the Australian Stocks and Flows Framework (ASFF) and the OzEcco embodied energy flows model. The first (ASFF) is a set of thirty-two linked calculators that follow, and account for, the important physical transactions that underpin our everyday life. The second (OzEcco) is based on the concept of embodied energy, the chain of energy flows from oil wells and coal mines that eventually are included or embodied in every good and service in both the domestic and export components of our economy. Both analytical frameworks are based on systems theory and implemented in a dynamic approach.

In chapter 7, Mannaerts presents STREAM, a partial equilibrium model for material flows in Europe, with emphasis on the Netherlands. The model provides a consistent framework for material use scenarios and related environmental policy analysis of dematerialization, recycling, input substitution, market and cost prices, and international allocation of production capacity. The chapter reports the effects of various environmental policy instruments for Western Europe and the Netherlands, including imposed energy taxation, taxation of primary materials, performance standards for energy and emissions, and deposit money for scrap. Chapter 7 shows that no absolute decline in material use can be found in member countries of the Organization for Economic Cooperation and Development (OECD), but that a relative decoupling of material use and GDP can be observed.

In chapter 8, Idenburg and Wilting present the DIMITRI model, a mesoeconomic model that operates at the level of production sectors, focusing on production and related environmental pressure in the Netherlands. The use of a multiregional input-output structure enables an analysis of changes among sectors and among regions resulting from technological changes. Because of the dynamic nature of the model, it allows analysis not only of the consequences of changes in direct or operational inputs, but also of shifts from operational inputs toward capital inputs and vice versa. Different analyses with the model are presented to show the benefits of a dynamic input-output framework for policy analysis.

1.4 Waste Management and Recycling

Chapters 9 and 10 address modeling of opportunities for waste management and recycling. In chapter 9, Bartelings, Dellink, and van Ierland present a general equilibrium model of the waste market to study market distortions and specifically flat-fee pricing. A numerical example is used to demonstrate the effects of flat-fee pricing on the generation of waste. The results show that introducing a unit-based price will stimulate both the prevention and recycling of waste and can improve welfare, even if implementation and enforcement costs are taken into account. Introducing an upstream tax can provide incentives for the prevention of waste but will not automatically stimulate recycling. A unit-based pricing scheme is therefore a more desirable policy option.

In chapter 10, van Beukering provides an overview of the internalization of waste flows. He focuses attention on the international trade of recyclable materials between developed countries and developing countries. Empirical facts indicate a high rate of growth of such trade. Moreover, a particular trade pattern has emerged, characterized by waste materials being recovered in developing countries and then exported to other developing countries, where they are recycled. Chapter 10 discusses the economic and environmental significance of the simultaneous increase in international trade and the recycling of materials.

1.5   Dynamics of Eco-Industrial Parks

Chapters 11 and 12 deal with industrial ecology dynamics at the lowest scale and consider the self-organization and stimulation of eco-industrial parks. A classic example in industrial ecology is the industrial symbiosis reported in Kalundborg. Both chapter 11 and chapter 12 use the Kalundborg case study as their starting point. The eco-industrial park in Kalundborg is one of the most internationally well-known examples of a local network for exchanging waste products among industrial producers. In chapter 11, Jacobsen and Anderberg offer an analysis of the evolution of the "Kalundborg symbiosis." This case has been studied from different viewpoints. For comparison and contrast, the authors discuss ongoing efforts to develop an eco-industrial park in Avedøre Holme, an industrial district around the major power plant in the Copenhagen area. The chapter closes by addressing the limitations of the Kalundborg symbiosis.

In similar vein, in chapter 12, Boons and Janssen use the Kalundborg example to analyze why there have been so many efforts to re-create Kalundborg in other locations. The main problem of creating an eco-industrial park is to overcome a collective action problem, which is not without costs. But such costs are generally neglected, since most studies look only at benefits from technical bottom-up and top-down designs. As research on collective action problems shows, top-down arrangements are often not effective in creating sustained cooperative arrangements. Furthermore, bureaucrats and designers may not have the required knowledge to see entrepreneurial opportunities for reducing waste flows. Boons and Janssen argue that more may be expected from incentives for self-organized interactions that stimulate repeated interactions and reduce investment costs, or from providing subsidies for investments in interfirm linkages.

1.6   Policy Implications

Finally, in chapter 13, van den Bergh, Verbruggen, and Janssen discuss the policy implications of the research reported here. In particular, the chapter considers the question of whether a combined policy specifically focused on materials and waste is needed. It concludes that a general dematerialization policy is meaningful for a number of theoretical and practical reasons. Dematerialization and waste policy support one another in the long run, even if, in the short run, they are often conflicting. The aim of a combined policy is to facilitate technological innovation and transitions toward a material-poor economy. Policy can attempt to stimulate the incorporation of certain physical requirements into production processes and products, even when from an economic point of view these requirements are second best.

References

Allen, D. T., and N. Behmanesh. (1994). Wastes as raw materials. In B. R. Allenby and D. J. Richards (eds.), *The Greening of Industrial Ecosystems*, 69–89. Washington, DC: National Academy Press.

Ayres, R. U., and L. W. Ayres (eds.). (2002). *A Handbook of Industrial Ecology*. Cheltenham, England: Edward Elgar.

Ayres, R. U., and A. V. Kneese. (1969). Production, consumption and externalities. *American Economic Review* 59: 282–297.

Bartone, C. (1990). Economic and policy issues in resource recovery from municipal wastes. *Resources Conservation and Recycling* 4: 7–23.

Boulding, K. E. (1966). The economics of the coming spaceship Earth. In H. Jarrett (ed.), *Environmental Quality in a Growing Economy*, 3–14. Baltimore: Johns Hopkins University Press for Resources for the Future.

Brunner, P. H. (2002). Beyond materials flow analysis. *Journal of Industrial Ecology* 6(1): 8–10.

Daly, H. E. (1968). On economics as a life science. *Journal of Political Economy* 76(3): 392–406.

Daly, H. E. (1977). *Steady-State Economics*. San Francisco: Freeman.

Daly, H. E., and K. N. Townsend (eds.). (1993). *Valuing the Earth: Economic, Ecology and Ethics.* Cambridge, MA: MIT Press.

Daniels, P. L. (2002). Approaches for quantifying the metabolism of physical economies: A comparative survey. Part II: Review of individual approaches. *Journal of Industrial Ecology* 6(1): 65–88.

Daniels, P. L., and S. Moore. (2002). Approaches for quantifying the metabolism of physical economies: A comparative survey. Part I: Methodological overview. *Journal of Industrial Ecology* 5(4): 69–93.

Factor Ten Club. (1994). *Carnoules Declaration.* Wuppertal, Germany: Wuppertal Institute.

Fischhoff, B., and M. J. Small. (2000). Human behavior in industrial ecology modeling. *Journal of Industrial Ecology* 3(2–3): 4–7.

Georgescu-Roegen, N. (1971). *The Entropy Law and the Economic Process.* Cambridge, MA: Harvard University Press.

Graedel, T. E., and B. R. Allenby. (2003). *Industrial Ecology.* 2nd edition (1st ed., 1995). Upper Saddle River, NJ: Pearson Education, Prentice Hall.

James, D. (1985). Environmental economics, industrial process models, and regional-residuals management models. In A. V. Kneese and J. L. Sweeney (eds.), *Handbook of Natural Resource and Energy Economics,* vol. 1, 271–324. Amsterdam: North-Holland.

Kneese, A. V., R. U. Ayres, and R. C. D'Arge. (1970). *Economics and the Environment: A Materials Balance Approach.* Baltimore: Johns Hopkins University Press.

Koenig, H. E., and J. E. Cantlon. (2000). Quantitative industrial ecology and ecological economics. *Journal of Industrial Ecology* 3(2–3): 63–83.

Socolow, R., C. Andrews, F. Berkhout, and V. Thomas (eds.). (1994). *Industrial Ecology and Global Change.* Cambridge: Cambridge University Press.

van den Bergh, J. C. J. M. (ed.). (1999). *Handbook of Environmental and Resource Economics.* Cheltenham, England: Edward Elgar.

von Weizsäcker, E., A. B. Lovins, and L. H. Lovins. (1997). *Factor Four: Doubling Wealth—Halving Resource Use: A Report to the Club of Rome.* London: Earthscan.

2 The Interface between Economics and Industrial Ecology: A Survey

Jeroen C. J. M. van den Bergh and Marco A. Janssen

2.1 Introduction

The Western consumer seems to be more and more enamored by material things. This has resulted in the availability of a wide variety of products, in turn giving rise to increasing pressure on the environment. Nearly all important local and global environmental problems as well as almost all environmentally related human health risks can be reduced to the flows and the accumulation of substances and materials in the economy. This is illustrated by such different problems as climate change (fossil fuel use), acid rain (fertilizer use, animal fodder, meat consumption), toxicity (metal use), water pollution (paper use), desiccation (water use), and the depletion of fisheries (fish consumption).

In this chapter we will survey the field of research that is concerned with the physical dimension of economic activities and products. This will entail a discussion of the concepts, theories, and methods that economists have used to study physical flows. An effective environmental policy analysis of an economy is enhanced by paying close attention to physical dimensions. Four reasons can be given.

First, the "physical economy" approach enables research to be conducted on the coherence of environmental problems by adding a realistic physical context to the monetary economy. For example, by adding "end-of-pipe technologies" to the production process, the emission of certain substances to air or water can be reduced. Without a reduction of material input, however, production will inevitably generate solid waste, which is also associated with negative environmental effects (Ayres 1998). In the best case this will result in a salable product, like gypsum with flue gas desulfurization.

A second reason for an explicit description of the physical dimension of the economy is that it creates a basis for dealing with notions like "sustainability" and "sustainable development." Elaboration of these concepts usually involves referring to physical and biological stocks and the feedback of changes therein to the physical economy (van den Bergh and Nijkamp 1994). Sustainability can be considered strongly connected to physical constraints, determined by exhaustion of raw material supplies and the accumulation of substances in the environment.

A third reason for a physical economy approach is that it facilitates multidisciplinary use of economic models and insights, in particular, the linking of these to physical, chemical, and biological models of environmental processes and compartments. This is especially valuable in integrated modeling aimed at the study of long-term effects and risks of substance flows through the environment and economy (see, e.g., Rotmans and de Vries 1997).

A final reason for a physical economy or economics-of-industrial-ecology approach is that it allows robust policy suggestions to be arrived at. The combination of the various dimensions, integrated in formal models, means that policies almost automatically are tested for physical, technical, and economic limits and opportunities.

The interface between economics and industrial ecology has its roots in the late 1960s and early 1970s. Ayres and Kneese (1969) is generally regarded as the first article that presents a theoretical, formalized framework to combine economic modeling, based on the general equilibrium format, with physical flow accounting. This was regarded by the authors as a general approach for dealing in a fundamental and correct way with externalities caused by the extraction, production, use, and waste of materials or commodities containing materials. An extended version of their work is contained in Kneese, Ayres, and D'Arge (1970). Other early work combining economics and material flow analysis is Georgescu-Roegen (1971a), which is almost philosophical in nature, and Ayres (1978), which emphasizes the use of input-output techniques. Both Ayres and Georgescu-Roegen address the implications of thermodynamics for the specification of production functions in economic models (see also Georgescu-Roegen [1971b] 1976, and for an evaluation of his work, Cleveland and Ruth 1997). Contributions in Daly and Umaña (1981) and Faber, Niemes, and Stephan (1987) can also be considered early discussions of the integration of economics with industrial ecology themes.

This chapter will illustrate the wide array of model approaches that allow integration of economics and industrial ecology as well as try to provide a comprehensive survey of relevant studies that have been performed thus far. The structure of the chapter is as follows. Section 2.2

discusses in more detail the relationship between physical flows through the economy and environmental problems. This includes a typology of materials as well as a short review of the main implications of thermodynamics for the economic analysis of physical flows. In section 2.3, the policy context is sketched. This involves discussing the traditional hierarchy in waste management, recycling and reuse, and dematerialization at various scales. Next, section 2.4 surveys the wide range of concepts, theories, and methods that are employed in environmental-economic analysis of physical flows. These cover, among other things, materials and resource accounting, mass balance, material-product chains, mass-balance production functions, recycling models, input-output (I/O) modeling, equilibrium theory and externalities, and economic growth theories. This is followed in section 2.5 by a closely linked discussion of related perspectives on concepts and methods in industrial ecology, with particular attention to material cycles, international trade, and qualitative network analyses. Section 2.6 presents conclusions.[1]

2.2　Materials, Substances, and the Environment

2.2.1　A Typology

This section discusses the relationship between materials (and substances) and environmental problems. Substances are amounts of atoms or molecules, for example, metals, sand, and water. Materials are physically bonded substances, such as paper, wood, plastic, metal alloys, and fossil fuels. The impact on the natural environment of physical flows and the accumulation of substances and materials in the economy covers many categories of environmental problems. A first category concerns "resource problems," connected to scarcity and exhaustion of natural resources and supplies of raw materials. A second category concerns "pollution problems," connected to waste flows and the emission of substances into the environment. These flows can be harmful to the health of humans or other living organisms because of their character (quality) or amount (quantity). Toxic and artificial substances are in the short term the most alarming in this regard, whereas solid waste and emissions of acidifying substances and greenhouse gases create long-term risks.

Several specific problems are associated with material flows in the economy. Substances accumulate in the economy and cause numerous indirect and delayed problems. Therefore, measuring and predicting the ultimate environmental effects of material flows and determining the causes of these problems is not straightforward.

In addition, substance flows are connected in two essential ways to energy use and related environmental effects. First, there is a strong coherence between the use of materials and energy in production processes, because energy is needed to transform and modify materials. Second, energy supply is dominated by fossil fuels in most countries and is linked to flows of several substances, particularly carbon compounds and nitrogen and sulfur oxides. For instance, energy use based on fossil fuels in the United States leads to a share of almost 40 percent in the total input of substances and materials in the economy (Wernick and Ausubel 1995). The enormous quantity of several substance flows additionally creates large amounts of movements of freight. This causes specific environmental problems, notably related to the use of space and energy.

Materials and substances can be classified on the basis of several characteristics, for example, physical, biological, and economic. The most important substances for biological processes on earth and for human beings are the nutrients carbon, nitrogen, sulfur, and phosphor us. These elements take part in the nutrient cycles of living and nonliving systems. Such cycles are characterized by natural recycling. The four substances occur in the biosphere—the living earth—in larger concentrations than in the physical, abiotic (nonliving) natural systems. This is caused by biotic processes and is indirectly the result of a long and slow process of organic evolution. In this context the term "Gaia" is sometimes used, based on the theory of a living earth as a system of several complex chemical and biological feedbacks that have tended toward natural balance by evolution (Lovelock 1979). However, the nutrient cycle is nowadays severely disturbed by human activities, in particular by the combustion of coal, oil, and gas. This has caused an increase in the concentrations of carbon dioxide, nitrogen oxides, and sulfur oxides in the atmosphere, which in turn creates the risk of a serious disturbance of the natural balance.

In addition to these nutrients, six other (partly overlapping) categories of substance and material flows can be distinguished that mix physical and economic considerations. Some of these categories will receive detailed attention in later chapters. (For a more detailed discussion, see, for example, Ayres 1999a.)

1. *Metals* These are used in numerous products. Some of them are toxic, like cadmium, copper, lead, and zinc. These are recycled only in small amounts by natural processes, so that they accumulate in the environment (Guinée et al. 1999; van der Voet, Guinée, and Udo de Haes

1999). This creates long-term risks for the health of humans and ecosystems. Moreover, the extraction of metals produces huge amounts of waste flows, because many ores contain low concentrations of metals. This holds for copper, lead, and nickel, and especially for gold, platinum, and uranium (Ayres and Ayres 1996). In addition, many metals are rather scarce anyway. Recycling by humans is possible but is hampered by the fact that metals are regularly a part of alloys.

2. *Plastics* Our direct living and working environment is increasingly dominated by this category of materials. It covers a variety of synthetic substances, with polypropene, polystyrene, polyethylene, and polyvinyl chloride (PVC) the most important. Production and waste treatment of some plastics is extremely environmentally damaging. Nowadays plastics are mainly manufactured from oil products. In a "postoil" and "postmetal" world, plastics produced on the basis of plant material might contribute to a sustainable economy (Ackerman 1997). Interestingly, the first plastic, cellulose, was already made of plant material. All knowledge about plastics that has accumulated in the last few decades—concerning material characteristics like stiffness, hardness, tolerance for different temperatures, and transparency—can be applied usefully to develop materials that satisfy the requirements set by modern life. Plastics are suitable for recycling, which often involves a process of moving through lower quality grades.

3. *Chemical products* These include products that can be toxic, carcinogenic, or persistent, and some are completely artificial, that is, not found in nature. Examples associated with considerable physical flows are chemical compounds of phosphate, sulfur, nitrogen, and chlorine (Ayres and Ayres 1996). The chemical industry and agriculture are the sectors most directly involved in these flows. To lower the environmental load in the chemistry sector, the following development directions have been identified (DTO 1997): production of methanol based on photovoltaic solar energy; hydrocarbon conversion in which power stations produce raw material the chemical industry in addition to energy; and integrated plant conversion, which uses biomass as raw material for the chemical industry. The use of pesticides in agriculture creates a specific problem, and although the situation is improving, it is far from positive.

4. *Minerals* This covers stone, gravel, clay, and sand, which are extracted in large amounts from the earth's crust. Although they are relatively harmless per unit of weight, they cause much transport, disturbance, water pollution, and damage to landscapes, through erosion by opencast mining.

5. *Packaging material* This type of material makes up a considerable part of the total weight and volume of waste generated by human activities. Its extremely short life duration is an important characteristic. The most essential types of materials encountered are glass, paper, cardboard, steel, aluminum, and plastics. Some authors claim that the lightest packing is the best packing material in terms of minimal environmental load, with the exception of materials with toxic components (Ackerman 1997). Note that this type of material, that is, packaging, differs from the other types in that waste resulting from it relates especially to the (intermediate and final) use stage, whereas environmental waste related to the other material types is generated in various stages, including production and use.

6. *Organic products* These are connected to food production, the use of paper and wood, and other uses of biomass. Paper making and wood production in particular generate toxic substances and cause eutrophication of surface water. Organic substances can be reused or biologically decomposed. Composting of such substances is possible, though this requires separation of organic and other waste, which is not so easily achieved. A specific problem is that metals accumulate in the soil via composting and eutrophication, resulting in increased metal concentrations in cultivated vegetables, which can after many feedback cycles exceed health risk thresholds (see Molenaar 1998). In addition, the biologist Vitousek and his colleagues (1997) have estimated that over 40 percent of the total biomass on earth is being used by humans or is severely threatened.

Traditionally, environmental economics aims to make environmental effects comparable in measurement terms: namely, through monetary valuation and the notion of "external costs."[2] This is an alternative to using weights, which is being done in life cycle analysis. Table 2.1 shows a comparison of external costs for a number of packaging materials. It is based upon a large survey by the Tellus Institute in Boston, which has performed many studies of alternatives for waste management and recycling within the United States. The table shows that PVC, new aluminum, and the plastic type polyethylene terephthalate (PET) are the most environmentally damaging substances per unit of weight. It must be kept in mind that for a complete picture, it is also necessary to take into account the weight of the packaging material needed per unit of packed product. From this point of view, glass packing is especially unattractive. Its undesirability on this criterion, however, is compensated for by its rela-

Table 2.1
External costs of packaging materials

Packaging material	Estimation of external costs*
Plastics	
HDPE (high-density polyethylene)	128
LDPE (low-density polyethylene)	158
PET (polyethylene terephthalate)	331
Polypropylene	148
Polystyrene	162
PVC (polyvinyl chloride)	1714
Paper	
Bleached kraft paperboard	121
Unbleached coated boxboard	94
Linerboard	95
Corrugating medium	101
Unbleached kraft paper	96
Boxboard from wastepaper	76
Linerboard from wastepaper	77
Corrugating medium from wastepaper	109
Glass	
Virgin glass	70
Recycled glass	48
Metal	
Virgin aluminum	928
Recycled aluminum	76
Steel	79

Source: Ackerman 1997, 102.
* In $US 1993 per ton packing material.

tively good performance in terms of external costs per unit of packaging material.

2.2.2   Fundamental Physical Backgrounds: Thermodynamics

Next, we briefly describe the main insights from physics, as these define constraints on the physical and technological processes that occur in the economy. The main relevant discipline is thermodynamics, or the science of energy and mass (for accessible treatments, see, among others, Ayres 1978 and Ruth 1993). The first and second laws of thermodynamics are of particular importance here. The first law, that of energy conservation, states that physical processes always involve conservation of energy/mass.[3] In other words, energy can neither be created nor destroyed. The second law, that of entropy, states that any physical process—biological or technological—leads to a loss of useful or concentrated energy (known as "exergy"). Entropy is defined as the distance to a thermodynamic equilibrium. It is sometimes used in a loose manner to describe dimensions other than the energy dimensions of physical processes, using the analogy with energy entropy. It is then interpreted as a measure of structure, information, or development. The term "material entropy" has been proposed to point to the diffusion of substances, the wastage of materials, and the erosion of material structures. This term, however, has no formal background and cannot easily be transformed into a quantitative standard.

From the first law, conservation of energy/mass, the mass-balance principle has been derived. It states that mass is preserved, so that inflow of substances in a system leads to accumulation or outflow of those substances. Although this principle is an approximation, the error margin is minimal under terrestrial conditions, in which materials and energy are almost perfectly separated categories. Creation or elimination of matter is very rare, unlike the transformation of materials and substances in a chemical sense.[4]

What do these laws and principles mean for environmental economics? First, they emphasize that economic activities do not take place in an independently operating system, but in an open system that exchanges energy and matter with its environment, that is, the natural environment. Second, processes in the physical economy lead to a degradation of the environment and natural resources, which can be compensated for only by a continuous inflow of solar energy. Third, even if many economists and technological optimists claim that substitution can solve most, if not all, environmental problems, there is no such thing as substitution of other production factors for energy. In other words, all physical, biological, and technical processes require a continuous input of energy (or more precisely: exergy). Sources of potential energy are the basis of all processes, and their functions are unique and irreplaceable. However, substitution among different forms of energy is possible. Fourth, thermodynamics determines the absolute boundary conditions of technical efficiency of the most advanced machinery, even machinery that has not yet been invented. This sets limits to the substitution of materials, to the amount of exergy needed for waste separation, to the amount of exergy required for extracting metals from ores, and so on. Some of these limits are conditional; that is, they can be determined on the basis of specific process characteristics, like required power, work per time unit, or temperature of combustion (see Peet 1992). Fifth, production and consumption are processes in which not mass, but shape, changes. This implies, for instance, that waste management, including incineration, will not reduce the total physical amount of waste (and emissions), but just its shape, form, or medium. Waste can be gaseous, liquid, or solid, but its state nevertheless highly influences spatial or temporal features of environmental damage. A particular case is the delay of environmental effects through the storage of toxic waste, which can cause the phenomenon known as a "chemical time bomb."

2.3 Principles for Waste Management, Recycling, and Dematerialization

2.3.1 The Hierarchy in Waste Management

The economic and environmental literature gives several suggestions for policy and strategies to influence substance and material flows. They can be categorized according to whether they pertain directly to waste management, recycling, or dematerialization. We now briefly review each of these, illustrated with some examples.

Many countries consider the dumping of waste as problematic because it shifts environmental risks to the future, or because dumping space is scarce. Instead, within the Dutch waste material policy, for example, prevention has the highest priority, followed by reuse, recycling, and then end processing, such as incineration and dumping. The U.S. Environmental Protection Agency (EPA) uses a similar reduce-reuse-recycle hierarchy but does not prefer incineration over landfilling. Since every option for waste management is associated with a variety of environmental effects, the most desirable choice from an environmental perspective is not immediately or generally clear. For example, if the contribution to an increased greenhouse effect by methane and carbon dioxide is considered, a comparison between dumping and incineration of waste results in a preference for dumping, as long as not more than 4.5 percent of the carbon in the dumped waste is released as methane. This preference is a consequence of the fact that one methane molecule is twenty-four times more effective than a carbon dioxide molecule in terms of warming-up potential (Ackerman 1997).

There is a fairly large economic literature on waste management policy by environmental and policy scientists (see Powell, Turner, and Bateman 2001) as well as by economists. So far, no serious economic policy in this regard has been implemented, which means that economic growth trends easily translate into trends of growing waste generation, both by industries and by households. Economic theory emphasizes that markets for waste either do not exist or are distorted, which economists frame as market failures (Wertz 1976; Choe and Fraser 1999; Ferrara 2003;

Fullerton and Kinnaman 1995, 1996; Morris 1994; Palmer and Walls 1997; Shinkuma 2003). An important factor of distortion is flat-fee pricing, in which the price paid to the (local) government for waste collection and treatment is unrelated to the amount of waste generated or supplied. But although theory and empirical work shows that user fees can reduce waste generation considerably, illicit dumping and burning may be unintended consequences (Hong, Adams, and Love 1993; Linderhof et al. 2001; Mirada and Aldy 1998; Sterner and Bartelings 1999). See chapter 9 for further discussion of this issue.

Waste management is best analyzed in an explicitly spatial dimension. For instance, the Netherlands is currently split into four waste regions, even if the primary responsibility for waste policy design and implementation is with the national government. The Dutch Waste Management Association organizes cooperation among the waste regions, on the one hand, and the provinces and municipalities, on the other. The municipalities take care of the collection of waste and encourage prevention through licenses. The provinces are responsible for the removal of waste material and the granting of licenses for processing. The national government is attempting to promote, through legislation, the prevention and recycling of products and waste oil. This division of tasks is a historical legacy that does not necessarily reflect a spatially optimal configuration. Only a carefully undertaken spatial analysis can provide information about such a configuration.

Although the waste problem is serious and is regarded as such by policymakers, it should be pointed out that the environmental effects of using materials in production processes are generally much more significant than the effects of waste processing after consumption (with the exception of nuclear and toxic wastes). This can be explained by the fact that extraction as well as chemical and physical processes in production generate a high level of emissions to air and water per ton of material and use much more energy than end processing of the same materials in a later waste phase. In addition, the environmental load of producing new materials is generally much higher than that of recycling materials.

2.3.2 Closing of Cycles

Closing cycles of substances or materials are generally considered an important strategy for reducing and preventing material waste, with the ultimate aim of achieving "zero emissions." Recycling can apply to substances, materials, and products. Products require repairing, whereas separation and physical and chemical processing are necessary for distilling substances and materials from waste. Certain materials are already reasonably well collected separately, particularly paper and glass (Ackerman 1997; van Beukering 2001). In addition, a distinction can be made among primary, secondary, and tertiary recycling. Primary recycling concerns reuse within a production process. Secondary recycling points at the processing and reuse of materials and substances obtained from the waste flow after consumption. Tertiary recycling refers to the combustion of waste to release stored chemical energy (Kandelaars 1999).

Reuse of a product is usually more attractive than separate reuse of the substances it contains, because the former option generally requires less physical and chemical processing and therefore causes less energy use and environmental pressure. Nevertheless, old-generation products may consume more energy than new products, which can undo energy and material savings in production. In addition, the reuse of products requires maintenance and repairing. Repairing of products has become less attractive in recent decades for several reasons. In the first place, products have generally become more complex in terms of their three-dimensional structure and the number of different materials used. Increased purchasing power has resulted in the faster replacement of products by consumers, partly because of the availability of new designs and added functions as well as shifting fashions. However, supply factors have also shortened the economic life of products, as a result of an endless search for product innovations stimulated by "Schumpeterian competition." This is clearly illustrated by the rapid changes characterizing current computer and telecommunications (mobile phone) markets. Even though this can be evaluated positively from an economic perspective, many disadvantages can be identified from an environmental perspective. Only a combined analysis can provide a definite answer with regard to how to define incentives for competition, innovation, and management of materials and substances.

Cascading of materials is a useful and an underrated strategy for saving energy and virgin materials. It follows from the previous point, that is, reuse of products being generally more attractive from an environmental perspective than separate reuse of the materials and substances of which they are made. Cascading means that recycling consists of different stages. The aim in any particular stage is to try, where possible, to achieve the same quality of use as in the preceding stage. Otherwise, recycling involves applications of waste materials and substances to uses of lower (but still the highest possible) quality. In other words, high-quality applications of energy and materials have priority, but as soon as the quality of materials decreases, new applications of lower quality open up. From a cascading perspective, using natural gas to heat buildings represents an enormous waste of high-quality (low-entropy) energy (exergy). Similarly, using virgin wood to produce packing material and high-grade paper for magazines with a short life is an enormous waste of high-quality material. To promote cascading of materials, physical flows from companies with very different and preferably "complementary" processes and material input-output characteristics must be coordinated. This should start as early as the planning phase of commercial areas, as it includes elements of spatial design and proximity. (This will be further discussed in section 2.5 on industrial ecology). At a higher aggregation level, incentives can be given to potential suppliers and buyers of waste material to stimulate their cooperation. Although successful examples of cooperation of this type among firms exist, the best-known perhaps being the Kalundborg site in Denmark (see chapter 11), most policies aimed at stimulating eco-industrial parks do not go beyond investment projects with relatively short payback times (see chapter 12).

2.3.3   Dematerialization

An additional important aim in environmental policy focused on materials and substances is dematerialization, which refers to a reduction of material throughput of the economy. Dematerialization can be interpreted in several ways: namely, as a reduction of substance and material weight on the level of a product, of a company, or of the whole economy. Dematerialization at a micro level means that the same service can be given with less direct and indirect input of substances and materials, or a shift to other, lighter materials. At the product level, dematerialization simply means that products become lighter. At a firm level, it can also imply that production processes are more materials-efficient and thus use less resource input as well as generate less material waste. In addition, dematerialization at the firm level can involve a shift to other products, or a change in the mix of products supplied. Cleveland and Ruth (1999) offer a more detailed discussion of definitions, indicators, and methods of analyzing dematerialization.

Dematerialization will reduce the environmental load, just like recycling, at the beginning and end of activity chains, that is, in the extraction of raw materials and waste processing. A main difference from recycling is that dematerialization is generally associated with less energy use and transport. Other terms that reflect elements of dematerialization are "eco-efficiency" and "Factor Four" (see section 2.5).

One can also consider dematerialization at the macro level, that is, of the whole national or even global economy. This creates a much more complex system of factors and impacts. Factors influencing dematerialization at this level are economic growth, changes in the structure of supply (sectors) and demand, changes in import and export, and technological innovations and substitution among materials in companies. The relation between, for example, the gross national product (GNP) per capita and material use has received some attention in the literature. Jänicke et al. (1989) performed an empirical cross-sectional analysis for thirty-one countries and concluded that dematerialization occurred at the macro level during part of the 1970s and 1980s. A follow-up study by de Bruyn and Opschoor (1997), however, shows that this type of dematerialization came to a halt around 1990 and that recently GNP and material use at a macro level have started to move in the same direction again. These types of studies are obviously rather sensitive to the choice of macroindicators for material use. In the aforementioned studies, attention was focused on steel, energy, cement production, and freight transport.

The relationship between energy use and dematerialization in U.S. metals sectors has received attention in the context of analysis of carbon dioxide (CO_2) emissions (Ruth 1995a, 1998). Ruth's studies focus on copper, lead, zinc, aluminum, and iron and steel sectors and describes the dynamic interrelationships among resource extraction, materials processing, fuel use, and technological change using time series data and engineering information. Subsequently, projections of material, energy, and CO_2 emissions are made for the period 1990 to 2020.

From a commercial point of view, dematerialization can be considered the result of a decision process that involves a number of considerations and trade-offs. Attention can be given to the size of a product ("miniaturization"), the use of light materials, the complexity of a product and its production, the physical and economic lifetime of a product, the available range of reuse options, and the safety of transporting and using a product (Herman, Ardekani, and Ausubel 1989). The expected costs and profits of alternative investments in new processes and products will, of course, be most influential on any decision. In addition to the supply perspective, a demand perspective can be adopted. This can provide a longer time perspective. For example, a higher income can lead to a saturation of certain types of material consumption, closely linked to shifts toward services and leisure activities. It is not easy to make predictions in this area, because of the large number of factors involved. This in turn means that predicting patterns of dematerialization at a macro scale will be extremely difficult, as is illustrated by the paradox that the electronic information technology revolution did not—as hoped and expected—result in less use of paper, but quite the opposite. Of course, one can argue that this revolution has only started and that in a future phase, paper use per unit of output will finally decrease.

In sum, a reduction of waste through dematerialization of production and products is in principle preferred to the recycling of products, materials and substances, mainly because it involves less indirect activities and related energy use, transport, and space use, and associated environmental impacts. Recycling, however, is much easier than the control of dematerialization, because the latter has many more different dimensions and is influenced by numerous factors at micro- to macro-levels. Recycling prevents waste, but it does not reduce the size of material flows through the economy, that is, if waste disposal and extraction phases are excluded. It is clear that a task is waiting for economists to systematically study the relations and considerations between the aforementioned strategies, as well as support choices among them with combined economic-environmental evaluation.

2.4 Themes and Methods in Environmental-Economic Analysis of Physical Flows

2.4.1 Introduction
In this section, a brief overview is given of how economists include physical flows in their analyses. This overview will involve conceptual, theoretical, and methodological points of view. Studying the relationship between physical changes in the economy and physical changes in the environment, and the influence of environmental policy on this relationship, is an important task of environmental economics. Changes in the economy can occur at different levels, which can be studied separately or in combination. Changes in inputs can be studied by formulating decision models with production functions; changes in intermediate products and indirect activities caused by, for instance, recycling can be examined with I/O models; changes in production and investment at a national scale can be studied by constructing growth models with material flows; and consumption and international trade can be linked to material flows in microeconomic market models.

2.4.2 Mass-Balance and Economic Analysis of Material-Product Chains
Many studies of material flows use the mass-balance principle mentioned earlier. Use of this principle allows us to make consistent statements about substances flowing in and out of a process at any level: machine,

factory, firm, industry, region, country, etc. With mass-balance conditions, the scarcity of resources can be coupled with problems of emissions and waste flows. For this purpose, models of substance and material flow have been developed (Moll 1993; van der Voet 1996).

For a thorough environmental or economic analysis, the concept of "material-product chain" (M-P chain) can serve as a starting point (Opschoor 1994). An M-P chain is a system or network of coupled flows of at least one material and one product. These flows connect activities or phases in the chain, for example, extraction, production, consumption, collection, reuse, dumping, and combustion of waste. Chains are usually not isolated, even if such isolation is often assumed in order to make analyses tractable or to restrict the amount of data required for an empirical analysis. This type of isolation involves, among other things, focusing on a specific substance, material, sector, region, or product.

An M-P chain can be used as a starting point for a specific analysis. It might be useful to make a distinction between an economic and an environmental M-P chain analysis. An example of the latter is life cycle analysis (LCA), which gives an overview of the environmental effects of a product for the whole chain. This is mainly suitable for discrete decision problems (Guinée 1995; van den Berg, Dutilh, and Huppes 1995). For example, the choice between milk packed in glass or that packed in cardboard can be based on the comparison of the most important environmental effects during the life cycle.[5]

By adding specific economic aspects to such an M-P chain, an economic M-P chain can be created. Such aspects include the description of decision-making agents in the chain; the allocation of scarce factors (work, capital); the substitution of inputs in production processes along the chain; changes in sector or firm structure that affect the structure of a chain; the dynamic aspects of investment and technological innovation; and the impact of various environmental policy instruments. Economic analysis can then be directed at questions of cost-effectiveness of chain control, consequences of economic growth and development of substance flows, or the influence of environmental policy on physical flows (Ayres 1978; Ruth 1999). The implementation of economic M-P chain analysis requires that material balance or substance flow models be coupled with economic models. Kandelaars (1999) gives a good overview of this type of work and also presents her own studies of M-P chains.

It is possible to combine environmental and economic analyses of M-P chains. This means, for instance, that a life cycle analysis can be combined with an economic evaluation of environmental effects (van

Beukering, Spaninks, and Oosterhuis 1998) or with an allocation mechanism and chain optimization (Weaver et al. 1997). The M-P chain approach makes it possible to illustrate problem shifts in environmental as well as economic dimensions, which is a central focus of industrial ecology. A choice can be made between an evaluation based on cost-effectiveness under given environmental and economic conditions (Starreveld and van Ierland 1994; Kandelaars and van den Bergh 1996a, 1996b), or on a multicriterion analysis (Kandelaars and van den Bergh 1997). The latter points to a difficult problem associated with environmental life cycle analysis: namely, the comparison of different environmental effects. Ayres (1995) states that economic monetary evaluation is, despite its limitations, the only consistent approach for the comparison of different environmental effects. van Beukering (2001) presents various models that operationalize this idea. In addition, he introduces the notion of an international material product chain (I-M-P chain), which combines material flow models with descriptions of interactions among various economic agents of activities or countries. The I-M-P chain tries to describe the spatial dimensions of "industrial metabolism" and will be discussed in more detail in section 2.5.

What are the essential differences among the life cycle analysis, economic M-P chain analysis, and substance flow analysis (Bouman et al. 2000)? Unlike economic M-P chain analysis, life cycle analysis and substance or material flow analysis do not address economic questions, since they lack a description of economic decision and market processes. Moreover, M-P chain analysis explicitly distinguishes the dimension "products," whereas neither substance nor material flow analysis does. In contrast to life cycle analysis, economic M-P chain analysis describes interactions among flows of different products. Finally, a major strength of economic M-P chain analysis is that it connects the essence of life cycle analysis with the essence of substance flow analysis, albeit in manner that simplifies both approaches, so as to control the amount of complexity involved in description and analysis (Kandelaars and van den Bergh 1997).

Bouman et al. (2000) offer a methodological comparison of how material flows are dealt with in three types of modeling approaches—namely material flow analysis (MFA), LCA, and partial equilibrium economic analysis—by examining the same numerical example with each of the three approaches. The main finding is that the approaches are largely complementary, rather than contradictory. In particular, the analysis illustrates that MFA and LCA tend to focus on technical solutions to environmental pressure, whereas economic analysis focuses on price mechanisms that stimulate substitution on both the demand and supply sides of the market. It is suggested that a sequential application of the three approaches is both effective and less tedious than a fully integrated modeling approach.

Unlike in traditional economics, M-P chain analysis does not consider consumption to be merely a process in which consumption goods enter and utility or welfare is the only output. Products have a life cycle and can best be regarded as capital goods, in that they exist and are used over a considerable period of time to render services (Noorman and Schoot Uiterkamp 1998). During this entire time, they retain materials and substances. From a capital perspective, the fact that many varieties of a certain product, produced at different times, are in use at the same time suggests the use of a vintage model. This means that changes over time in the material composition of a product are taken into account. A vintage model for consumer goods in use can provide an accurate picture of accumulation of substances in the economy, as well as the time delay between the extraction of raw materials and waste processing (Kandelaars and van den Bergh 1997). For illustrative case studies, see chapters 5 and 6.

2.4.3 Direct and Indirect Substitution of Input and Mass-Balance Production Functions

In what way should production processes be described when material flows are studied? Georgescu-Roegen (1971a, 1971b [1976]) made an important contribution to environmental economics through publications on the relationship between economics and thermodynamics. He emphasized that we should distinguish among four aspects of production systems: "supplies," "flows," "stocks," and "services." This division into qualitatively different inputs in the production process creates several views on substitution and complementarity. Stocks, such as machinery and work, generate services, which transform flows (such as energy), substances, and semimanufactured products. The term "substitution" is cryptically used in many environmental-economic analyses, especially in the context of the growth debate. A distinction is made between direct and indirect substitution (van den Bergh 1999). Direct substitution refers to changes within a category of relatively homogeneous production factors that occupy the same function in a production process. An example within the input category "materials" is the replacement of steel with plastics and aluminum to lower product weight. Indirect substitution refers to the relation among production factors that play a different role within a production process. This "different role" means that they are to a certain degree complementary, but that some changes in the complementarity relation are possible, and these changes are identified as substitution. For example, through more input of work or machines in processing a material into a product, production waste (given a certain production level) can be avoided or primary recycling can be increased, all resulting in a reduction in material use. Direct substitution can be considered as the "replacement" of one input with another, for example, with different materials. Indirect substitution is closer to "saving" or "increase in efficiency and productivity."

Thermodynamics teaches us that indirect substitution of materials or energy and other inputs in production can occur only within certain limits. By using more work or machinery in production, the amount of material input required can be reduced, but not to zero, at least if the production output is a physical good. What is the influence of technical change on these aforementioned relationships among production inputs? No fundamental alterations occur in these relationships as a result of technical change, although a technological or thermodynamic optimum can be approached. This is, however, limited by certain characteristics of a production process (Berry, Salamon, and Heal 1978).

The standard production functions used by economists, also called "neoclassical production functions"—for example, those of the Cobb-Douglas type—at first sight seem inconsistent with the lessons of thermodynamics, especially the derived material balance principle. Cleveland and Ruth (1997) and Daly (1997) argue that such production functions describe inputs usually in a symmetric or identical way, so that qualitative differences among the distinct categories of inputs remain unclear. Characteristics of flows, supplies, and stocks are therefore not explicitly distinguished, which makes the models rather unsuitable for studying substitution problems. To achieve a minimal level of realism in material flow studies, a separate variable "material inputs" should be part of any production function, or should be "essential."

According to the definition of "essential" by Dasgupta and Heal (1979), in order to have a positive output of production, a strictly positive amount of material input is required. The amount, however, does not need to have a positive lower limit, as long as the marginal productivity of production inputs is assumed to go to infinity when the production output approaches zero. Although this assumption seems unrealistic at first, it cannot be proved to be so, because it involves a translation of physical units (substances, materials) into functional (product) units or monetary (value) units. It has turned out to be impossible to determine an absolute upper limit for the amount of value to be derived from a given amount of material inputs, even if it is thought that such a limit should exist (see Stern 1997). This leaves room for different, subjective opinions, leading, for instance, to the coexistence of growth optimists and pessimists. A more explicit approach to studying substitution in production, as previously proposed, could contribute to a better understanding of the differences between positions in the growth debate.[6]

2.4.4 Environmental Economic Analysis of Recycling

The economic analysis of recycling is still underdeveloped within environmental economics, possibly because recycling and processing of waste are not considered standard categories of economic production.[7] The economic study of recycling can start with an interaction between new and reused materials (or products), which can be regarded as imperfect substitutes. The price of each good will depend on the prices of other goods and the cross-price elasticity of the demand, on the one hand, and the costs of new and reused materials (or products), on the other hand. The costs of reused materials can include several categories of costs, such as those of collection, cleaning, separation, and chemical processing. As long as these costs are relatively high compared with the costs of new materials, the recycling of some substances and materials will remain a small-scale activity. To equalize the two types of costs, the scarcity of raw materials as well as the environmental consequences of waste flows and emissions need to be translated into adequate levies on materials.

Generally, the prices of waste materials suitable for reuse are subject to much fluctuation as compared with product prices in general. In the United States, this is the case for old paper, aluminum waste, and iron. This fluctuation is probably caused by a combination of economic business cycles and scarcity of and price developments regarding new materials. In addition, an important factor may be that industrial activities often prefer new substances and materials to used ones (see Ackerman 1997).

A striking feature of reuse and recycling is that the collection of some used products and materials seems to take place even when the cost savings are small and economic incentives are missing. Ackerman (1997) states that consumers seem to realize that their high level of material consumption has negative consequences for the environment. They try to compensate for this by regularly visiting the paper and glass recycling containers. Probably this apparently altruistically behavior occurs only for their own peace of mind, since material consumption does not decrease. This view on recycling and reuse as the most logic practical contribution that individual consumers can make to a "better environment" corresponds to the recycling ideology as propagated by several environmental organizations. But perhaps a more credible explanation for "collection behavior" is that recycling has become a social norm, at least one that has spread through a part of society. Whatever the explanation, economists should ask themselves whether their models are correct in explaining collection behavior, and which insights alternative behavioral theories suggest as to what can be expected from price instruments of material policy (van den Bergh, Ferrer-i-Carbonell, and Munda 2000).

A trend of the last few decades has been that recycling is increasingly linked to international trade, following the general pattern of globalization. To judge whether such a trend is desirable from economic, environmental, and developmental perspectives, there is a need for careful analysis that takes into account the wide range of externalities in the material-product chain, from extraction through production to consumption and back through recycling to production. Van Beukering (2001) presents a number of statistical and optimization analyses to examine these issues (see also section 2.5.3).

2.4.5 Input-Output Analysis of Economic Structure Changes

Let us consider a higher level: namely, the structure of the economy. By relating changes in material flows to changes in economic structure, one can learn which changes are more or less desirable from a combined environmental-economic perspective. In particular, a structural decomposition analysis can be performed. This provides information about the relationship between changes in physical flows and changes in factors like volume, technology, sector structure, input structure of sectors, final consumption and export. Such an analysis uses, among other things, information contained in detailed I/O tables of the economy. Such tables describe the economic structure through information about the mutual delivery of materials and semimanufactured materials within economic sectors. By comparing two I/O tables for different years, the indirect effects of changes in one sector on other sectors can be taken into account. This technique is known as structural decomposition analysis (SDA). By linking such an analysis to environmental indicators, a detailed insight is obtained about which changes have gone along with, and possibly have caused, certain changes in environmental pressure and material flows (Rose and Casler 1996). Rose (1999) and Hoekstra and van den Bergh (2002, 2003) present the state of the art of this technique as well as applications to material flow analysis. Rose, Chen, and Adams (1996) is the first application of SDA to material flow indicators. Wier and Hasler (1999) use SDA in a study of nitrogen in Denmark. Hoekstra (2003) presents a detailed analysis with SDA of material flows (iron and metals, and plastics) for the Netherlands. This study produces two hybrid-unit I/O tables, one for 1990 and one for 1997, and analyzes these using SDA. Subsequently, the results serve as an input to backcasting and forecasting scenario analyses (see chapter 4 for more details).

When the relationship between economic structure and physical flows is studied, the purpose is usually to create long-period views about economic development. Input-output modeling can be useful, then, because not only does it focus on structure, but it also avoids excessive assumptions regarding substitution, market characteristics, and individual behavior. In addition, it allows integration of data on monetary flows with data on substance flows, which in turn can be linked to natural resources, waste, and emissions in national accounts. Moreover, material balance conditions can easily be added, since they fit the linear I/O structure seamlessly. Next, an I/O model, however detailed it might be, can be extended with an optimization module without losing operationality. The Dutch study "Space for Growth" (WRR 1987), and the study "Sustainable Economic Development Structures" (Dellink, Bennis, and Verbruggen 1996) are examples of such an extension. Finally, the coupling of I/O models with market models or applied (general) equilibrium models is possible, although this requires much work. The most important advantage of a combined I/O-equilibrium model is that it allows more refined policy analysis in which the influence of specific policy instruments on the behavior of producers and consumers is taken into account.[8]

Duchin has argued in favor of using dynamic I/O analysis in environmental economics (1996) and in industrial ecology (1992). The implementation of this technique means that a great deal of attention needs to be given to the design of logical and detailed future scenarios that gather information about new technologies with respect to energy saving, reduction measures, dematerialization, and recycling. Duchin and Lange, supported by Thonstad and Idenburg (1994), made an empirical I/O analysis to test a hypothesis given in the well-known report *Our Common Future* of the Brundtland Committee (WCED 1987): that economic growth and ecological sustainability at a world scale can be united. The I/O model through which this hypothesis was tested is a sort of updated version of the Leontief I/O world model from the 1970s. The analysis relates to the period 1980 to 2020. The model describes sixteen regions, fifty sectors, and changes in the international trade of goods, capital flows, and economic and development aid. Calculations with the model of the effects of detailed scenarios, mainly on energy and material use of metals, cement, paper, and chemicals, show rising trends for the world as a whole. Environmental pressure indicators are calculated as well: CO_2 emissions double worldwide in the studied period; emissions of sulfur dioxide (SO_2) remain nearly constant; and nitrous oxide (NO_x) emissions more or less double. These and other results lead to a clear rejection of the hypothesis of the Brundtland Committee. Strictly speaking, this implies that we should think about how development, rather than growth in GDP terms, can be combined with sustainable environment quality at a world scale. For a similar study with a dynamic I/O model, but at a national level, see the study of DIMITRI in chapter 8.[9]

All together, scenarios and an I/O model can assemble an incredible amount of information in a clear and consistent way in a prospective dynamical analysis. "Natural resource accounting" and "social-accounting matrices" offer a basis for this. Disaggregation of households is thus possible with respect to income, education, labor, family size, or surroundings (for example, city or countryside). This allows, for example, the study of changes in lifestyle and demographic developments. Evidently, both have wide-ranging consequences for the use of products, and indirectly for material and substance flows. In addition, the use of physical flow information at the level of economic sectors, leading to physical I/O tables, can provide much insight into the physical structure of the economy. The construction of these, however, is a very time-consuming task that has not often been performed (on this issue, see chapter 4). Conclusively, we can say that dynamic I/O models in combination with scenarios are able to translate a detailed description of reality into possible future patterns of physical flows and economic structure. Operationalizing them is, however, a very laborious task.

2.4.6   Other Macro Models

Endogenous substitution within sectors and price mechanisms does not play a role in input-output analysis. That would require market and price balance models, which can describe the total economy, or part of it. An example is the STREAM model, which has been developed by the

Netherlands' Central Planning Bureau in cooperation with RIVM (National Institute for Public Health and the Environment), both of which are Dutch research institutes concerned, respectively, with economics and the environment (see chapter 7). STREAM is an empirical partial equilibrium model with sectors as units that was developed to study the size and causes of environmental problems related to the flows of seven bulk goods through the Dutch and European economy: iron and steel, aluminum, artificial manure, chlorine, plastics, phosphor, and paper. Each of these is characterized by strategies containing specific options and costs coupled with dematerialization, reuse, and waste management. Because of the aggregate character of the model, cost curves are constructed that classify the options technically possible within each of those strategies in terms of marginal cost-effectiveness. This allows, for example, the study of recycling at a macro level.

STREAM employs a macro level of aggregation. Other macro-level models have followed a resource-accounting framework, which aims to include information on the use of flow and stock resources from the environment in an economic framework. Resource accounting explicitly links information on processes in the economy with that on processes in the environment. Energy is often used as the general resource factor. An interesting framework used to study the physical dimensions of economic systems is the ECCO approach: Enhancement of Capital Creation Options (Slesser 1990). This has seen a number of applications (Noorman 1995; Ryan 1995; Slesser, King, and Crane 1997; Battjes 1999; and chapter 6). The ECCO modeling approach can be characterized as a dynamic energy-accounting approach that links the production of human-made capital to the natural capital that physically enables a given production level. The ECCO methodology determines the system-wide, long-term effects of implementing policy options at the national or regional level. It does this by assessing the growth potential of the economy in the context of the existing economic structure and user-defined policies, technology options, and environmental objectives. In turn, changes in growth potential alter a wide range of demand and supply terms and so reflect many other aspects of the evolving economy.

ECCO models emphasize the physical fixed capital requirement associated with a particular policy and the ability of the economic system to deliver that capital, either by direct manufacture or trade. The wider impacts of policies are realized through their effect on the overall allocation of fixed capital among sectors, and therefore on the rate of growth of the system, which is endogenously computed.

2.4.7 Growth Debate and Material Flows

The relation between economic growth and material flows in the longer run remains a difficult topic. The thermodynamical insights discussed earlier do not easily and directly lead to absolute physical limits at the macro level, because of the aforementioned separation between value (welfare, use, monetary value) and physical amount (kilograms, joules). This is one of the reasons for the growth debate, which can be characterized best by the following three main questions: Is economic growth desired? Is economic growth possible? And can we control or regulate economic growth? (van den Bergh and de Mooij 1999). The growth-optimistic view is disseminated by, among others, the economists Julian Simon (e.g., Simon and Kahn 1984) and Wilfred Beckerman (1999). They state that growth is good and maybe even necessary for both a good environmental policy and the maintenance or recovery of environmental quality. The most recent support for this, according to some scholars, is "environmental or green Kuznets curves (EKCs)." These are empirical assessments of a "delinking" of growth (income per capita) and certain environmental pressure indicators. A possible explanation for such a delinking is that a higher income is associated with a more advanced technology, allowing, for instance, resources to be used in production more efficiently. Moreover, income growth leads to an increased interest in the quality of nature and the environment, which is known to economists as the "environment as a luxury good." However, the hypothesis is valid for only a limited number of environmental indicators, which moreover have a weak relationship with global or even local sustainability issues (de Bruyn and Heintz 1999). In addition, the EKC research reflects that environmental policy has emphasized urgent and local environmental problems threatening human health. Other problems, of a worldwide character, or waste problems, are postponed in space or time. However, because of the partial character of the EKC studies, these problems are neglected.

What role do materials and thermodynamics play in the growth debate? Obviously, growth of all physical flows through the economy inevitably leads to more environmental pressure, ceteris paribus. But what is the effect of GDP growth, which is not necessarily equivalent to an increase in the "physical economy"? Maybe a substitution among inputs in production away from material resources, or from consumption of goods to services, can succeed in combining such a growth with dematerialization at the macro level. For example, an increase in "clean" services in the GDP is noticeable for most countries over the recent decades. The question is, of course, whether such a trend can continue. The connection between activities suggests that it is likely that service sector growth implies growth of environmental pressure caused by intermediate sectors that are more material intensive and pollutive. Such intermediate relations would imply a limit to dematerialization at the macro level. D'Arge and Kogiku (1973), Gross and Veendorp (1990), van den Bergh (1993), and van den Bergh and Nijkamp (1994) have included mass balance in economic growth models to deal with this question at a theoretical level. Structural decomposition analysis, discussed in section 2.4.5, can get a grip on this issue from empirical and policy-relevant angles.

Next, an important question is whether recycling of substances, materials, and products offers a solution for important environmental problems. This comes down to the question of whether 100 percent recycling is possible. Many individuals have studied this, but it is has turned out to be difficult, if not impossible, to come up with a resolute "yes" or "no." The diffusion of substances is an argument for a "no"—think of rubber particles lost from a tire being used on the road, or the peeling off of metal-containing paint through weathering. Practically seen, it seems impossible to collect all component substances of a product for recycling. Complete recycling might be approximated when energy is available without limit and cost. But then other problems related to energy will of course create a bottleneck. On the other hand, Ayres has stated that perhaps not all substances and materials are totally reusable at the same time, but each time a considerable part can be reused. For this, he has proposed the terminology "waste mining" (see, e.g., Ayres and Ayres 1996).

2.5 From Extended Material Flow Analysis to Industrial Ecology

2.5.1 Introduction

Since the late 1980s, an integrated perspective on physical flow industrial systems has been called "industrial ecology" (Allenby and Richards 1994; Socolow et al. 1994; Graedel and Allenby 2003; Ayres and Ayres 2002). A central metaphor of this field is the "industrial ecosystem," which reflects the fact that an objective in an industrial system can be that "the consumption of energy and materials is optimized and effluents of one process ... serve as the raw materials for another process" (Frosch and Galopoulos 1989, 94), much like nutrient flows in biological ecosystems. An ecosystem is "the living community and the nonliving environment functioning together" (Odum 1963, 4). The boundaries of ecosystems are not always clearly geographically defined, except for islands or lakes, and are often linked to cycles of energy, water, nutrients, and carbon. Analogously, industrial production units that are linked to one another through fluxes of materials and energy can be thought of as comprising an industrial ecosystem. An industrial ecosystem is seen as a network of mutually dependent transformation processes, which form part of a larger whole, analogous to the function of a local community or ecosystem in relation to its global environment.

In what follows, we examine a number of themes in industrial ecology from an economic perspective.

2.5.2 Industrial Metabolism

Related to the notion of industrial ecology is that of "industrial metabolism." This refers to flows of materials and energy within and among industrial and ecological systems, as well as their transformation into products, by-products, and effluents (Ayres and Simonis 1994; Ayres 1999b). Both the metabolism of economic systems and that of organisms change over time through natural or social-economic evolution as well as through coevolution of the environment-economy system (van den Bergh and Gowdy 2000). Nevertheless, the change in the metabolism mechanism of economic systems is much more rapid: Witness the history of mankind, from the Stone Age through to the Industrial Revolution on to the current Information Age. The diversity of products, materials, and substances, as well as of production processes, human labor, and interactions among economic agents, has increased tremendously over time. Moreover, direct connections among agents now extend throughout the globe, which is the fundamental feature of the often misused concept of "globalization." Materials and substances that come in large volumes or are exotic to the natural environment have a disturbing influence on global biogeochemical cycles. The world economy can no longer be regarded as a minor influence on the world's environment.

One of the insights from industrial ecology and analyses metabolism is that partial policies focused on one part of a system may lead to shifting problems to another part of the system. To avoid this shifting, an integrated, system-wide approach is required. It is fair to say, however, that all applied studies in the area of industrial ecology unavoidably have some element of incompleteness or partiality as well. The boundaries of the relevant system must then be decided on the basis of interactions with a wider system that are considered negligible.

Research on industrial metabolism has focused on the description of material flows in economic systems (see, e.g., the case studies in Ayres and Simonis 1994). Studies along these lines provide insights about the size of material flows and the identification of stocks in which certain materials accumulate. (For a study on heavy metals, see, for instance, Guinée et al. 1999.) This leads to concepts like "chemical time bombs" (think of the accumulation of chemicals in the river soils) and "waste mining" (materials accumulating that can become an important resource in the future) (Stigliani et al. 1991; Rohatgi, Rohatgi, and Ayres 1998). However, a number of issues cannot be addressed properly with such a descriptive approach, especially the (economic) motivations underlying the actions of the economic agents.

Ayres (1994) noted that the economic system is a metabolic regulatory mechanism that balances supply and demand for both products and labor through price mechanisms. The economic system as a whole is essentially a collection of firms, together with institutions and worker-consumers. A manufacturing firm converts material inputs into marketable products and waste materials. Like species in biological systems, firms specialize in certain types of activity. From a material flow perspective, the following products can be distinguished:

- *Primary commodities* or *virgin materials:* raw materials that have been extracted from natural resources.
- *Secondary commodities* or *recyclable waste materials:* raw materials that have been recovered after production or consumption.
- *Final commodities:* intermediary products suitable to be converted directly into consumer goods.
- *Consumer products* or *final goods:* products generated in the final production (manufacturing) stage before consumption.
- *Waste materials:* residue materials from various stages in the industrial material cycle that can no longer be converted into useful materials or products in an economically feasible way.

Products of one firm can be used as an input by another. Firms are part of some sort of material cycle. This cycle is related to the general scheme of industrial metabolism as defined by Ayres (1994). We have adapted the terminology in this scheme and added an additional stock, waste, to explicitly include the waste treatment sector (figure 2.1). Materials are extracted from the environment and used to produce final commodities, which in turn are used to produce consumer products. Waste is generated in every step of the material cycle. This waste disappears into the environment or is collected for reuse or recycling.

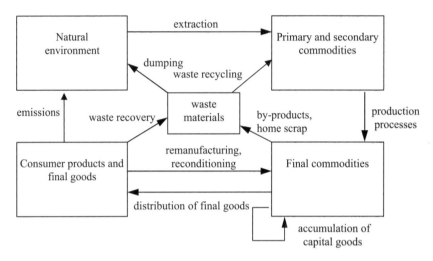

Figure 2.1
Five-box scheme for industrial material cycles.
Source: Adapted from Ayres 1998.

2.5.3 International Material-Product Chains

The relationship between material flows and international trade has received some attention from economists. An early study is Grace, Turner, and Walter (1978). Van Beukering et al. (2001) discuss a multi-regional version of international material cycles, also known as the I-M-P chain, which combines descriptions of material flows with those of inter-actions among various economic agents or activities in different countries. The I-M-P chain describes the physical dimension of economic systems in a setting of international interactions.

Figure 2.2 shows a possible I-M-P chain for two interacting or trad-ing countries. Country A represents a developed country that is well endowed with high-tech capital, skilled labor, and recyclable waste mate-rials. Country B represents a developing country that is poorly endowed with capital, recyclable waste, and know-how and well endowed with unskilled labor and primary raw materials. The arrows between country A and B represent international trade flows of raw materials and prod-ucts. Note that trade can relate to all stages of the chain. Materials and products flow horizontally, diagonally, and vertically from one segment to another within the I-M-P chain. Horizontal flows reflect intraindustry trade, and diagonal flows indicate interindustry trade. The traditional M-P chain is represented by the autarchic vertical material flows within

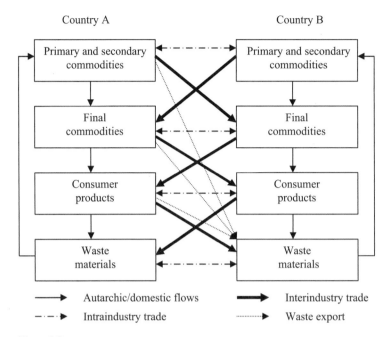

Figure 2.2
An example of the international material-product chain.
Source: van Beukering et al. 2001.

each country, given that borders are closed and countries are fully dependent on their own resources.

Traditionally, the immobility of labor and capital has caused natural resource availability to be an important driving factor of interindustry trade. With the mobilization of capital, production centers have become less dependent on the local availability of material resources. Technological knowledge, scale effects, and vicinity to consumer markets have become decisive factors, causing intraindustry trade. Materials and products have been traded horizontally among segments in the I-M-P chain. Differences in strictness of environmental policies have caused polluted materials and products to flow to developing countries.

Empirical studies confirm changes in the I-M-P chain (e.g., van Beukering and Bouman 2001). A pattern that links developed and developing countries has emerged. In particular, developing countries have become more important as importers of primary and secondary commodities and as exporters of commodities to their own region (van Beukering 2001). A number of other studies also address empirical issues of trade in materials (Byström and Lönnstedt 1995; Michael 1998; van Beukering and Duraiappah 1998; van Beukering and Janssen 2001).

Finally, Janssen and van den Bergh (2004) link international trade in materials to the research on environmental Kuznets curves, which addresses the relationship between economic growth and environmental pressure (see section 2.4.7). They formulate a numerical optimal growth model with material resource use in two trading countries and then simulate a number of scenarios. The results can explain some of the patterns obtained by empirical EKC studies.

2.5.4 Industrial Symbiosis

One of the most noticeable topics in industrial ecology is the analysis of "industrial parks." Such analysis relates to the concept of "industrial symbiosis," that is, the idea that the negative ecological impact of industrial activities can be reduced efficiently and effectively by stimulating the spatial proximity of location and cooperation among firms. By stimulating a larger, organized system of industrial relations, it is possible to prevent spillover effects, which occur when efforts to reduce negative ecological impact in one part of the system create additional or worse impacts in other parts of the system. In a collective approach, firms achieve competitive advantage by physical exchange of materials, energy, water and by-products (Chertow 2000). A prominent example of a successful industrial symbiosis is the industrial park at Kalundborg in Denmark (see chapter 11). Even though the term "park" suggests a planned or artificial approach, it can be the outcome of a slow process of self-organization, assisted by a policy environment that is favorable to cooperation and symbiosis. Desrochers (2002) broadens the industrial symbiosis concept from an industrial park to an entire city.

Sager and Frosch (1997), studying the metals sector in New England (figure 2.3), found that this sector has developed into a well-functioning industrial ecosystem, indicated by very low percentages of material loss from the system: 0.5 percent for copper and 4.5 percent for lead. They also found that larger firms are more efficient, but concluded from a survey of thirty-five metal-processing firms that none of the firms employs an industrial symbiosis perspective. Industrial symbiosis can be regarded as an emergent property of the system of local interactions of individual firms. A critical role in the functioning of the whole system is played by the secondary processors, such as scrap dealers and melters. Sager and Frosch also discovered that a transparent governmental waste policy in terms of prices and regulation stimulates the performance of the indus-

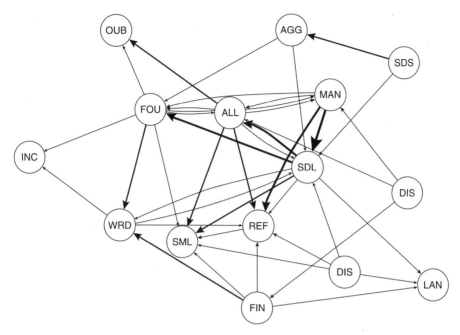

Figure 2.3
Flows of metals among metal processors in New England, with DIS = distributors/
virgin metal suppliers; AGG = agglomerators/brokers; SDS = scrap dealers (small);
SDL = scrap dealers (large); ALL = alloyers; FOU = foundries; FIN = finishers;
DIS = dismantlers; INC = incinerators; SML = smelters; MAN = manufacturers; WRD =
waste reclaimers/disposers; REF = refiners; OUB = other users of by-products; and
LAN = landfills.
Source: Frosch et al. 1996.

trial ecosystem. In contrast with Kalundborg (which is characterized by
fixed interactions among firms associated with particular physical flows),
in New England, the metals sector is more flexible as a result of mobile
links among production firms, scrap dealers, and melters. Because of
these mobile links, the industrial ecosystem probably has a larger capacity
to adapt to external stimuli than one characterized by fixed links (the latter being more fragile).

Industrial ecosystems can be analyzed at various levels of aggregation,
depending on whether the core questions deal with the context or the underlying mechanisms. At the lowest level, concrete people and products
come into view. Andrews (2001) argues that the agency level needs to be
studied to explain the human considerations and motivations—possibly
involving profits, welfare, ethics and social norms—that ultimately drive
industrial ecology.

An important area of micro-level analysis is consumer behavior. Current economies are very much focused on the consumption of material goods rather than on maximizing the utilization of these goods (Stahel 1994). Axtell, Andrews, and Small (2002) recommend the application of (multi)agent models in industrial ecology, which they argue are capable of analyzing innovation, cooperation, and diffusion processes.

2.5.5 Services and Dematerialization

A transition from a consumption economy toward a service economy would emphasize the function or service that products may provide instead of the ownership of the product itself. Examples are laundry services, carpooling or car-sharing services, teleshopping, and maintenance services. The choice is thus between buying a good or buying its services through leasing: for example, buying a washing machine or taking your clothes to a laundry service center. The latter means a more professional approach, economies of scale, better maintenance of machines, and more control of material and wastewater flows. Stimulation of services will require a behavioral change of both consumers and firms. When consumers share or lease products, companies will seriously and systematically consider the repair or refurbishing of products, thus considering all consumer products as valuable capital goods. To allow for such a transition, product designs and product services need to change. Products need to be developed in such a way that components can be taken back for remanufacturing. Remanufacturing is different from recycling in that it is a form of product prolongation, not a simple material loop closing (Stahel 1994). Innovative producers may perceive a possible new market share for alternative dematerialized products and hence may adopt machines that are more robust, more reparable, and more adjustable qua function. Innovative consumers, that is, the first individuals who start using alternative products, provide a behavioral example to other consumers.

It is expected that a service economy would be much less material intensive, relative to the satisfaction of human needs, than the present consumption economy (Ehrenfeld 1997). Technical options like changing product design can significantly reduce material and energy use. The market entrance of products with designs charged to accomplish such reductions, however, often fails. One reason is that current production and consumption patterns are "locked in" (Arthur 1989), indicating that the behavior of agents is interdependent, as a result of network externalities, social imitation, switching costs, or simply economies of scale. A misunderstanding is the idea that lock-in is permanent. History shows that all kinds of technologies eventually become replaced by a new technology. In the case of text processors, WordPerfect replaced WordStar, and nowadays Microsoft Word is the dominating text processor. Even the QWERTY keyboard, a popular lock-in example, may be replaced one day by another technology: for example, technology related to speech recognition. Janssen and Jager (2002) present a simulation model of coevolution of product choice by consumers and product development by firms. They find that successful introduction of alternative products needs to be accompanied by adequate price shocks to get out of the locked-in situation.

2.5.6   Factor Four and Rebound Effects

Many authors have focused attention on improving "eco-efficiency," that is, the environmental performance per unit of useful economic output. Solutions comprise technical design of products and processes, organization, and logistics. The "Factor Four" approach promoted by the German Wuppertal Institute is perhaps the best-known example of this. Factor Four denotes that a doubling of welfare is feasible while halving material and energy use (von Weizsäcker, Lovins, and Lovins 1997). There is, however, a risk that all concrete eco-efficiency suggestions at a micro level give rise to overly optimistic estimations of net gains on materials at a macro level, because indirect effects that run through various markets (resource, product, capital, labor, and financial) may be significant. This can stimulate substitution patterns in production and consumption that ultimately may undo part of the first-order eco-efficiency gains. Such indirect effects are now known under various names: general equilibrium, macroeconomic, or rebound effects. For example, through less use of a particular material resource as a result of a new production technology or product design, prices of the respective resource will fall, which in turn can increase the use of the original products or processes or stimulate entirely new uses of the resource. Only an integration of industrial ecology and economic models (i.e., technical information, material flows, and economic mechanisms) will make it possible to resolve such difficult issues. Reijnders (1998) in addition argues that in striving for a Factor X improvement, achievable values of X will widely vary among economic activities, depending on the specific technological, economic, and institutional conditions prevailing. The relevance of institutions for eco-efficiency is stressed by Bleischwitz (2003). Again, models representing a good integration of economics, technology, and institutions are needed to address these issues.

2.6 Conclusions

This chapter has shown that there is an overlap of interest and expertise between industrial ecology and environmental economics. Each covers a wide range of approaches to studying physical flows through firms, sectors, and the economy as a whole. More important, however, is that their methods and insights are largely complementary. This suggests a need for further linking and integration.

Adopting a physical economy or economics-of-industrial-ecology approach allows policy proposals to be tested for robustness in the widest possible sense. This includes taking account of various dimensions: physical, technical, environmental, and economic. Integrated models with a firm economic basis, usually of a numerical format, provide a good starting point for synthesizing these dimensions, as the complexity of the analysis rapidly increases. A variety of model types are available: partial and general equilibrium models, dynamic input-output models, structural decomposition models, etc. They can best be used so as to generate complementary insights. This reflects a pluralistic approach, which is consistent with the remainder of this book.

Physical flows cover a variety of specific materials and substances, each with unique potential environmental impacts. Especially when addressing larger material flows, materials accounting, using the mass-balance principle, is indispensable for arriving at accurate prediction and scenario analyses that can provide useful information for policymaking. Dynamic analysis of the relationship between growth and material flows suggests a number of critical factors: dematerialization at product and production technology levels; recycling of substances, materials, and products; and changes in the structure of demand and production.

Environmental and ecological economics offer various conceptual frameworks and techniques for studying the physical dimensions of economic systems. They can be linked to ideas coming from the young field of industrial ecology. Thus, the risks of transferring environmental problems in systems, space, and time can be better understood. This can be followed by studying the design of innovative policies, such as those aimed at chain management, cascading of substances and materials, waste mining, substance deposit-refund systems, and producer lifetime responsibility for, and ownership of, final products.

Acknowledgements

We are grateful to Heleen Bartelings and Pieter van Beukering for helpful comments.

Notes

1. Other surveys of economic analysis of material flows are offered by Ruth (1993, 1999) and Kandelaars and van den Bergh (2001).

2. The concept "external effect" or "externality" is part of microeconomic welfare theory. It is defined as an unplanned physical effect, outside the market, of a decision made by one individual on the welfare, health, or production of someone else, without any compensation's taking place. Environmental economics studies, in particular, negative external effects, or "external costs." The welfare theory approach focuses attention on the optimal level of external costs, to be achieved through the implementation of adequate policy instruments (Baumol and Oates 1988).

3. The equivalence of mass and energy is demonstrated by the famous law $E = mc^2$, first formulated by Albert Einstein.

4. Both nuclear fission and fusion processes can be neglected in terms of the mass of matter being transformed.

5. Heijungs (1997) presents a methodological comparison of the various formal methods of analysis commonly used within environmental science, notably material flow analysis, LCA, and I/O modeling.

6. Production functions combined with the mass-balance principle have been formulated by Georgescu-Roegen ([1971b] 1976), Gross and Veendorp (1990), van den Bergh and Nijkamp (1994), Ruth (1995b), and van den Bergh (1999).

7. Numerous publications offer theoretical economic considerations of waste control and recycling. Turner (1995) uses a partial static equilibrium framework. Dynamic models of recycling are studied in Lusky (1975) and van den Bergh and Nijkamp (1994). Dinan (1993) and Fullerton and Kinnaman (1995) study recycling with theoretical general equilibrium models. Recent overviews of economic aspects and models of recycling can be found in McClain (1995), Kandelaars (1999), Kandelaars and van den Bergh (2001), and van Beukering (2001).

8. Dellink and Kandelaars (2000) performed an analysis in which an applied general balance model is combined with a material flow model. A similar type of dynamic model has been developed by Ibenholt (2003), who applied it to Norway.

9. Perrings (1986, 1987) presents a very abstract and general mass-balance I/O approach that covers interactions not only among sectors within the economy, but also between the economy and the environment.

References

Ackerman, F. (1997). *Why Do We Recycle? Markets, Values, and Public Policy*. Washington, DC: Island Press.

Allenby, B. R., and D. A. Richards (eds.). (1994). *The Greening of Industrial Ecosystems*. Washington, DC: National Academy Press.

Andrews, C. J. (2001). Building a micro foundation for industrial ecology. *Journal of Industrial Ecology* 4(3): 35–51.

Arthur, B. (1989). Competing technologies, increasing returns, and lock-in by historical events. *Economic Journal* 99: 116–131.

Axtell, R., C. J. Andrews, and M. J. Small. (2002). Agent-based modeling and industrial ecology. *Journal of Industrial Ecology* 5(4): 10–13.

Ayres, R. U. (1978). *Resources, Environment and Economics: Applications of the Materials/ Energy Balance Principle.* New York: Wiley-Interscience.

Ayres, R. U. (1994). Industrial metabolism. Chap. 1 in R. U. Ayres and U. E. Simonis (eds.), *Industrial Metabolism—Restructuring for Sustainable Development.* Tokyo United: Nations University Press.

Ayres, R. U. (1995). Life cycle analysis: A critique. *Resources, Conservation and Recycling* 14: 199–223.

Ayres, R. U. (1998). Industrial metabolism: Work in progress. In J. C. J. M. van den Bergh and M. W. Hofkes (eds.), *Theory and Implementation of Economic Models for Sustainable Development,* 195–228. Dordrecht, the Netherlands: Kluwer Academic.

Ayres, R. U. (1999a). Materials, economics and the environment. In J. C. J. M. van den Bergh (ed.), *Handbook of Environmental and Resource Economics,* 867–894. Cheltenham, England: Elgar.

Ayres, R. U. (1999b). Industrial metabolism and the grand nutrient cycles. In J. C. J. M. van den Bergh (ed.), *Handbook of Environmental and Resource Economics,* 912–945. Cheltenham, England: Elgar.

Ayres, R. U., and L. W. Ayres. (1996). *Industrial Ecology: Towards Closing the Materials Cycle.* Cheltenham, England: Elgar.

Ayres, R. U., and L. W. Ayres (eds.). (2002). *A Handbook of Industrial Ecology.* Cheltenham, England: Elgar.

Ayres, R. U., and A. V. Kneese. (1969). Production, consumption and externalities. *American Economic Review* 59: 282–297.

Ayres, R. U., and U. E. Simonis (eds.). (1994). *Industrial metabolism: Restructuring for Sustainable Development.* Tokyo: United Nations University Press.

Battjes, J. J. (1999). Dynamic Modelling of Energy Stocks and Flows in the Economy (An Energy Accounting Approach). Ph.D. diss., University of Groningen, Groningen, the Netherlands.

Baumol, W. J., and W. E. Oates. (1988). *The Theory of Environmental Policy.* 2nd ed. Cambridge: Cambridge University Press.

Beckerman, W. (1999). A pro-growth perspective. In J. C. J. M. van den Bergh (ed.), *Handbook of Environmental and Resource Economics,* 622–634. Cheltenham: Elgar.

Berry, R. S., P. Salamon, and G. Heal. (1978). On a relation between economic and thermodynamic optima. *Resources and Energy* 1: 125–137.

Bleischwitz, R. (2003). Cognitive and institutional perspectives of eco-efficiency. *Ecological Economics* 46: 453–467.

Bouman, M., R. Heijungs, E. van der Voet, J. van den Bergh, and G. Huppes. (2000). Material flows and economic models: An analytical comparison of SFA, LCA and partial equilibrium models. *Ecological Economics* 32(2): 195–216.

Byström, S., and L. Lönnstedt. (1995). Waste paper usage and fibre flow in Western Europe. *Resources, Conservation and Recycling* 15: 111–121.

Chertow, M. R. (2000). Industrial symbiosis: Literature and taxonomy. *Annual Review of Energy and the Environment* 25: 313–337.

Choe, C., and I. Fraser. (1999). An economics analysis of household waste management. *Journal of Environmental Economics and Management* 38: 234–246.

Cleveland, C. J., and M. Ruth. (1997). When, where, and by how much do biophysical limits constrain the economic process? A survey of Nicholas Georgescu-Roegen's contribution to ecological economics. *Ecological Economics* 22: 203–223.

Cleveland, C. J., and M. Ruth. (1999). Indicators of dematerialization and the materials intensity of use. *Journal of Industrial Ecology* 2(3): 15–50.

Daly, H. E. (1997). Georgescu-Roegen versus Solow/Stiglitz. *Ecological Economics* 22: 261–266.

Daly, H. E., and A. F. Umaña (eds.). (1981). *Energy, Economics and the Environment.* AAAS Selected Symposia Series. Boulder, Co: Westview.

d'Arge, R. C., and K. C. Kogiku. (1973). Economic growth and the environment. *Review of Economic Studies* 40(1): 61–77.

Dasgupta, P. S., and G. M. Heal. (1979). *Economic Theory and Exhaustible Resources.* Cambridge: Cambridge University Press.

de Bruyn, S. M., and R. J. Heintz. (1999). The environmental Kuznets curve hypothesis. In J. C. J. M. van den Bergh (ed.), *Handbook of Environmental and Resource Economics*, 656–677. Cheltenham, England: Elgar.

de Bruyn, S. M., and J. B. Opschoor. (1997). Developments in the throughput-income relationship: Theoretical and empirical observations. *Ecological Economics* 20: 255–268.

Dellink, R. B., M. Bennis, and H. Verbruggen. (1996). Sustainable Economic Development Structures: Scenarios for Sustainability in the Netherlands. IVM report W96/27, Institute for Environmental Studies, Free University, Amsterdam.

Dellink, R. B., and P. P. A. A. H. Kandelaars. (2000). An empirical analysis of dematerialisation: Application to metal policies in the Netherlands. *Ecological Economics* 33: 205–218.

Desrochers, P. (2002). Cities and industrial symbiosis. *Journal of Industrial Ecology* 5(4): 29–44.

Dinan, T. M. (1993). Economic efficiency effects of alternative policies for reducing waste disposal, *Journal of Environmental Economics and Management* 25: 242–256.

DTO (Duurzame Technologische Ontwikkeling). (1997). *DTO Sleutel Chemie—Zon en Biomassa: Bronnen van de Toekomst* (Sustainable Technological Development: Key Chemistry—Sun and Biomass: Resources for the Future). Interdepartementaal Onderzoeksprogramma Duurzame Technologische Ontwikkeling. The Hague: Ten Hagen & Stam Publishers.

Duchin, F. (1992). Industrial input-output analysis: Implications for industrial ecology. *Proceedings of the National Academy of Sciences* 89: 851–855.

Duchin, F. (1996). Ecological economics: The second stage. Chapter 14 in R. Costanza, O. Segura, and J. Martinez-Alier (eds.), *Getting Down to Earth: Practical Applications of Ecological Economics*, 285–299. Washington, DC: Island Press.

Duchin, F., and G. M. Lange, in association with K. Thonstad and A. Idenburg. (1994). *The Future of the Environment: Ecological Economics and Technical Change.* Oxford: Oxford University Press.

Ehrenfeld, J. R. (1997). Industrial ecology: A framework for product and process design. *Journal of Cleaner Production* 5(1–2): 87–95.

Faber, M., H. Niemes, and G. Stephan. (1987). *Entropy, Environment and Resources: An Essay in Physico-Economics.* Heidelberg: Springer-Verlag.

Ferrara, I. (2003). Differential provision of solid waste collection services in the presence of heterogeneous households. *Environment and Resource Economics* 26(2): 211–226.

Frosch, R. A., W. C. Clark, J. Crawford, T. T. Tschang, and A. Weber. (1996). The Industrial Ecology of Metals: A Reconnaissance. From a lecture delivered at the Royal Society/Royal Academy of Engineering meeting, May 29–30, London. Reprinted in *Philosophical Transactions of the Royal Society* 355: 1335–1347.

Frosch, R. A., and N. E. Gallopoulos. (1989). Strategies for manufacturing. *Scientific American* 261(3): 94–102.

Fullerton, D., and T. C. Kinnaman. (1995). Garbage, recycling, and illicit burning or dumping. *Journal of Environmental Economics and Management* 29: 78–91.

Fullerton, D., and T. C. Kinnaman. (1996). Household demand for garbage and recycling collection with the start of a price per bag. *American Economic Review* 86: 971–984.

Georgescu-Roegen, N. (1971a). *The Entropy Law and the Economic Process.* Cambridge, MA: Harvard University Press.

Georgescu-Roegen, N. ([1971b] 1976). Process analysis and the neoclassical theory of production. In N. Georgescu-Roegen, *Energy and Economic Myths*, 37–52. New York: Pergamon.

Georgescu-Roegen, N. (1976). *Energy and Economic Myths.* New York: Pergamon.

Grace, R., R. K. Turner, and I. Walter. (1978). Secondary materials and international trade. *Journal of Environmental Economics and Management* 5: 172–186.

Graedel, T. E., and B. R. Allenby. (2003). *Industrial Ecology.* 2nd ed. (1st ed., 1995). Upper Saddle River, NJ: Pearson Education, Prentice Hall.

Gross, L. S., and E. C. H. Veendorp. (1990). Growth with exhaustible resources and a materials balance production function. *Natural Resource Modeling* 4: 77–94.

Guinée, J. B. (1995). Development of a Methodology for the Environmental Life-Cycle Assessment of Products—with a Case Study on Margarines. Ph.D. diss., Centre for Environmental Science, Leiden University.

Guinée, J. B., J. C. J. M. van den Bergh, G. Huppes, P. Kandelaars, T. M. Lexmond, S. W. Moolenaar, E. van der Voet, and E. Verkuijlen. (1999). Evaluation of risks of metal flows and accumulation in economy and environment. *Ecological Economics* 30(1): 47–65.

Heijungs, R. (1997). Economic drama and the Environmental Stage: Formal Derivation of Algorithmic Tools for Environmental Analysis and Decision-Support from a Unified Epistemological Principle. Ph.D. diss., Leiden University.

Herman, R., S. A. Ardekani, and J. H. Ausubel. (1989). Dematerialization. In J. H. Ausubel and H. E. Sladovich (eds.), *Technology and Environment*, 50–69. Washington, DC: National Academy Press.

Hoekstra, R. (2003). Structural Change of the Physical Economy: Decomposition Analysis of Physical and Hybrid-Unit Input-Output Tables. Tinbergen Institute Research Series no. 315. Ph.D. diss., Thela Thesis Academic Publishing Services, Amsterdam.

Hoekstra, R., and J. C. J. M. van den Bergh. (2002). Structural decomposition analysis of physical flows in the economy. *Environmental and Resource Economics* 23: 357–378.

Hoekstra, R., and J. C. J. M. van den Bergh. (2003). Comparing structural and index decomposition analysis. *Energy Economics* 25: 39–64.

Hong, S. R., M. Adams, and H. A. Love. (1993). An economic analysis of household recycling of solid waste: The case of Portland, Oregon. *Journal of Environmental Economics and Management* 25: 136–146.

Ibenholt, K. (2003). Material accounting in a macroeconomic framework: Forecast of waste generated in manufacturing industries in Norway. *Environmental and Resource Economics* 26(2): 227–248.

Jänicke, M., M. Mönch, T. Ranneberg, and U. E. Simonis. (1989). *Structural change and environmental impact: Empirical evidence on thirty-one countries in the East and West. Environmental Monitoring and Assessment* 12(2): 99–144.

Janssen, M. A., and J. C. J. M. van den Bergh. (2004). Into the black box of environmental Kuznets curves: Optimal growth and material resource use in two trading countries. *The Annals of Regional Science* 38: 93–112.

Janssen, M. A., and W. Jager. (2002). Stimulating diffusion of green products: Co-evolution between firms and consumers. *Journal of Evolutionary Economics* 12: 283–306.

Kandelaars, P. P. A. A. H. (1999). *Economic Models of Material-Product Chains for Environmental Policy Analysis.* Dordrecht, the Netherlands: Kluwer Academic.

Kandelaars, P. P. A. A. H., and J. C. J. M. van den Bergh. (1996a). Materials-product chain analysis: Theory and an application to zinc and PVC gutters. *Environmental and Resource Economics* 8: 97–118.

Kandelaars, P. P. A. A. H., and J. C. J. M. van den Bergh. (1996b). A dynamic simulation model for materials-product chains: An application to gutters. *Journal of Environmental Systems* 24: 345–371.

Kandelaars, P. P. A. A. H., and J. C. J. M. van den Bergh. (1997). Dynamic analysis of materials-product chains: An application to window frames. *Ecological Economics* 22: 41–61.

Kandelaars, P. P. A. A. H., and J. C. J. M. van den Bergh. (2001). A survey of material flows in economic models. *International Journal of Sustainable Development* 4(3): 282–303.

Kneese, A. V., R. U. Ayres, and R. C. D'Arge. (1970). *Economics and the Environment: A Materials Balance Approach.* Baltimore: Johns Hopkins University Press.

Linderhof, V., P. Kooreman, M. Allers, and D. Wiersma. (2001). Weight-based pricing in the collection of household waste: The Oostzaan case. *Resource and Energy Economics* 23: 359–371.

Lovelock, J. E. (1979). *Gaia: A New Look at Life on Earth.* Oxford: Oxford University Press.

Lusky, R. (1975). Optimal taxation policies for conservation and recycling. *Journal of Economic Theory* 11: 315–328.

McClain, K. T. (1995). Recycling programs. In D. W. Bromley (ed.), *The Handbook of Environmental Economics,* 222–239. Oxford: Blackwell.

Michael, J. A. (1998). Recycling, international trade, and the distribution of pollution: The effect of increased U.S. paper recycling on U.S. import demand for Canadian paper. *Journal of Agricultural and Applied Economics* 30(1): 217–223.

Mirada, M. L., and J. E. Aldy. (1998). Unit pricing of residential municipal solid waste: Lessons from nine case study communities. *Journal of Environmental Management* 52(1): 79–93.

Moll, H. C. (1993). Energy Counts and Materials Matter in Models for Sustainable Development. Ph.D. diss., Groningen, the Netherlands: STYX Publications.

Molenaar, S. W. (1998). Sustainable Management of Heavy Metals in Agro-ecosystems. Ph.D. diss., Wageningen University.

Morris, G. (1994). The economics of household solid waste generation and disposal. *Journal of Environmental Economics and Management* 26: 215–234.

Noorman, K. J. (1995). Exploring Futures From an Energy Perspective: A Natural Capital Accounting Model to Study into the Long-Term Economic Development of the Netherlands. Ph.D diss., University of Groningen.

Noorman, K. J., and A. J. M. Schoot Uiterkamp (eds.). (1998). *Green Households? Domestic Consumers, Environment and Sustainability.* London: Earthscan.

Odum, E. P. (1963). *Ecology.* New York: Holt, Rinehalt and Winston.

Opschoor, J. B. (1994). Chain management in environmental policy: Analytical and evaluative concepts. In J. B. Opschoor and R. K. Turner (eds.), *Economic Incentives and Environmental Policies,* 197–227. Dordrecht, the Netherlands: Kluwer Academic.

Palmer, K., and M. Walls. (1997). Optimal policies for solid waste disposal taxes, subsidies, and standards. *Journal of Public Economics* 65: 193–205.

Peet, J. (1992). *Energy and the Ecological Economics of Sustainability*. Washington, DC: Island Press.

Perrings, C. (1986). Conservation of mass and instability in a dynamic economy-environment system. *Journal of Environmental Economics and Management* 13: 199–211.

Perrings, C. (1987). *Economy and Environment*. New York: Cambridge University Press.

Powell, J. C., R. K. Turner, and I. J. Bateman. (2001). *Waste Management and Planning*. Vol. 5 of Managing the Environment for Sustainable Development. Cheltenham, England: Elgar.

Reijnders, L. (1998). The factor X debate: Setting targets for eco-efficiency. *Journal of Industrial Ecology* 2(1): 13–22.

Rohatgi, P., K. Rohatgi, and R. U. Ayres. (1998). Materials futures: Pollution prevention, recycling, and improved functionality. In R. U. Ayres (ed.), *Eco-Restructuring: Implications for Sustainable Development*, 109–148. Tokyo: United Nations University Press.

Rose, A. (1999). Input-output decomposition analysis of energy and the environment. In J. C. J. M. van den Bergh (ed.), *Handbook of Environmental and Resource Economics*, 1164–1179. Cheltenham, England: Elgar.

Rose, A., and S. Casler. (1996). Input-output structural decomposition analysis: A critical appraisal. *Economic Systems Research* 8: 33–62.

Rose, A., C. Y. Chen, and G. Adams. (1996). Structural Decomposition Analysis of Changes in Material Demand. Working paper, Department of Mineral Resources, Pennsylvania State University.

Rotmans, J., and B. de Vries. (1997). *Perspectives on Global Change: The Targets Approach*. Cambridge: Cambridge University Press.

Ruth, M. (1993). *Integrating Economics, Ecology and Thermodynamics*. Dordrecht, the Netherlands: Kluwer Academic.

Ruth, M. (1995a). Technology Change in US Iron and Steel Production: Implications for Material and Energy Use, and CO_2 Emissions. *Resources Policy* 21(3): 199–214.

Ruth, M. (1995b). Thermodynamic implications for natural resource extraction and technical change in US copper mining. *Environmental and Resource Economics* 6: 187–206.

Ruth, M. (1998). Dematerialization in five US metals sectors: Implications for energy use and CO_2 emissions. *Resources Policy* 24(1): 1–18.

Ruth, M. (1999). Physical principles and environmental economic analysis. In J. C. J. M. van den Bergh (ed.), *Handbook of Environmental and Resource Economics*, 855–866. Cheltenham, England: Elgar.

Ryan, G. J. (1995). Dynamic Physical Analysis of Long-Term Economy-Environment Options. Ph.D. diss., University of Canterbury.

Sager, A. D., and R. A. Frosch. (1997). A pespective on industrial ecology and its application to a metals-industry ecosystem. *Journal of the Cleaner Production* 5: 39–45.

Shinkuma, T. (2003). On the second-best policy of household's waste recycling. *Environmental and Resource Economics* 24: 77–95.

Simon, J., and H. Kahn (eds.). (1984). *The Resourceful Earth: A Response to the Global 2000 Report*. New York: Basil Blackwell.

Slesser, M. (1990). *ECCO: Simulation Software for Assessing National Sustainable Development. Part II, User's Manual*. Edinburgh, Scotland: Resource Use Institute.

Slesser, M., J. King, and D. C. Crane. (1997). The Management of Greed: A Bio-Physical Appraisal of Environment and Economic Potential. Dunblane, Scotland: Resource Use Institute.

Socolow, R., C. Andrews, F. Berkhout, and V. Thomas (eds.). (1994). *Industrial Ecology and Global Change*. Cambridge: Cambridge University Press.

Stahel, W. (1994). The Utilization-Focused Service Economy: Resource Efficiency and Product-Life Extension. In B. R. Allenby and D. J. Richards (eds.), *Greening of Industrial Ecosystems*, 178–190. Washington, DC: National Academy of Engineering.

Starreveld, P. F., and E. C. van Ierland. (1994). Recycling of plastics: A material balance optimisation model. *Environmental and Resource Economics* 4: 251–264.

Stern, D. I. (1997). Limits to substitution and irreversibility in production and consumption: A neoclassical interpretation of ecological economics. *Ecological Economics* 21: 197–215.

Sterner, T., and H. Bartelings. (1999). Household waste management in a Swedish municipality: Determinants of waste disposal, recycling and composting. *Environmental and Resource Economics* 13: 473–491.

Stigliani, W. M., P. Doelman, W. Salomons, R. Schulin, G. R. B. Smidt, and S. W. A. T. M. van der Zee. (1991). Chemical time bomb: Predicting the unpredictable. *Environment* 33: 26–30.

Turner, R. K. (1995). Waste management. In H. Folmer, H. L. Gabel, and J. B. Opschoor, *Principles of Environmental and Resource Economics: A Guide for Students and Decision-Makers*, 440–466. Aldershot, England: Elgar.

van Beukering, P. J. H. (2001). *Recycling, International Trade and the Environment: An Empirical Analysis*. Dordrecht, the Netherlands: Kluwer Academic.

van Beukering, P. J. H., and M. Bouman. (2001). Empirical evidence on recycling and trade of paper and lead in developed and developing countries. *World Development* 29: 1717–1737.

van Beukering, P. J. H., and A. Duraiappah. (1998). The economic and environmental impact of wastepaper trade and recycling in India: A material balance approach. *Journal of Industrial Ecology* 2(2): 23–42.

van Beukering, P. J. M., and M. A. Janssen. (2001). Trade and recycling of used tyres in Western and Eastern Europe. *Resources, Conservation and Recycling* 33: 235–265.

van Beukering, P. J. H., F. A. Spaninks, and F. H. Oosterhuis. (1998). Economic Valuation in Life Cycle Assessment: Applied to Recycling and Solid Waste Management. Report W98/02, Institute for Environmental Studies. Amsterdam.

van Beukering, P. J. H., J. C. J. M. van den Bergh, M. A. Janssen, and H. Verbruggen. (2001). International Material-Product Chains: An Alternative Perspective on International Trade and Trade Theories. Discussion paper TI 2000-034/3, Tinbergen Institute, Amsterdam.

van den Berg, N. W., C. E. Dutilh, and G. Huppes. (1995). *LCA voor beginners—Handleiding milieugerichte levenscyclusanalyse* (Introduction to LCA: A tutorial on environmentally oriented life-cycle analysis). Centre for Environmental Science, Leiden University, Leiden, the Netherlands.

van den Bergh, J. C. J. M. (1993). A framework for modelling economy-environment-development relationships based on dynamic carrying capacity, materials balance and sustainable development feedback. *Environmental and Resource Economics* 3: 395–412.

van den Bergh, J. C. J. M. (1999). Materials, capital, direct/indirect substitution, and materials balance production functions. *Land Economics* 75(4): 547–561.

van den Bergh, J. C. J. M., and R. de Mooij. (1999). An assessment of the growth debate. In J. C. J. M. van den Bergh (ed.), *Handbook of Environmental and Resource Economics*, 643–655. Cheltenham, England: Elgar.

van den Bergh, J. C. J. M., A. Ferrer-i-Carbonell, and G. Munda. (2000). Alternative models of individual behaviour and implications for environmental policy. *Ecological Economics* 32(1): 43–61.

van den Bergh, J. C. J. M., and J. M. Gowdy. (2000). Evolutionary theories in environmental and resource economics: Approaches and applications. *Environmental and Resource Economics* 17(1): 37–57.

van den Bergh, J. C. J. M., and P. Nijkamp. (1994). Dynamic macro modelling and materials balance. *Economic Modelling* 11: 283–307.

van der Voet, E. (1996). Substances from Cradle to Grave. Ph.D. diss., Moolenaarsgraaf: Optima Druk.

van der Voet, E., J. B. Guinée, and H. A. Udo de Haes (eds.). (1999). *Heavy Metals: A Problem Solved? Methods and Models to Evaluate Policy Strategies for Heavy Metals.* Dordrecht, the Netherlands: Kluwer Academic.

Vitousek, P. M., H. A. Mooney, J. Lubchenco, and J. M. Melillo. (1997). Human domination of Earth's ecosystems. *Science* 277: 494–499.

von Weizsäcker, E., A. B. Lovins, and L. H. Lovins. (1997). *Factor Four: Doubling Wealth—Halving Resource Use. A Report to the Club of Rome.* London: Earthscan.

WCED (World Commission on Environment and Development). (1987). *Our Common Future.* Oxford and New York: Oxford University Press.

Weaver, P. M., H. Landis Gabel, J. M. Bloemhof-Ruwaard, and L. N. van Wassenhove. (1997). Optimising environmental product life cycles: A case study of the European pulp and paper sector. FEEM Paper (Nota di Lavoro) 88.97, Fondazione Eni Enrico Mattei, Milaan.

Wernick, I. K., and J. H. Ausubel. (1995). National materials flows and the environment. *Annual Review of Energy and Environment* 20: 493–573.

Wertz, K. L. (1976). Economic factors influencing household production of refuse. *Journal of Environmental Economics and Management* 2: 263–272.

Wier, M., and B. Hasler. (1999). Accounting for nitrogen in Denmark: A structural decomposition analysis. *Ecological Economics* 30: 317–331.

WRR (Wetenschappelijke Raad voor het Regeringsbeleid [Netherlands Council for Government Policy]). (1987). Ruimte voor Groei [Space for Growth]. Rapporten aan de regering 1987/29, Staatsuitgeverij, Den Haag.

II Historical Analysis of Structural Change

3 Is the U.S. Economy Dematerializing? Main Indicators and Drivers

Robert U. Ayres, Leslie W. Ayres, and Benjamin Warr

3.1 Background

Dematerialization is a popular environmental slogan these days. The notion that dematerialization is desirable follows directly from the idea that large mass flows are environmentally harmful as such (Schmidt-Bleek 1993a, 1993b, 1994; Adriaanse et al. 1997). It would follow almost automatically that less harm must follow from reduced mass flow. We argue that this is a misleading and essentially false proposition.

There is a widespread impression, even among sophisticated economists, that the economy of industrialized countries is well advanced along an inevitable process of dematerialization. Many examples have been presented to demonstrate the supposed trend toward dematerialization in industrial societies. The primary example, of course, is the computer chip. Undoubtedly the size and weight of computer chips, and products containing them, from radios and portable telephones to computers, are much less massive—and the products use less electric power—than their less powerful counterparts ten, twenty, or thirty years ago. We will have more to say about this subsequently.

To a lesser extent, the same trend can be observed in a variety of other products. Bicycles made from aluminum or carbon-fiber frames are lighter than the older steel-framed units. Refrigerators are smaller and lighter, thanks to improved thermal insulation. Automobiles became lighter from 1975 to 1990, thanks to the CAFE (corporate average fuel economy) standards (but prior to the great popularity of sport utility vehicles [SUVs]). Cans are thinner and lighter. Cold-weather clothing is lighter than in the past. Many products formerly made of metal are now made of plastic. And so forth.

Consider the following two quotes:

Fiber-optics has replaced huge tonnages of copper wire, and advances in archi-
tectural and engineering design have made possible the construction of buildings
with much greater floor space but significantly less physical material than the
buildings erected just after World War II. *Accordingly, while the weight of current
economic output is probably only modestly higher than it was a half century ago,
value added, adjusted for price change, has risen well over threefold.* (Alan Greenspan, from speech at The Conference Board, NY, October 16, 1996; quoted in
Cairncross 1997, 212; italics ours)

or (based on the above)

America's output, measured in tons, remains about as heavy as it was a century
ago, even though real GDP, measured in value, is twenty times greater. The main
reason for this striking shift from material goods to intangibles, described in a
speech in 1996 by Alan Greenspan, chairman of the Federal Reserve Board, can
be identified as the rising proportion of total cost of the 'knowledge' content of
goods and services, relative to materials and energy. (Cairncross 1997, 212)

In fact, neither of these two eminent authorities has done sufficient home-
work. However, that may be forgivable: The subject is much more complicated than it appears to be at first glance.[1] Some of the complexities
are discussed in the next section. Subsequently we present the historical
data on mass flows from three different perspectives: namely, mass per
capita and exergy per capita, then mass and exergy per unit of GDP, then
embodied exergy per unit of mass. The material flows themselves are
considered in five groupings: namely, fossil fuels, metals, agricultural
products, construction materials, and chemicals. Finally, we present some
interesting conclusions.

3.2   Conventions and Complications

Since GDP is a measure of economic activity in monetary units, it is lit-
erally impossible to weigh the GDP, as such. However, it is possible to
do some other things. For instance, one might weigh the "final" products
sold to end users. However, this is not as simple as it might seem at first
glance. Does the gasoline purchased by a farmer for his pickup truck
count as an end use or an intermediate use? Is a commuter's rail pass an
end use? What about a business suit? There are many self-employed persons and corresponding ambiguities, and no statistics on the allocation of

physical flows between intermediate and end uses. (We return to the discussion of final versus intermediate products later.)

It might be easier to weigh raw material inputs. But here, too, we need to decide what is meant by a "raw material." The usual definition refers to primary commodities exchanged in markets (but obviously not all commodities). However, this definition would exclude firewood self-gathered by the end user, as well as vegetation grazed by sheep or cows and manure produced by those animals. (In India these would be major omissions.)

If air and water are included as raw materials, as they should be from a purely physical perspective, the mass involved will be far larger than that of all other inputs combined. But figures for air and water are unavailable in official statistics and would, in any case, be ambiguous. We would need to decide whether to include irrigation water and cooling water among these inputs. We would also need to decide whether atmospheric nitrogen that does not participate in the combustion process should be counted as an input. In practice, these numbers on nitrogen flows can be estimated roughly, but only for recent years, and probably not more precisely than plus or minus 50 percent.

A more reasonable convention might be to include oxygen from the air insofar as it participates in combustion processes and animal respiration, but to subtract oxygen generated by photosynthesis in agriculture and forestry. The same convention, applied to air, would rule out nitrogen that is nonreactive. (Only a tiny percentage of atmospheric nitrogen throughput is "fixed" by natural biological processes or combustion processes, and this fraction can generally be neglected in a large-scale mass balance.) Similarly, we might try to include water only when it participates actively in a reaction, as in photosynthesis or when it combines with plaster of paris or Portland cement. But then water driven off by dehydration processes should be subtracted. Water that participates *only* as a passive carrier of heat or a diluent of waste should probably not be counted, if only because there are no good statistics on it. This reasoning suggests that air and water should both be excluded from consideration.[2]

This convention would also exclude some of the "hidden flows," such as earth—topsoil or subsoil—that is simply moved from one place to another by road-building or construction projects or surface erosion. These flows have been given considerable weight by Schmidt-Bleek and others (Schmidt-Bleek 1993a, 1993b, 1994; Adriaanse et al. 1997). We think that displacement alone is not a sufficient argument for inclusion, although (if data were available) an argument could be made for counting erosion because of its double impact on agricultural land, on the one hand, and on siltation of lakes, river valleys, and estuaries, on the other. Despite the poor data, the first major international study of mass flows did include erosion (Adriaanse et al. 1997). However, an argument for not including it is that there is no corresponding inflow, and hence no way to use the mass-balance condition for verification.

The case of material displaced by surface mining (overburden) is not quite so simple, however, since mine waste normally contains some chemically active minerals, such as pyrites that can oxidize when exposed to air and water. Although the concentration of these contaminant minerals is low, the quantities are vast. Moreover, they are commonly piled in upland valleys where they result in acidification and mobilization of toxic metals and drastically alter topography and stream flows. It seems defensible to include mine overburden both as part of an input to the economic system and as an unwanted output as well.

If air and water are excluded, but mining overburden, other minerals (like sand, gravel, clay, and limestone), and crude metal ores are still included, one obtains a smaller but more justifiable set of numbers. A further convention is needed to decide how to deal with crude ores associated with imported metals. For instance, a hundred tons of platinum used each year in the United States for industrial and automotive catalysts requires that a hundred million tons of crude ore (more or less) must be dug up and processed in South Africa. In principle, we think that this crude ore should probably be assigned to the consuming country.

The materials embodied in imported products, especially automobiles and electronic goods, raise a similar question. In principle, we think that not only the material content of the finished goods but the indirect material flows associated with their manufacture in other countries should be assigned to the consuming country. On the other hand, revenues earned by manufacturing goods for export are likely to be spent (on average, over time) on consumable imports. It is probably unavoidable, if theoretically unjustified, to assume that these inflows and outflows compensate, at least roughly, for one another. Ad hoc corrections might be possible in specific cases, such as South Africa, Chile, and some other African countries, where the indirect flows associated with exported mineral products, such as diamonds, gold, and copper concentrates, vastly exceed the indirect flows associated with imports. A detailed analysis based on international trade data would obviously complicate the analysis enormously and probably be beyond any practical possibility of implementation in the near future.

There is another set of ambiguities, especially related to agriculture. For instance, should we include the grass consumed by cattle among the inputs? How should we treat crop residues and manure that are recycled to the soil? How should we treat manure that is not recycled? The most consistent rule, albeit not the easiest one to apply, is to draw the line at photosynthetic products harvested directly by humans or harvested by animals whose products are consumed by humans. By this rule, plant parts that are physically removed, either by human farmers or by grazing animals, are counted as "inputs," regardless of what happens to them subsequently. On the other hand, plant parts remaining in or on the ground (roots, fallen leaves) are not counted. According to this rule, all plant parts that are harvested but not utilized, as well as animal wastes, are "waste outputs."

We cannot state a general, all-encompassing rule for deciding what is in and what is out, but the two primary criteria in making the determination are common sense and data availability.

3.3   Exergy as an Alternative Measure

Exergy is an unfamiliar term, except to chemists or physicists. But it is really what nontechnical people usually mean when they speak of *energy*. When people speak of energy consumption or energy production, it is really exergy that they mean. Notice that exergy can be used up, whereas energy is always conserved. The exergy embodied in a fuel can be equated approximately to the *heat of combustion* of that fuel.

However, exergy does have a formal definition in thermodynamics that is somewhat broader. Exergy is defined as the *maximum* amount of work that can be extracted from a subsystem as it approaches equilibrium reversibly with its surroundings. Thus a substance with little or no exergy content is almost indistinguishable from its surroundings, and conversely.

Combustion is an example of *chemical* exergy in which a substance reacts with oxygen rapidly and generates combustion products that subsequently diffuse and thus equilibrate with the atmosphere. Combustion processes generate heat, which can do work via a Carnot cycle heat engine. Of course, oxidation need not be rapid. Rusting of iron is an example of slow oxidation. Heat is generated, but so slowly that it is not noticeable. But iron (like most other metals) will burn rapidly and liberate heat rapidly at a high enough temperature. Similarly, the respiration process in animals is another form of oxidation (which is why the energy—actually exergy—content of food is expressed in Calories). There are several other kinds of exergy, including physical exergy (kinetic energy) and thermal exergy (heat). However, for our purposes, only chemical exergy need be considered.

There are some economically important processes that are essentially the reverse of combustion, in the sense that chemical exergy is concentrated and embodied in a target substance. Photosynthesis is an example. Carbo-thermic reduction of metal ores and ammonia synthesis are other examples. In the first case, exergy from the sun is captured and embodied in carbohydrates, which are combustible chemical substances. In the metals case, a metal oxide in contact with red hot carbon is converted into a pure metal plus carbon dioxide. The latter reaction is disguised combustion. In the ammonia case, natural gas plus air is converted to ammonia plus carbon dioxide by a series of catalytic processes at high temperatures and pressures, which also amount to disguised combustion.

However, there are other processes that can do work (in principle). So when salt is dissolved in water, some heat is generated, and work could be done using that heat if it were not rapidly diffused away. Desalination is the reverse of this diffusion process, and quite a lot of heat is required for the purpose. It follows that any useful material that is present in concentrations above the average in the air (if it is a gas) or the ocean (if it is soluble) or the earth's crust (if it is neither a gas or soluble) also embodies some exergy. Thus, pure rainwater contains some exergy as compared with seawater, which has zero exergy by definition. Pure salt also contains some exergy for the same reason. Similarly pure oxygen or pure nitrogen contain some exergy, whereas the mixture that is air has zero exergy content, by definition. Finally, mine overburden has little or no exergy if it is chemically indistinguishable from the surrounding earth or rock.

The point of this rather long explanation is that all natural raw materials contain some exergy, insofar as they are chemically different from the air, water, or earth, depending on whether the end product of all chemical reactions is a gas, a liquid or soluble solid, or an insoluble solid. By this test, the exergy embodied in a ton of sand is extremely small compared with the exergy embodied in a ton of coal. In effect, exergy is a measure of distance from equilibrium. This means that some chemical exergy is embodied in all material wastes. But of course the most dangerous wastes, environmentally speaking, are the ones with the highest exergy content, meaning that they are chemically reactive.

Fuels, hydropower, nuclear heat, and products of photosynthesis (phytomass) are the major sources of exergy input into the economy. Most other materials have very little exergy in their original form but gain exergy from fuels, as in metal reduction or ammonia synthesis. Nevertheless, the exergy content of materials is an interesting measure, especially in contrast to the traditional measure (mass).

3.4 Mass Flows in the U.S. Economy, 1993

Based roughly on the above conventions, a summary of the major mass flows in the U.S. economy for 1993 is shown in figure 3.1.[3] (We also included overburden and erosion, in this case, since estimates of these mass flows were available.) The mass-balance principle was used to estimate a number of flows that could not be measured directly. In particular, we used the mass balance to calculate the amount of oxygen generated by photosynthesis in agriculture and forestry, the amount of atmospheric oxygen required to burn all the fossil fuels (and wood) and the amount of water vapor generated by the combustion process. (We used official estimates of carbon dioxide production from fuel combustion and calculated the others as ratios, based on chemical reaction formulas.)

It goes without saying that "useful" outputs weigh a lot less than inputs. The difference constitutes production-related waste flows. But, the word "useful" is also ambiguous. In economic terms, useful outputs would presumably be those with a well-defined market and market price. In general, lots of outputs are inputs for other downstream products. Yet some of the physical outputs of the system—as defined by the foregoing conventions—must clearly be useful without having market prices. An obvious example of this would be forage and silage fed to animals on the farm. These are unpriced, but not unvalued, intermediates. Manure generated and recycled by grazing animals on the farm is another obvious example; it is unpriced, but it would clearly be inappropriate to regard it as a waste. (In India this material is harvested, dried, and used as domestic fuel.) On the other hand, animal manure generated in large industrialized feedlots *is* a waste. Finally, oxygen—a by-product of photosynthesis—is clearly valuable, albeit unpriced. The same is true of water vapor.

Raw agricultural products harvested in the United States in 1993 amounted to 868 million metric tons (MMT), of which 457 MMT was crops and the rest was silage, hay, and grass. Of this, 83 MMT (net) was exported, mostly for animal feed. Animal products amounted to 119.5 MMT. The food-processing sector converted 374 MMT of harvested inputs (dry weight) to 286 MMT of salable products, of which 203 MMT was food consumed by Americans, 66 MMT was by-products, animal feeds, and food exports, and 14 MMT was a variety of nonfood

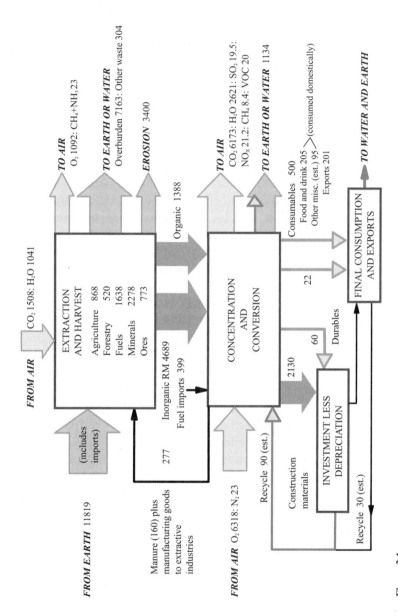

Figure 3.1
The U.S. economic system as a whole from a mass flow perspective, 1993 (in millions of metric tons).

products including natural fibers, leather, tobacco, and alcohol. Evidently 500 MMT, more or less, was "lost" en route to the consumers, mostly as water vapor and CO_2, though other wastes were also significant (Ayres and Ayres 1998).

Consider forest products. Inputs (raw wood harvested) amounted to 520 MMT in 1993, not counting timber residues left in the forests (about 145 MMT). About 200 MMT of this weight was moisture. Finished dry wood products (lumber, plywood, particle board) weighed about 61 MMT. Finished paper products amounted to 83 MMT, which included some paper made from imported wood pulp from Canada and some recycled wastepaper. The output weight also included 3.7 MMT of fillers and other chemicals embodied in the paper. Again, the difference between input and output weights was very large. Quite a lot of the difference was lignin wastes from the paper mills, which are burned on site for energy recovery, but some of which still end up as pollution. A great deal of harvested wood (about 168 MMT, including paper mill wastes) was burned as fuel, producing about 230 MMT of CO_2 as a by-product. By our convention, this must be counted as part of the weight of outputs (Ayres and Ayres 1998).

Economists have tried to finesse these awkward complexities by creating a classification known as "final" goods, namely, goods sold to "final" consumers in markets. This class of goods is reasonably well-defined. It is, presumably, what Greenspan and Cairncross had in mind when they spoke of "the weight" of the GNP. However, even so-called final goods (except for food, beverages, and medicinals) are not physically consumed. They are, in a sense, producers of immaterial metabolic services. By this test, *all final outputs (not excepting food and beverages) are immaterial services and therefore weightless*.[4] However, Greenspan and Cairncross were clearly talking about finished products, not ultimate services provided by those products. Products do have mass, as well as monetary value (counted in the GNP). The question, which we mentioned briefly at the beginning of section 3.2, is: What products should be counted as "final" and "useful" for purposes of weighing the GNP?

It might seem at first sight that intermediate products should also be weighed, inasmuch as they have economic value. However, on reflection, the value of intermediate products is already included in the value of the final goods to which they contribute. Similarly, the mass of final goods intermediates incorporates (part of) the mass of intermediates. The difference is the mass of wastes generated by the final stage of production. It can be argued that the mass of wastes is also, in some sense, part of the mass of the GNP, as produced, on the grounds that the wastes are direct consequences of the production process. However, we insist that the useful materials should be separated conceptually (and in the statistics) from the wastes. Nonetheless, we do agree with one of this volume's reviewer that with this convention, the mass of the final products does not really correspond to the GDP and should not be defined as the mass of the GDP.

As a practical matter, it seems reasonable to start by counting the weight of finished materials, which is to say materials that are embodied in products, or otherwise used, without further chemical transformation. Steel is an example. There is relatively little difference between the weight of raw steel produced (89 MMT in the United States in 1993 [Ayres and Ayres 1998]) and the weight of "finished" steel products. The small losses of steel in the rolling, casting, and machining stages of production are almost entirely recycled within the steel industry.[5] The same can be said of other "finished materials," from paper and plastics to glass and Portland cement: Very little or none of the finished material is lost after the last stage of production, except as consumption or demolition wastes.

What of fuels and intermediate goods like ammonia, caustic soda, chlorine, and sulfuric acid? Raw fuels are refined, of course, with some losses (such as ash) and some fuel consumption (around 10 percent) to drive the refineries. But refined fuels are converted, in the course of use, mainly to heat, mechanical power, and combustion wastes. Fuels cannot be recycled. The mass of raw hydrocarbon fuel *inputs* was a little over 1,600 MMT in 1993 (Ayres and Ayres 1998). It was mostly combined with atmospheric oxygen. The combustion of hydrocarbon fuels in the United States, in 1993, generated around 5,200 MMT of CO_2, the most important greenhouse gas (OECD/IEA 1995, 39). This may be a slight underestimate, since some of the hydrocarbons produced by refineries do not oxidize immediately (asphalt and plastics, for instance), but except for what is buried in landfills, all hydrocarbons oxidize eventually.

Minerals such as salt, soda ash, and phosphate rock, as well as petrochemical feedstocks, are converted into other chemicals. Some of these (mainly polymers) end in finished goods (like tires, carpets, packaging materials, and pipes). Others are converted into wastes in the course of use. Examples include acids and alkalis, cleaning agents, detergents and solvents, pesticides, and fertilizers. A rational accounting system should distinguish between dissipative intermediates such as these and nondissipative materials embodied in finished durable goods that might (in principle) be repaired, reused, or remanufactured and thus kept in service for a longer period.

Another important distinction must be made, namely, one between "potentially reactive" and "inherently inert" materials. Most metals, paper, plastics, and so on are in the "reactive" category, insofar as they can oxidize or react with other environmental components. (Most of these, especially paper and plastics, can be burned for energy recovery). However, as a practical matter, these energy-rich materials are considerably outweighed by the inert materials utilized in structures, such as glass, brick and tile, concrete, plaster, gravel, and stone. All of the latter group of materials are chemically inert, even though some of the processes for manufacturing them involve heating.[6]

Portland cement (made from limestone, clay, and shale) is the single most important component of construction materials in terms of energy inputs and dollar value, but not in terms of weight (81.7 MMT) (U.S. Bureau of Mines 1993, "Cement," table 1). Calcined gypsum products consumed in the United States in 1993 amounted to 15.2 MMT (U.S. Bureau of Mines 1993, "Gypsum," table 1). Brick and tile production accounted for 12.3 MMT of clay in 1993. Refractory products accounted for an additional 2.5 MMT of clay (U.S. Bureau of Mines 1993, "Clay"). Glass production in the United States was approximately 14 MMT, of which only 4 MMT was flat (window) glass (U.S. Bureau of Mines 1993, "Soda Ash," figure 1). The total mass of thermally processed building materials consumed in the United States in 1993 was 125 MMT.

On the other hand, inert materials such as construction sand and gravel consumed in the United States in 1993 (mainly aggregate for concrete) amounted to 870 MMT (U.S. Bureau of Mines 1993, "Sand and Gravel," table 1), and the total quantity of crushed stone—including other sand and gravel—produced and consumed was 1,120 MMT (U.S. Bureau of Mines 1993, "Crushed Stone," table 1). Of this, about 110 MMT of raw limestone and dolomite was used as raw material in cement and quicklime (CaO) manufacturing. The rest of the sand, gravel, and stone, including 1.2 MMT of "dimension stone" (such as marble and granite) was used as such, without further processing. Altogether, chemically inert structural materials consumed in the United States in 1993 without thermal processing amounted to about 1,870 MMT.

In contrast, the weight of all metals produced (and consumed) in the United States in 1993 was less than 100 MMT. By far the greater part, especially of steel, was used for construction purposes or motor vehicles.

The biggest consumer of copper (mostly as wire) is the construction industry, which accounted for 1.03 MMT in 1993; the motor vehicle industry consumed 270,000 metric tons (MT), and consumer durables (home appliances) used 250,000 MT. The rest of the metals were mainly embodied in durable goods such as transportation equipment and other machines and appliances.

The total weight of materials consumed in the production of motor vehicles in 1993 in the United States amounted to about 27.8 MMT. Exact figures for other categories are difficult to obtain, however. The weight of all machinery, including electrical machinery and electronics, produced in the United States can only be estimated on the basis of inputs (which themselves are approximate). We can account for about 12.3 MMT, mostly steel plus some aluminum; copper accounts for only 0.32 MMT.

Against these numbers, the weight of other consumer products is modest. For example, the weight of all textiles produced, including cotton, wool and all synthetics, amounts to around 5 MMT. Products of textiles, partly clothing and partly furnishings (including carpets), must be of the same order of magnitude.

3.5   Factors Driving or Inhibiting Dematerialization

It is appropriate now to consider changes over time. Factors driving materials consumption per capita trends include declining ore grades (for metals), increasing efficiency of discovery, separation and recovery of valuable materials from ores, improved material properties (such as strength at high temperatures), and design improvements (such as miniaturization and longer lifetimes). Other factors inhibiting dematerialization include increasing purification requirements, shorter product life, and greater complexity of products, especially as it affects recycling.

Declining ore grades obviously generate more process waste (as well as overburden), other factors remaining equal. It might seem that this would lead to higher prices and thus, less use. However, so far, ore grades have decreased, but prices have fallen, not risen (Barnett and Morse 1962; Barnett 1979). Increased efficiency of discovery and process technology probably account for this. In fact, a major message of this chapter is that falling prices (together with rising personal income) have stimulated rising demand for most material products.

The history of iron and steel production over the past two centuries or more is a good illustration. In the mid-eighteenth century, iron was reduced from ore (an oxide) by contact with red-hot charcoal, derived

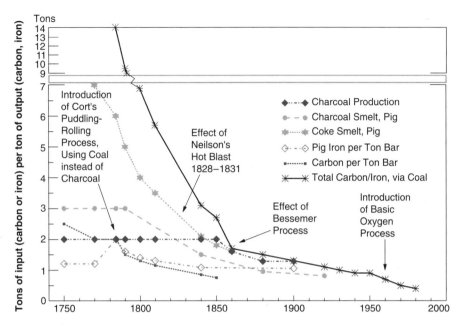

Figure 3.2
Iron and steel production (carbon and iron input per ton of carbon and iron output).

from controlled partial oxidation of wood. About three tons of charcoal were needed to produce a ton of iron, and as much as five tons of carbon in wood were required per ton of charcoal used. When coal first replaced wood, the coal/iron ratio was about eight tons/ton, as a result of inefficient coking and low-temperature blast furnaces. The introduction of air preheating (hot blast) in the 1830s cut the fuel requirements threefold. Today, the average carbon/iron ratio for the industry is less than 0.6 tons of carbon (as coke) per ton of iron, which is approaching the theoretical limit of 0.44 tons/ton (figure 3.2). The quality of the pig iron produced in 1800 was essentially the same as that of the pig iron produced today, but the quantity of materials processed per ton of pig iron was vastly greater, as was the cost.

As costs fell, over time, the demand for iron and steel, per capita, grew even faster, largely as a result of new uses. This is a prime example of what is called the "rebound effect." Iron and steel consumption per capita rose, in mass terms (discontinuously, to be sure) from the beginning of the Industrial Revolution until the 1950s and did not decline significantly until the late 1970s (figure 3.3). Aluminum is another example. Electric

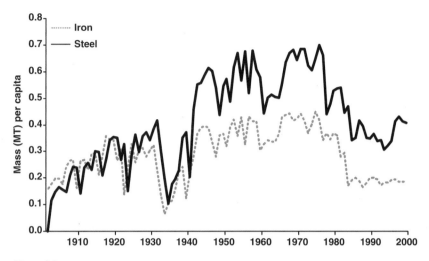

Figure 3.3
Iron and steel consumption (mass per capita), United States, 1900–1998.

power requirements for electrolytic smelting have fallen by a factor of nearly four since the Hall-Heroult process was introduced in 1880 and by a factor of two since 1900. Meanwhile, per capita consumption grew steadily until around 1980 and has declined only slightly since then (figure 3.4).

Copper offers another case in point, complicated by significantly declining ore grades, as shown in figure 3.5. Nevertheless, prices have fallen in real terms since the nineteenth century, albeit very irregularly. From 1870 through 1918, copper prices averaged 50 cents/lb and ranged between 40 and 60 cents/lb (in 1967 dollars). From 1919 through 1952 the average price was about 25 cents/lb and ranged between roughly 20 cents and 30 cents/lb, except for a peak in 1929 and a trough in 1932. From 1932 to 1970, copper prices increased fairly steadily, from 15 cents/lb to 50 cents/lb, with an intervening peak in 1955–1956. Since 1970, prices have declined again quite sharply to below 25 cents/lb in 1984–1986. There was another surge in the late 1980s and early 1990s, followed by an even lower bottom in 1998–2000, with prices similar to the those in the mid-1930s. The long-term trend appears to be down, although there was a long period (1920 through 1970) of generally rising prices (figure 3.6). Meanwhile, consumption per capita has increased fourfold since 1900, ignoring wartime peaks, despite periods of rising prices, driven by steady electrification of the economy.

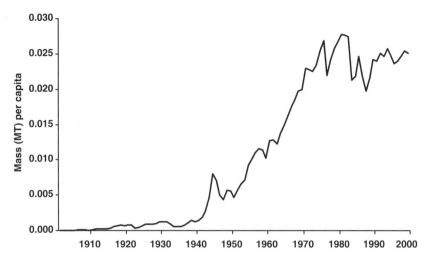

Figure 3.4
Aluminum consumption (mass per capita), United States, 1900–1998.

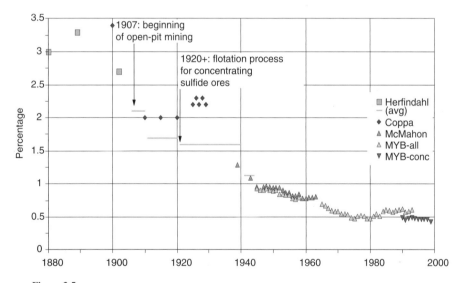

Figure 3.5
U.S. copper ore grade percentage, 1880–2000.
Sources: U.S. Bureau of Mines (MYB) (various years); McMahon (1965); Herfindahl (1967); Coppa (1984).

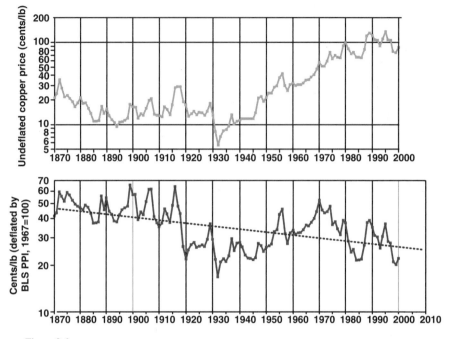

Figure 3.6
Price of copper on the New York market, 1870–2000 (cents per pound in current and constant 1987 U.S. dollars). BLS PPI: Bureau of Labor Statistics Producer Price Index.

Other examples of dramatically increasing efficiency of chemical synthesis processes are illustrated in figure 3.7. However, thanks to price
declines, per capita consumption of the major industrial chemical species,
as a group, increased quite steadily until 1969 or so and has since fluctuated but increased slightly in aggregate terms (figure 3.8).

In the case of fossil fuels, significant gains in thermodynamic (as opposed to economic) efficiency on the production and processing side are
largely confined to petroleum refining, as indicated by figure 3.9, which
plots the historical trends in crude oil cracking and refining technology.
The net result has been increased utilization of crude oil for purposes of
propulsion versus declining usage for other purposes such as illumination
(kerosine lamps), cooking, and space heating. Use for lubrication and
petrochemicals has also increased, although use for these purposes is not
plotted in the graph.

Per capita fuel usage is also affected by efficiency gains in energy services (work) performed, including those resulting from use of steam
engines and turbines (for electricity production) and internal combustion

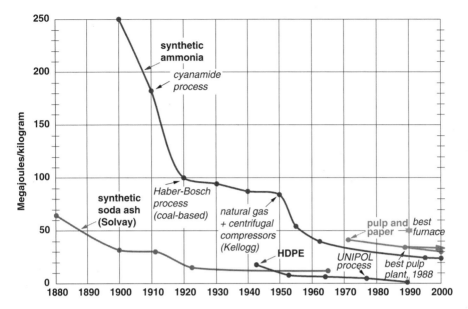

Figure 3.7
Exergy consumption by industrial processes, United States, 1880–2000. HDPE: high-density polyethylene.

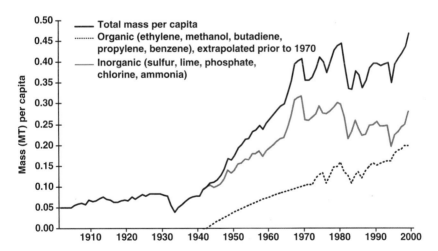

Figure 3.8
Organic and inorganic chemical consumption (mass per capita), United States, 1900–1998.

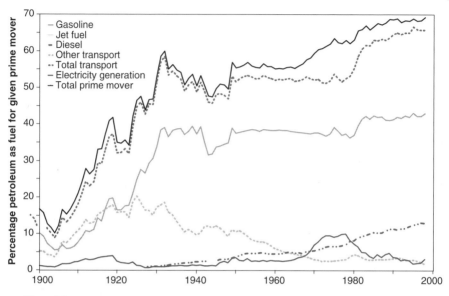

Figure 3.9
Petroleum utilization efficiency (percentage as fuel for prime movers).

engines for vehicular propulsion as well as heat utilization (figure 3.10). Nevertheless, despite efficiency gains, per capita fossil fuel consumption in mass terms has more than doubled since 1900 (see right-hand scale in figure).

As regards inert construction materials, including cement, gypsum, clay, stone, sand, and gravel, thermodynamic efficiency gains on the construction side have been minor and largely limited to the cement industry. Gains—if any—on the consumption (end use) side are difficult to measure, especially since there have been significant changes in the mix of materials (more steel, glass, and plastics, less brick and plaster) and more prefabrication. Wood products are now used more efficiently than in 1900, as plywood and particle board have replaced simple lumber for many purposes such as siding. Nevertheless, the mass of construction materials consumed per capita rose nearly sixfold from 1900 to its postwar peak in the mid-1960s. The bulge of construction material was a specific consequence of President Dwight D. Eisenhower's forty-one-thousand-mile interstate highway program. The program, created in 1956, consumed an enormous amount of concrete and aggregate ("enough to build 80 Hoover dams or six sidewalks to the moon" [Yergin 1991, p. 553]). The program was substantially complete by 1970. Yet there has been only

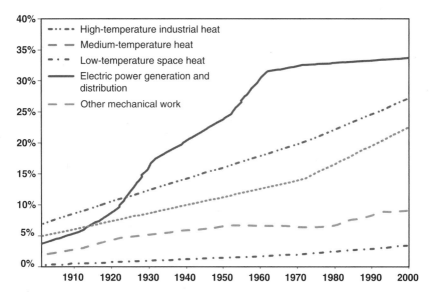

Figure 3.10
Energy (exergy) conversion efficiencies, United States, 1900–1998.

a modest (less than 20 percent) decline in consumption of construction materials since then.

To complete the picture, gross biomass consumption per capita declined significantly (about 40 percent) from 1900 to its low point in the middle of the Depression (figure 3.11). This probably reflects the declining use of fuelwood for home heating and cooking and less need for feed for horses and mules on farms, resulting from the increasing use of tractors after the 1930s. Since then, biomass consumption per capita has trended slightly upward, reflecting increasing meat consumption, though it remains less than the 1900 level.

Although generalizations based on incomplete data are always open to question, it seems fairly clear that the most important factor opposing dematerialization is demand for new products and services, induced by declining prices and rising incomes. In fact, efficiency gains, in general, have been more than compensated for by increasing demand (the "rebound effect").

Improvements in materials properties, such as strength/weight ratios, have resulted in some dematerialization. Figure 3.12 indicates the extent of technical improvements, for the strongest materials, since historical times. However, the very strong new lightweight materials, notably

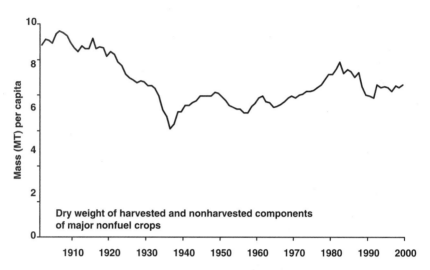

Figure 3.11
Biomass consumption (mass per capita), United States, 1900–1998.

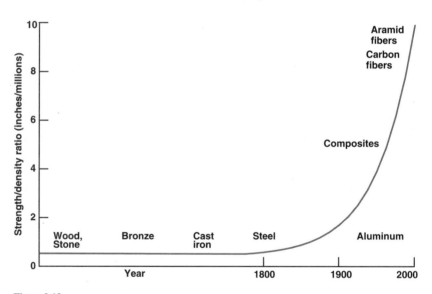

Figure 3.12
Progress in materials strength/density ratio (showing a fifty-fold increase).
Source: National Research Council, National Academy of Sciences (1989).

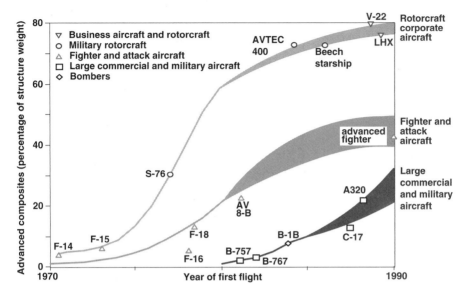

Figure 3.13
Timetable for use of composites in aircraft.

aluminum, titanium, and their composites, have found uses mainly in aircraft (figure 3.13) and for extremely limited-consumption items such as tennis rackets and golf clubs. Power/weight ratios of the most advanced prime movers have increased by four orders of magnitude since 1800 and by tenfold since 1900 (figure 3.14). However, these gains have resulted mainly from design innovations (such as the diesel engine and the gas turbine) made possible by the availability of better alloys and precision manufacturing techniques. They have made bigger-power units possible and contributed mainly to increased use of electricity generation and propulsion power, as already indicated.

Similarly, very dramatic improvements in telecommunications technology increased telecommunications channel capacity by at least eight orders of magnitude from 1800 to 1980 (Martin 1971), when glass fibers began to replace copper wires. Channel capacity has increased by several more orders since then. The mass of a single equivalent telephone circuit has fallen from 100 kg/km in 1914 to around .005 kg/km now (figure 3.15).

The substitution glass optical fibers for copper cannot be detected in the aggregate data. The weight of refined copper consumed in the United States in 1993 was about 1.8 MMT, of which just 504,000 MT

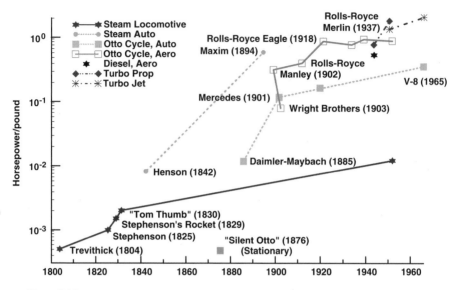

Figure 3.14
Power/weight ratios (mobile).

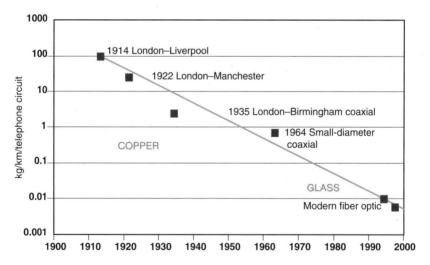

Figure 3.15
Reducing material consumption in long-distance telecommunication cables.

(0.5 MMT) was destined for wire manufacturing (Ayres and Ayres 1998). Thus, although Greenspan may be right that fiber optics have replaced "huge tonnages" of copper wire, the adjective "huge" must be understood in a comparative sense. Optical fiber usage for telecommunications has increased sharply, displacing some copper wire usage. However, although glass fibers can carry far more information per unit mass than copper, the amount of copper displaced cannot be more than a few hundred thousand tons at the very most. So glass fiber has replaced a small part of the demand for copper wire, but probably not (yet) enough to affect the tonnage per capita or tonnage/GDP trend for copper, and certainly not enough to affect the overall trend.

The final example of (apparent) dematerialization resulting from technological improvements in design is the computer chip. Since Moore's law was first stated in 1965, the number of circuit elements per square centimeter of chip surface has doubled every eighteen months, with astonishing regularity. However, demand for chips—and computers to contain them—has increased even faster.

The weight of an electronic computer circa 1955 (as now) consisted mostly of peripherals. The first commercial computer made by IBM, the model 701 (1953), occupied—with its peripherals, such as DC power supply, tape drives, card readers, and printers—a floor space of at least forty square meters, not counting space required for special air-conditioning units. The machine itself occupied a volume of at least six cubic meters and contained several thousand vacuum tubes (for logic circuitry) and cathode ray tubes (for memory), not to mention steel racks. By 1955 there were probably twenty such machines in the United States (including the competing Universal Automatic Computer, or UNIVAC, units), and including peripherals, each one probably weighed as much as a car, or around one metric ton, if not more.

Now jump to the present. Domestic shipments of personal computers (PCs) in 2000 were about fifty million units. These were mostly laptop and desktop units. The average weight of each desktop unit is over 10 kg (of which the monitor accounts for something like half), whereas laptops average something like 3–4 kg. So, at 6–7 kg per unit, the overall mass of small computers shipped in the United States in the year 2000—disregarding workstations, mainframes, and supercomputers—must have been around 300,000,000 kg or 1 kg per capita. (The mass of silicon chips contained therein was much smaller, probably less than a thousand metric tons.) Other types of computers, such as servers, though fewer in number, are also bigger and heavier than PCs. The mass of PCs and

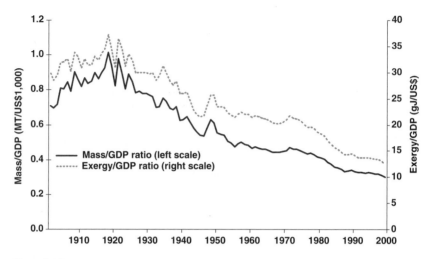

Figure 3.16
Fossil fuel consumption (mass/GDP versus exergy/GDP), United States, 1900–1998.

associated hardware produced and sold annually in the United States is of the order of 2 kg per capita, and this does not include the rather large indirect use of materials for processing ultrapure silicon, wafers, and chips.[7] The point is that the alleged "dematerialization" in this case is mostly mythical: Computers are individually smaller, but their numbers are enormously greater than was the case forty years ago, and aggregate material consumption is up, not down.

To summarize, there is little evidence of per capita dematerialization of the U.S. economy. On the contrary, increased demand seems to over-compensate for efficiency gains in every case we have investigated.

3.6   Trends in Exergy per Capita, Mass/GDP, and Exergy/GDP

Up to now we have considered mass of materials consumed per capita. There are three other measures of possible interest, however: namely, exergy per capita, mass/GDP, and exergy/GDP. For instance, though exergy consumption is proportional to mass, for a given material, this is not necessarily true for groups of materials (e.g., construction materials or fuels) because of changes in the mix or composition of the group. The differences, as shown in figures 3.16–3.19, for fuels, metals, construction materials, and chemicals, offer some interesting insights. (In the case of biomass, exergy and mass are essentially proportional.) Figure 3.20 shows

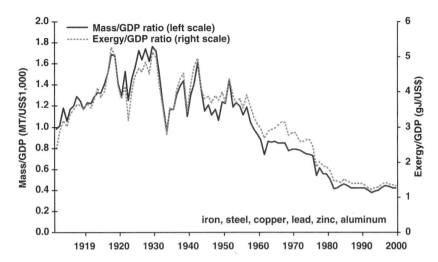

Figure 3.17
Metals consumption (mass/GDP versus exergy/GDP), United States, 1900–1998.

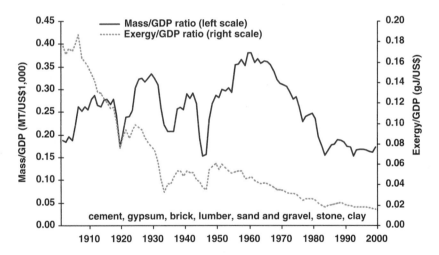

Figure 3.18
Construction materials consumption (mass/GDP versus exergy/GDP), United States, 1900–1998.

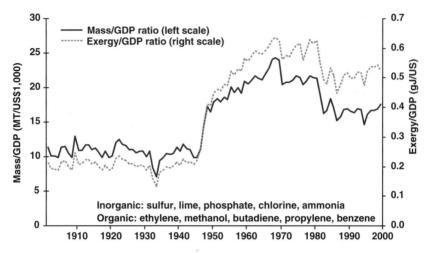

Figure 3.19
Chemicals consumption (mass/GDP versus exergy/GDP), United States, 1900–1998.

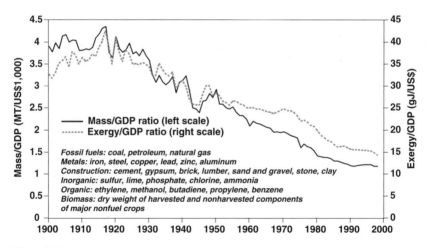

Figure 3.20
Total fuels, metals, construction, chemicals, and biomass (mass/GDP versus exergy/GDP), United States, 1900–1998.

mass and exergy per unit GDP for the total of all the materials considered, including biomass.

It is noteworthy that there has been a modest dematerialization of energy carriers since 1900. That is to say, the mass/exergy ratio for primary fuels consumed in the United States has declined from 0.042 metric tons per terajoule (tJ) in 1900 to 0.03 MT/tJ in 1995. This is an overall decline of 28 percent, due primarily to the increased use of natural gas and decreased use of coal. But actually the minimum point was reached in the decade 1965–1975 (0.028 MT/tJ). The trend has been rising since that time, as coal has increased its share of the electric power generation market since the energy crisis of 1973–1974.

The pattern for steel is quite similar to that for fuels. The peak in production per unit of GDP occurred in 1920, at about 75 MT/$million. There were subsequent (lower) peaks in 1940, 1950, and 1955, followed by a sustained decline. Steel has lost ground to many other structural materials, including aluminum and concrete. Aluminum production, in contrast, exhibits a sustained and monotonic *increase* in the tonnage/ GDP ratio from 1900 through 1970, largely as a result of the increasing importance of air transportation and the use of aluminum cans for beverages. There was a small decrease in 1975 followed by another increase in 1980. Both peaks were about 1.5 MT/$million. However, since 1980, the mass/GDP ratio for aluminum has fallen to 1.2 MT/$million, probably because two of the main existing aluminum markets were saturated by then and because of competition from plastics, mainly PVC, in the building sector.

As might be expected, plastics exhibit the most dramatic increase, from virtually zero in 1900 to about 6.6 MT/$million in 1995. For the most part, plastics have been substituting for other materials, including steel, aluminum, lumber, and paper.

Trends in mass/GDP and exergy/GDP diverge significantly from trends in mass per capita and exergy per capita. The per GDP trends tend to exhibit declines (albeit with some exceptions for specific materials during certain periods), but most of the per capita trends show increases, (again, with a few limited exceptions) as shown in figures 3.21–3.24 and summarized in figure 3.25. This reflects the fact that GDP has increased faster than mass (or exergy) consumption, which follows, in turn, from the overall shift from products to services in the economy.

For instance, there has been a steady long-term decline in fossil fuel exergy/GDP (tJ/$million) in 1987 dollars. This indicator was about 27 tJ/ $million in 1900, peaked at 33 tJ/$million around 1920, and has declined

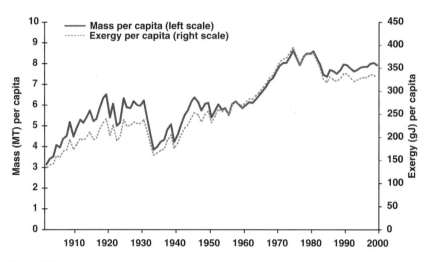

Figure 3.21
Fossil fuel consumption (mass per capita versus exergy per capita), United States, 1900–
1998.

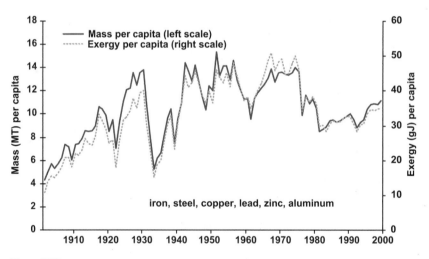

Figure 3.22
Metals consumption (mass per capita versus exergy per capita), United States, 1900–1998.

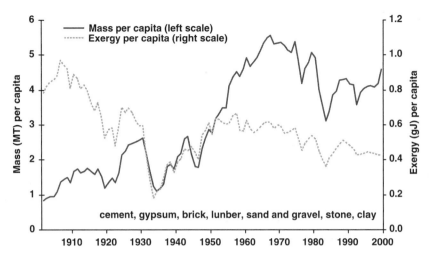

Figure 3.23
Construction materials consumption (mass per capita versus exergy per capita), United States, 1900–1998.

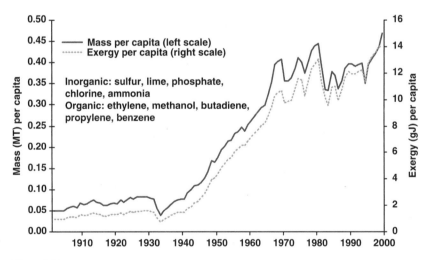

Figure 3.24
Chemicals consumption (mass per capita versus exergy per capita), United States, 1900–1998.

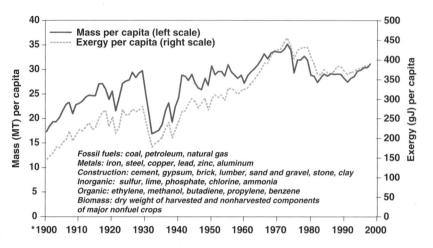

Figure 3.25
Total mass per capita and exergy per capita, United States, 1900–1998.

more or less monotonically since then to 13 tJ/$million in 1995, except
for a sharp but temporary drop below trend for World War II and a
slight rise during the early 1970s because of the U.S. embargo on Middle
Eastern oil and its aftermath. Combining mass/exergy and exergy/GDP,
the maximum occurred in 1920 (1.25 MT/$million). This indicator had
dropped to approximately 0.55 MT/$million by 1960 and around 0.4
MT/$million by 1990 and was slightly lower in 1995. The recent rate of
decline has been around 0.8 percent per annum.

Cement consumption per unit GDP has fallen monotonically since
1955 (38 MT/$million) to 18 MT/$million in 1995. Curiously, the
gypsum/GDP ratio appears to have reached a postwar minimum in 1975,
since which time it has risen slowly but steadily. The tonnage/GDP ratio
for bricks has declined more or less continuously, on average, despite
temporary fluctuations.

Lumber and paper exhibit opposite trends in relation to GDP. For
lumber the tonnage/GDP trend has been steadily downward since 1905,
except for a slight rise from 1945 to 1950. This clearly reflects a substitution of other building materials for wood. For paper, however, the trend
was steadily upward from 1900 through 1960 (except for 1945), but since
then it has fallen slightly, with a recent increase. The pattern for rubber is
somewhat similar to that for paper, with the tonnage/GDP ratio rising to
a peak in 1960. Since then, the ratio has declined quite sharply, however,

from a maximum of 0.9 MT/$million to a recent level of around 0.42 MT/$million.

Two examples of nonstructural materials, as opposed to the structural materials we've been considering, are sulfur and copper. Sulfur, which is almost entirely used to produce sulfuric acid, is a well-known indicator of the chemical industry, but roughly half of the total output currently is used for the processing of phosphate rock to make superphosphate. In this case, the tonnage/GDP ratio rose sharply to a high in 1930. Thereafter, the ratio fluctuated, with a lower peak in 1940, a higher one in 1950, and another in 1970. Since then, the ratio has declined fairly steadily from 3.5 MT/$million to around 2 MT/$million. This decline probably reflects saturation in the phosphate fertilizer market.

The tonnage/GDP ratio for copper production rose modestly from 1.3 MT/$million in 1900 to about 1.9 MT/$million in 1925 and 1930, and after a sharp dip in the 1930s, it reached almost the same level in 1940. After that, the trend was sharply down to about 0.5 MT/$million in 1980. Since then, the ratio has fluctuated at, or slightly below, that level. The major decline from 1940 through 1980 reflects reduced consumption of brass (replaced by other cheaper materials), substitution of PVC pipe for copper pipe, and the temporary replacement of some copper wire by aluminum wire. The latter substitution proved abortive (for safety reasons), and copper has made a comeback.

Obviously, some substitution has occurred, especially of plastics for metals and some structural materials. This tendency has reduced the mass of finished goods, to some degree, but not the potential hazards of process wastes. Chemical wastes are, on average, more reactive than other wastes and hence more likely to be hazardous.

3.7 Further Remarks on Dematerialization

Referring again to larger question raised by the comments of Greenspan and Cairncross, it is difficult to decide exactly what categories of "final" products should be included in a meaningful comparison between (say) 1900, 1950, and 2000. To mention a few contrary examples that show the opposite of dematerialization, cars and houses are much bigger than they were then, and there are more of each. Moreover, current trends in car size and house size are still up.

Similarly, per capita consumption of rail transport in 1900 was considerably higher than today, but the reverse is true for motor vehicles and air transport. More important, the fuel consumed per capita for all transportation purposes has increased enormously in the past century and in the last half century. Similarly, the fuel consumed to generate electric power has increased enormously. A century ago, electric power was available only in big cities, and it was still an expensive luxury. Today it is a necessity. Obviously, there have been substantial improvements in the efficiency and power/weight ratios of gasoline engines, gas turbines, and electric power generators. But these and similar technological contributions to dematerialization have been far outstripped by increased demand.[8]

Of course, apart from fuels and food, the total mass of materials consumed by the economy was—and still is—dominated by construction materials, especially sand, gravel, and crushed stone. It is certainly true, as Greenspan notes, that technological improvements in the building sector have permitted some increase in the available floor space per unit of mass input. But again, these efficiency gains are minor compared to the increased demand for floor space. So even in this case, the quantity of building materials consumed per capita in the United States has increased significantly in the last century, even if not per dollar of GDP.

In terms of mass per capita, the U.S. economy is not dematerializing at all. There are two countervailing trends that appear to explain this result. One is the phenomenon known as the "rebound effect," which is the fundamental growth mechanism that operates in the economic system. In brief, increased efficiency of production or usage leads to reduced costs and greater demand. The latter more than compensates for the efficiency gains. The other countervailing phenomenon is increased complexity. Simply stated, as functionality increases (e.g., computer chips become smaller and more powerful), the manufacturing process becomes more and more complex. The ratio of indirect material consumption to material actually embodied in the product is extremely large, which is, in itself, a new phenomenon.

This situation is in marked contrast with that of conventional metal or plastic products. Stone, glass, clay, and concrete products generate very little waste during the quarrying, mining, and processing of the materials, and even less is lost during construction. Forest products are also used quite efficiently: Wood that is not used directly as lumber is generally converted into fiberboard or paper. Even the wastes are burned as fuel. In the case of paper, most of the waste is at the consumption end. Only in the case of metals is the waste mass during mining, concentration, and smelting significantly greater than the mass of the refined product. (This is especially true for copper, and even more so for the precious metals.) However, from that point on in the manufacturing process, losses are small and mostly recovered. Again, it is mainly in the final consumption stage that significant unrecoverable losses do occur.

It is true that the economy is gradually dematerializing in terms of mass/GDP. The overall decline in this ratio from 1905 to 1995 was almost exactly a factor of three. After 1950, the decline was a factor of two. However, this measure is not particularly relevant, especially as long as actual mass consumption per capita is not decreasing.

3.8   Summary and Final Comments

The decline in aggregate tonnage of processed materials per unit of GDP, summarized in figure 3.20, is consistent with (modestly) increasing resource productivity. It is also consistent with the fact that some of the materials produced today are worth more, in real terms, than half a century or a century ago (although this would not necessarily be the case for most fuels or for tonnage materials like lumber, cement, bricks, or even structural steel). Quality and performance have improved in many areas. For instance, within the iron and steel category, the use of cast iron is now largely limited to automobile engines, and low-grade structural steels are being replaced in many applications by higher-grade steels that can be used in significantly smaller quantities. This is particularly evident in the case of rolled sheet steels for cans and automobile bodies, both of which are significantly thinner and lighter than they were a generation ago. It is fair to say that more "information" is embodied in (some) materials and products made nowadays than was the case in the past.

On the other hand, it is clear that the total mass of materials processed (and wasted), both per capita (figure 3.25) and in total (figure 3.26), has increased considerably over that same time span. The use of structural materials in housing has increased because, despite modest improvements in performance, each American occupies more floor space and consequently requires a greater quantity of walls, floors, roofing, plumbing, carpets, and furniture. This fact is reflected in the data.

The use of dissipative household consumables, such as fuels, detergents, bleaches, solvents, refrigerants, household pesticides, cosmetics, and packaging materials, has, if anything, increased even more dramatically. Within industry (and agriculture), the use of intermediates, such as explosives, fertilizers, herbicides and insecticides, drilling fluids, flotation agents, flocculents, industrial acids, catalysts, drying agents, and water

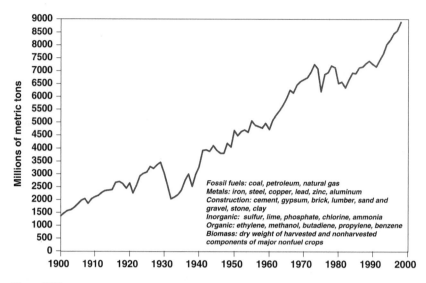

Figure 3.26
Total consumption mass, United States, 1900–1998.

and waste treatment chemicals, has undoubtedly grown faster than total throughput. However, the quantitative details are not easily documented.

Returning to the main theme of this chapter, recent discussions of the implications of sustainable development imply that per capita, and total, consumption of materials must be reduced in the industrialized world (e.g., by "Factor 10" [Factor Ten Club 1994, 1997] or "Factor Four" [von Weizsäcker, Lovins, and Lovins 1994, 1998]), in order to allow some room for absolute growth by the developing countries without overloading planetary waste assimilative capacity. The underlying imperative is clear.

However, the quoted comments of Greenspan and Cairncross (among many other economists) suggest that such a dematerialization process is already happening spontaneously. The intended implication is that governmental intervention to accelerate the process is unnecessary.

On the contrary, our data strongly suggest the opposite conclusion. In the first place, the optimists have overstated the extent of ongoing dematerialization. The weakly favorable trend in resource productivity in a few sectors is not nearly enough to compensate for increasing consumption per capita. It follows that major policy changes will be needed to bring about the needed reductions in materials and energy intensity.

Notes

1. A more academic source of background material can be found in an excellent literature survey in the *Journal of Industrial Ecology* (Cleveland and Ruth 1997). In particular, it listed and compared forty-one relevant studies, on topics including intensity of materials use (the EKC), long-wave theory, materials use decomposition analysis, and related I-O studies. These studies varied widely as to country or country grouping, time period covered, materials, degree of aggregation, data analysis method used, and implications drawn.

2. Other studies of mass flows have generally arrived at the same decision, that is, to exclude air and water flows (Adriaanse et al. 1997; Stahmer, Kuhn, and Braun 1998; Gravgård-Pedersen 1999; WRI 2000).

3. All of the detailed production and consumption statistics in the following paragraphs were derived by means of an extensive sectoral analysis of mass flows in the U.S. economy for 1993. The details are given in Ayres and Ayres (1998).

4. It can be argued that food and beverages are also service carriers, inasmuch as they pass through the body and become wastes almost immediately, except for the tiny fraction that is retained in body mass. Even that is returned to the environment at the end of life, except for the annual incremental increase in the mass of the human population.

5. Actually 51 MMT of the 89 MMT of steel produced in the United States in 1993 was recycled scrap. Domestic pig iron inputs were only 48 MMT. (The two input streams add up to 99 MMT; the weight difference consists mostly of slag and CO_2 [Ayres and Ayres 1998]).

6. Glass is manufactured by means of a thermal process from a mixture of silica (sand), magnesia, kaolin, and soda ash (sodium carbonate), plus traces of other metal oxides. Carbon dioxide is released. Portland cement is made by heating (calcining) a mixture of crushed limestone, clay, gypsum, and other silicate minerals. Carbon dioxide is released. Concrete is made from cement, sand, and other fillers, with added water. Brick and ceramic tiles are made from clay by heating to drive off water. Plaster is produced from natural gypsum by heating to drive off water, but the material is rehydrated (as in the case of Portland cement) to solidify.

7. The fabrication of computer chips is undoubtedly the most complex process in industry. Consider a day's production in a representative modern plant (circa 1998). The raw material input for this process consists of approximately seven hundred eight-inch wafers of ultrapure silicon metal, with a total mass of 14.2 kg. Each wafer has a thickness of 0.76 mm and a cubic volume of 8.55 cm^3.

It is easy to verify that for each kilogram of silicon wafer input, 161 kg of other chemicals, not including air or water, are used and discarded. The yield of the process is less than 100 percent, so for each kilogram of usable silicon chips from the process, at least 200 kg of chemicals (dry weight) are consumed and discarded.

The exergy embodied in pure silicon metal is 30.43 megaJoules (mJ)/kg, although the process of reduction from silica and purification to electronic grade (99.9999 percent pure) is very energy intensive (about 234 mJ/kg). The exergy embodied in fabrication process input chemicals, however, is approximately 788 mJ per kilogram of silicon processed, if water is not counted. If the exergy of ultrapure water is included, the total exergy embodied in the inputs is much larger: 2,100 mJ/kg, or roughly seventy times the exergy embodied in the silicon itself. The surprise is that because so much is required, ultrapure water accounts for almost two-thirds of the exergy inputs to the process. (Seawater is an environmental sink, with an exergy of zero, by definition, but pure fresh water has a small but nonzero exergy.)

But that is not the whole story. If, in the spirit of LCA, we also take into account the exergy inputs to the processes of manufacturing the major bulk chemicals, the totals above are significantly underestimated. Nevertheless, it is interesting to note that utility energy (roughly equivalent to exergy) used indirectly in manufacturing bulk chemicals amounts to at least a further 7,000 mJ/kg, mostly attributable to the energy (exergy) required to separate nitrogen from oxygen in air.

In short, the microelectronic products that are commonly cited as examples of dematerialization are really illustrations of quite a different and less favorable trend: namely, a sharply increasing ratio of process wastes to finished goods. See Williams, Heller, and Ayres (2003) for a full discussion.

8. The offsetting effects of economic growth, sometimes called the "rebound effect," have been documented by a number of authors (e.g., Key and Schlabach 1986; Wernick 1994; Jänicke, Binder, and Monch 1997).

References

Adriaanse, A., S. Bringezu, A. Hammond, Y. Moriguchi, E. Rodenburg, D. Rogich, and H. Schultz. (1997). *Resource Flows: The Material Basis of Industrial Economies.* Washington, DC: World Resources Institute.

Ayres, R. U., and L. W. Ayres. (1998). *Accounting for Resources 1: Economy-Wide Applications of Mass-Balance Principles to Materials and Waste.* Cheltenham, England: Elgar.

Barnett, H. J. (1979). Scarcity and growth revisited. In V. K. Smith (ed.), *Scarcity and Growth Reconsidered,* 163–217. Baltimore: Johns Hopkins University Press.

Barnett, H. J., and C. Morse. (1962). *Scarcity and Growth: The Economics of Resource Scarcity.* Baltimore: Johns Hopkins University Press.

Cairncross, F. (1997). *The Death of Distance.* Cambridge, MA: Harvard Business School Press.

Cleveland, C. J., and M. Ruth. (1997). When, where and by how much do biophysical limits constrain the economic process? A survey of Nicholas Georgescu-Roegen's contribution to ecological economics. *Ecological Economics* 22(3): 203–224.

Coppa, L. V. (1984). Copper, Lead, Zinc, Gold and Silver Waste Disposal Activities and Practices in the United States. Open file report, United States Bureau of Mines, Washington, DC.

Factor Ten Club. (1994, 1997). *Carnoules Declaration.* Wuppertal, Germany: Wuppertal Institute.

Gravgård-Pedersen, O. (1999). *Physical Input-Output Tables for Denmark: Products and Materials, 1990; Air Emissions, 1990–92.* Copenhagen: Statistics Denmark.

Herfindahl, O. (1967). Depletion and economic theory. In M. Gaffney (ed.), *Extractive Resources and Taxation.* 68–90. Madison: University of Wisconsin Press.

Jänicke, M., M. Binder, and H. Monch. (1997). Dirty industries: Patterns of change in industrial countries. *Environmental & Resource Economics* 9: 467–491.

Key, P. L., and T. D. Schlabach. (1986). Metals demand in telecommunications. *Materials & Society* 10: 433–451.

Martin, J. (1971). *Future Developments in Telecommunications.* London: Chapman and Hall.

McMahon, A. D. (1965). Copper: A Materials Survey. Information circular 8285, United States Bureau of Mines, Washington, DC.

National Research Council, National Academy of Sciences. (1989). *Materials Science and Engineering for the 1990s.* Washington, DC: National Academy Press.

OECD/IEA (Organization for Economic Cooperation and Development/International Energy Agency). (1995). *Energy Statistics of OECD Countries, 1992–1993.* Paris: Organization for Economic Cooperation and Development.

Schmidt-Bleek, F. B. (1993a). MIPS—A universal ecological measure? *Fresenius Environmental Bulletin* 2(6): 306–311.

Schmidt-Bleek, F. B. (1993b). MIPS revisited. *Fresenius Environmental Bulletin* 2(8): 485–490.

Schmidt-Bleek, F. B. (1994). *Wieviel Umwelt braucht der mensch? MIPS, Das Mass für Ökolologisches Wirtschaften* (How many resources do humans use? MIPS, the measure for ecology). Berlin: Birkhauser Verlag.

Stahmer, C., M. Kuhn, and N. Braun. (1998). *Physical Input-Output Tables for Germany, 1990.* Weisbaden, Germany: German Federal Statistical Office.

United States Bureau of Mines (annual). *Minerals Yearbook.* Washington, DC: Government Printing Office.

von Weizsäcker, E. U., A. B. Lovins, and L. H. Lovins. (1994). *Faktor Vier* [Factor Four]. Munich: Droemer Knaur.

von Weizsäcker, E. U., A. B. Lovins, and L. H. Lovins. (1998). *Factor Four: Doubling Wealth, Halving Resource Use.* London: Earthscan.

Wernick, I. K. (1994). Dematerialization and secondary materials recovery: A long-run perspective. *Journal of the Minerals, Metals and Materials Society* 46: 39–42.

Williams, E., M. Heller, and R. U. Ayres. (2003). The 1.7 kg microchip: Energy and material use in the production of semi-conductor devices. *Environmental Science and Technology* 36(24): 5504–5510.

WRI (World Resources Institute). (2000). *Weight of Nations: Materials Outflows from Industrial Economies.* Washington, DC: World Resources Institute.

Yergin, D. (1991). *The Prize: The Epic Quest for Oil, Money and Power.* New York: Simon and Schuster.

4 Structural Decomposition Analysis of Iron and Steel and of Plastics

Rutger Hoekstra and Jeroen C. J. M. van den Bergh

4.1 Introduction

Economic activities require a wide variety of material and energy inputs and generate waste by-products. The resulting physical flows are responsible for many environmental problems. Reducing material use while maintaining economic growth is desirable from both an environmental and an economic perspective. This is referred to as "dematerialization," "delinking," or "decoupling." There are several kinds of dematerialization. If an economy grows while its physical throughput decreases, this is referred to as "absolute dematerialization." If physical throughput increases but at a smaller rate than the economy, this is referred to as "relative dematerialization." Finally, if physical throughput increases at a greater rate than the economy, this is referred to as "rematerialization."

There is no consensus about the feasibility of absolute dematerialization. Optimists point to the environmental Kuznets curve (EKC) hypothesis, which suggests that as nations become richer, environmental pressure decreases. However, EKC studies are only partial views of individual pollutants that regard the economy as a black box without identifying the underlying mechanisms that cause changes in pollutant levels (de Bruijn and Heintz 1999).

Structural decomposition analysis (SDA) is capable of analyzing the driving forces that underlie the EKC. This method is used to analyze input-output (I/O) tables from two or more years, in order to identify which structural changes have caused a change in a particular variable (Rose and Casler 1996). It has been employed in the study of socioeconomic variables such as labor volumes, value added, and total output (Rose and Casler 1996).

Rose (1999) discusses the application of SDA to energy and environmental issues. Hoekstra and van den Bergh (2002) reviewed thirty

SDA studies of environmental issues. The great majority, twenty-six out of the thirty, dealt with energy and energy-related emissions. One study addressed nitrogen loading (Wier and Hasler 1999), and only three studies (Rose, Chen, and Adams 1996; Moll et al. 1998a, 1998b) examined bulk-material flows.

This chapter will present a new SDA of bulk materials in the Netherlands (1990–1997) as well as illustrate the use of its results in scenario analysis. As materials to be analyzed, iron and steel, and plastics have been chosen, because they are used in large quantities as well as in a wide variety of applications: They end up in consumer products, investment goods, and intermediate inputs. Moreover, the production and consumption of iron and steel and of plastic products are related to various environmental problems, such as global warming, toxic pollution, resource depletion, and waste disposal.

The data that are used in this study were constructed in conjunction with the Dutch National Statistical Bureau (CBS). A hybrid-unit I/O framework is used that integrates physical and monetary data (Miller and Blair 1985). The iron and steel and plastic products are measured in kilograms, and the other goods and services in the economy are measured in monetary units. Hybrid-unit I/O tables are constructed for 1990 and 1997 to assess the impact of structural changes, through SDA, on these physical products. This provides insights into the driving forces of environmental problems that are associated with the production and consumption of iron and steel and of plastic products.

The organization of this chapter is as follows. In section 4.2, structural decomposition analysis is introduced, and the empirical results of an SDA of bulk materials for the United States by Rose, Chen, and Adams (1996) are discussed. Section 4.3 describes the construction of the hybrid-unit I/O data for the Netherlands for 1990 and 1997. In section 4.4 SDAs of physical output and value added are presented, and the driving forces of these indicators are discussed. Section 4.5 illustrates the use of the SDA results in scenario analysis. Section 4.6 concludes.

4.2 SDA of Environmental Issues

Changes that occur in environment-related indicators, such as carbon dioxide emissions and energy use, are caused by underlying factors. For example, technological change, growth in exports, and changes in consumer demand all affect the use of energy. Decomposition analysis makes it possible to examine to what extent these factors have contributed to an observed change in energy use. In general, decomposition analysis assesses the influence of n determinant effects on a change that has occurred in a particular indicator:

$$\text{change in indicator} = \text{effect of determinant } 1 + \cdots + \text{effect of}$$

$$\text{determinant } n. \tag{4.1}$$

If the function that is decomposed is based on the I/O model, then the analysis is referred to as structural decomposition analysis (SDA). Rose and Casler (1996, 33) define SDA as an "analysis of economic change by means of a set of comparative static changes in key parameters in an I/O table." The strength of SDA is that it is capable of decomposing changes in the sector structure, technology, and demand side of an economy. Furthermore, since it is based on the I/O model, it is capable of distinguishing the direct and indirect effects of demand changes. The I/O model is given in equation (4.2) (see Miller and Blair 1985):

$$q = L \cdot y. \tag{4.2}$$

Here q is the output vector and y is the vector of final demand. The Leontief inverse $L = (I - A)^{-1}$ is calculated using matrix A, which denotes the intermediate input requirements per unit output of each sector. Decomposing this equation results in the "I/O coefficients effect" ($\Delta L \cdot y$), attributable to changes in the intermediate input structure, and a "final demand effect" ($L \cdot \Delta y$), attributable to changes in y:

$$\Delta q = \Delta L \cdot y + L \cdot \Delta y. \tag{4.3}$$

The I/O coefficient effect signifies the direct and indirect effects of technological changes in the intermediate input requirements. The final demand effect quantifies the change in the output that is attributable to changes in final demand. Both effects can be decomposed further. Decomposition of the I/O coefficient effect provides insight into the substitution and productivity changes of intermediate inputs that affect the total output. The final demand effect can be split into its various components, such as changes in exports, government expenditures, and private consumption. The influence of changes in final demand can be split into level and mix effects, which reflect, respectively, the influence of the change in the magnitude of final demand and the change in the distribution of commodities.

The introduction to this chapter noted that SDA has contributed mostly to the analysis of the driving forces of energy use and energy-related emissions. However, Rose, Chen, and Adams (1996) decompose individual bulk materials.[1] The development of plastics, rubber, glass, iron and steel, and nonferrous metals are decomposed for the United States for 1972–1982. The study uses a closed I/O model[2] in which the intermediates are divided into capital, labor, energy, materials, and other materials (KLEMO) aggregates. The results of this study, which are reported in table 4.1, depict the relative contribution, in percentages, of each of the determinant effects. The last row of the table indicates the percentage increase or decrease in material use over the period. Absolute dematerialization occurs for glass (−6 percent), iron and steel (−30 percent) and nonferrous metals (−7 percent), whereas the use of plastic (16 percent) and rubber (25 percent) increased over the period 1972–1982.

The first two effects depicted in the table are the final demand level (FDL) and final demand mix (FDM) effects, which together add up to the final demand effect. The FDL effect indicates the influence of overall economic growth on materials use. The FDM effect is negative for all materials in the table, except for rubber. This indicates that the shift in the final demand package has contributed to lower use of most materials through direct and indirect demand. The FDM effect for iron and steel is particularly large, almost large enough to counterbalance the positive FDL effect. However, the sum of FDL and FDM effects—the net final demand effect—is positive for all materials because of the large positive FDL effects.

A wide range of technological change (TC) effects is shown in the table. Significant results were obtained for technological changes in capital and labor, which contribute toward increasing material use. These technological changes in capital and labor are induced effects that arise because of the use of a closed I/O model. The dominant downward pressure on material use resulting from TC effects is caused by intermediate substitution, direct technological change in materials, and KLEMO substitution. The intermediate substitution effect indicates that the change in the intermediate input structure has led to less material use. The direct TC effects in materials reflect the conservation of materials. KLEMO substitution indicates that shifts in the intermediate inputs have led to less material use through direct and indirect demand. Note that the direct material substitution effect indicates a shift from iron and steel, and glass, to rubber, plastics, and, to a lesser extent, nonferrous metals. The aggregate results for all materials shows that this is a zero-sum effect.

Table 4.1
SDA of material use change, United States, 1972–1982 (percentages)

Decomposition effects	Plastics	Rubber	Glass	Iron and steel	Nonferrous metals	Weighted average
Final demand effects						
FDL	119	78	330	57	281	185
FDM	-25	2	-64	-47	-18	-72
Total final demand effects*	94	80	266	10	263	114
Technological effects						
Technological change in capital	54	36	145	26	124	84
Technological change in labor	56	34	149	20	92	70
Technological change in energy	15	10	39	6	29	21
Linkage technological change in materials	-6	-1	-1	-1	13	0
Direct technological change in materials	2	-100	-249	-45	-220	-147
Technological change in intermediates	2	6	48	7	3	15
Linkage material substitution	51	6	-2	-8	-13	-3
Direct material substitution	26	115	-117	-36	15	0
Intermediate substitution	-120	-66	-302	-76	-214	-195
Interfuel substitution	4	3	11	2	8	6
KLEMO substitution	-77	-23	-85	-4	-200	-64
Total TC*	7	20	-364	-109	-363	-214
Total	100	100	-100	-100	-100	-100
FDM/FDL*	-0.21	0.05	-0.21	-0.82	-0.05	-0.38
TC/FDL*	0.05	0.20	-1.05	-1.88	-1.32	-1.15
Percentage change	16	25	-6	-30	-7	-10

Source: Rose, Chen, and Adams (1996).
Note: Rows indicated by an asterisk have been added to the table reported in Rose, Chen, and Adams (1996). The percentages do not always sum to 100 (or −100) because of rounding errors.

Two ratios have been calculated for the table: FDM/FDL and TC/FDL. The FDL effect is a good numeraire because it is the same, in percentage terms, for all materials.[3] The FDM/FDL and TC/FDL ratios can therefore be used to assess the relative importance of the FDM and TC effects for each material and allow, as well, for comparison with other studies.

Absolute dematerialization seems to have arisen in regard to certain materials, because of the large negative TC effect, which is larger than the positive FDL effect, that is, the ratio of the two is smaller than -1. This is the case for glass, iron and steel, and nonferrous metals. Although the absolute dematerialization is also caused by sizable FDM effects, these are not large enough, in themselves, to outweigh the FDL effect.

Comparing the absolute values of the ratios, the importance of the decomposition effects becomes apparent. For the materials that experience absolute dematerialization, the ranking of the absolute size of the effects is TC, FDL, FDM. For plastics, the order is FDL, FDM, TC. For rubber, it is FDL, TC, FDM. The overall impression, which is bolstered by the weighted average results, is that for this period, the most important driving force was the TC effects, followed by the effect of economic growth (FDL). The FDM effect was the least important among the three effects examined in this study.

Are the driving forces of material flows different from those for other physical throughput variables? Rose and Chen (1991) and Casler and Rose (1998) decompose energy demand and carbon dioxide (CO_2) emissions, respectively. These studies use decomposition methods similar to that used in Rose, Chen, and Adams (1996) for the same period in the United States, which make these three studies ideally suited for a comparison. Table 4.2 is based on the results that are presented in Rose and Chen (1991) and Casler and Rose (1998).

There are a number of similarities and differences among the three studies. First, consider the total material use, energy use, or CO_2 emissions, denoted either by "total" or "weighted average" in tables 4.1 and 4.2. The FDM/FDL and TC/FDL ratios are all negative, indicating that these effects are helping to alleviate the upward pressure of the FDL effect. However, the relative importance of the determinant effects is different. The ratios indicate that the ranking of the absolute values of the determinant effects for total energy use is FDL, TC, FDM, whereas for total CO_2 emissions it is FDL, FDM, TC. For material demand, this ranking is TC, FDL, FDM. The TC effects therefore seem to be

Table 4.2
Aggregate SDA results for energy demand and CO_2 emissions for the United States, 1972–1982

Decomposition effect	Energy use					CO_2 emissions			
	Coal	Petroleum	Natural gas	Electricity	Weighted average	Coal	Natural gas	Petroleum	Total
FDL	66	510	88	65	1,846	51	53	156	441
FDM	−45	−159	−46	−10	−798	−33	−46	−94	−300
Total	22	350	43	55	1,048	18	7	62	141
TC	78	−450	−143	45	−948	82	−107	−162	−241
Total	100	−100	−100	100	100	100	−100	−100	−100
FDM/FDL	−0.67	−0.31	−0.52	−0.16	−0.43	−0.65	−0.87	−0.60	−0.68
TC/FDL	1.18	−0.88	−1.62	0.70	−0.51	1.60	−2.02	−1.04	−0.55
Percentage change	28	−4	−20	27	1	23	−19	−7	−2

Source: The results for energy demand are based on tables 3 and 4 of Rose and Chen (1991), and those for CO_2 emissions are based on tables II and III of Casler and Rose (1998).

more important in explaining changes in material use than those in energy use and in carbon dioxide emissions for this period in the United States.

Further analysis of the ratios of the individual materials and energy carriers within each study leads to number of conclusions, for this specific country and time period and these types of material:

• TC effects are nearly always more important than FDM effects, because the absolute value of the TC/FDL ratio is always larger than the FDM/FDL ratio (the only exception is rubber).
• The FDM effects help to reduce environmental pressures in all three studies, which is illustrated by the fact that the FDM/FDL ratio is nearly always negative (the only exception is rubber).
• TC effects are important driving forces for increases and decreases in environmental pressures, as indicated by the range of the TC/FDL ratio (−2.02 to 1.60). The FDM effects are smaller and nearly all negative. The range of the FDM/FDL ratio is −0.87 to 0.05.

For this period the physical economy of the United States seems to have experienced a shift from iron and steel, and glass, to plastics, rubber, and nonferrous metals, and a shift from petroleum and natural gas to coal. All three studies show the influence of the extraordinary oil crises experienced in the decade examined. The shifts in the environmental variables of the three studies could probably have been caused by changes in the (relative) energy prices and regulatory policies that stimulated lower energy and material use.

4.3   Construction of Hybrid-Unit Input-Output Tables

The SDA presented in section 4.4 uses a hybrid-unit I/O model, which is theoretically preferable to the standard I/O model (see Miller and Blair 1985, chap. 6). A hybrid-unit model uses multiple units, which in this case means that physical units are adopted for iron and steel and for plastic products (kilograms), whereas monetary values (€1997) are used for the other goods and services (see table 4.3). The tables for 1990 and 1997 used in section 4.4 were constructed in conjunction with the CBS, which involved cooperation with specialists in iron and steel products, plastics, trade, environmental national accounts, waste accounts, and national accounts integration.

Table 4.3
A hybrid-unit commodity-by-commodity I/O table

	Other commodities	Iron, steel, and plastic commodities	Final demand	Total output
Other commodities				
Iron, steel, and plastic commodities				
Primary inputs				

Note: The shaded area indicates the physical units (kilograms). The rest of the table is in monetary units (€1997).

I/O tables are produced from supply and use tables. In the study presented in this chapter, a hybrid-unit commodity-by-commodity I/O table is produced by going through six steps:

1. *Construction of monetary supply and use tables for 1990 and 1997 in 1997 prices* The data in the supply and use tables for 1997 in current prices were used without alteration. However, the data in the monetary supply and use tables for 1990 had to be inflated to 1997 prices. To do this, a time series of supply and use tables for 1990–1997 was constructed in basic prices of the current and previous year. This required that the classification scheme of the 1990–1995 supply and use tables be converted, since the CBS implemented a revision in 1995. Element-specific price indices were obtained from the time series to inflate the 1990 table.
2. *Construction of physical supply and use tables for iron and steel for 1990 and 1997*[4] Physical information from trade statistics, waste statistics, and recycling sources was used, and otherwise monetary information was converted to physical units by using a representative price. This yielded an initial physical supply and use table. However, the mass-balance principle dictates that the mass of commodities supplied must be equal to the mass used. Similarly, the industry totals of the supply and use tables should balance. In an iterative process, appropriate adjustments were made to ensure that the physical supply and use tables balance. The details of this process are described in CBS (2001) and chapter 5 of Hoekstra (2003).

3. *Integration of the monetary and physical supply and use tables* The physical information about iron, steel, and plastics commodities from step 2 was used to replace the corresponding monetary rows in the supply and use tables of step 1. The result is a set of hybrid-unit supply and use tables for 1990 and 1997.

4. *Making the intermediate portion of the supply and use tables square* The commodity technology assumption applied in step 5 requires square supply tables. By assigning each commodity to its primary producer, the number of commodities was therefore made equal to the number of industries.

5. *Applying the commodity technology assumption* Here it was assumed that a commodity is produced in the same way irrespective of the industry in which it is being produced. This resulted in a commodity-by-commodity I/O table, which adheres to I/O modeling assumptions. However, a main disadvantage is that the intermediate and primary input matrices of the resulting I/O table include negative elements.[5]

6. *Eliminating the negative values of the commodity-by-commodity I/O table* Theoretically, negative elements in step 5 can be caused by data errors, heterogeneous production processes, aggregation of commodities, and nonuniform prices (Konijn 1994; Hoekstra 2003). The negative elements were analyzed, and subsequently the supply and use tables were adjusted to eliminate them. The commodity technology assumption was then applied to the adjusted tables. Any remaining small negative elements were eliminated using a mathematical procedure (employed by the CBS) to balance the elements of the I/O table to the row and column totals.

The resulting nonnegative hybrid-unit I/O table has forty-six monetary and ten physical commodities. Table 4.4 summarizes the development of physical output and value added for the iron, steel, and plastic commodities. Furthermore, the last column indicates the type of dematerialization that has taken place. Note that our definition of (de)materialization is based on a comparison of the physical development and the value-added growth. Absolute dematerialization occurs if the physical output decreases. Relative dematerialization takes place if value added increases at a higher rate than physical output. Rematerialization occurs if physical growth exceeds the growth in value-added terms. Note also that absolute dematerialization is the only development that leads to lower material use, with an associated lower environmental pressure.

There is some debate about relevant indicators of dematerialization, notably about the extent of aggregation. It is not our aim to address this

Table 4.4
The development of physical output and value added of iron and steel and plastic products in the Netherlands (1990–1997)

Product group	Domestic output (10⁹ kg)			Value added (millions of €1997)			Type of dematerialization
	1990	1997	%	1990	1997	%	
Primary plastics	3.48	5.11	47%	736	1,403	91%	Relative dematerialization
Plastic products	1.09	1.42	30%	1,220	1,472	21%	Rematerialization
Primary iron and steel	4.94	6.33	28%	1,226	1,329	8%	Rematerialization
Iron and steel products	2.80	3.92	40%	1,965	2,259	15%	Rematerialization
Machines	0.96	1.18	22%	3,262	3,619	11%	Rematerialization
Office equipment and computers	0.07	0.15	124%	78	244	212%	Relative dematerialization
Electrical appliances	0.26	0.51	92%	934	899	–4%	Rematerialization
Electronics	0.19	0.51	169%	419	514	23%	Rematerialization
Cars and engines	0.56	0.63	14%	950	1367	44%	Relative dematerialization
Transportation equipment	0.41	0.29	–29%	399	425	7%	Absolute dematerialization

Note: The values for physical output reported here are slightly different from those in CBS (2001) because the diagonals of the I/O table have been set to zero. This alleviates the influence of changes in double counting (see appendix 8.A. of Hoekstra [2003] for details).

issue here. Nevertheless, it should be noted that the data set used here is fairly disaggregated: iron and steel, and plastics, are divided into ten commodity groups.

Overall, the table suggests that the period 1990 to 1997 in the Netherlands was characterized by growing physical output for nine out of the ten product groups. Six of the ten product groups experience rematerialization, which implies that the commodities produced less value added in 1997, in real terms, per kilogram of output than in 1990. Among these are primary iron and steel, iron and steel products, and plastic products, all of which are very large in terms of their physical output. Three activities, namely primary plastics, office equipment, and cars and engines, underwent relative dematerialization. Only transportation equipment experienced absolute dematerialization for this period.[6]

4.4 SDA of Iron and Steel and of Plastics in the Netherlands (1990–1997)

This chapter presents four SDAs that investigate the development of iron and steel, and plastics products, in the Netherlands for the period 1990 to 1997. Since dematerialization concerns both physical output and value added, both these variables are decomposed. The four SDAs are

1. *Decomposition of output* The change in output is decomposed into I/O coefficients (IOC), FDL, FDM, and import substitution (IS) effects.
2. *Decomposition of final demand categories* The final demand effect that is found in the previous decomposition is split into various category effects: exports, household consumption, government consumption, investments, and stock changes effects.
3. *Detailed decomposition of final demand categories* The final demand category effects that are found in the previous decomposition are split into level and mix effects.
4. *Decomposition of value added* The change in value added is decomposed into value-added coefficient, IOC, final demand, and IS effects.

These four SDAs are discussed in the following subsections.

4.4.1 Decomposition of Output

Equation (4.4) decomposes the change in the physical output (Δq) into four effects—the IOC ($\Delta L \cdot y$), FDL ($L \cdot \Delta y^{\text{lev}}$), FDM ($L \cdot \Delta y^{\text{mix}}$) and IS ($L \cdot \Delta b$) effects:

$$\Delta q = \Delta L \cdot y + L \cdot \Delta y^{\text{lev}} + L \cdot \Delta y^{\text{mix}} + L \cdot \Delta b, \qquad (4.4)$$

where $q = L \cdot y = L \cdot (y^{\text{lev}} \cdot y^{\text{mix}} - b)$ is the "I/O model including imports" and

q = Output
L = Leontief inverse
y = Final demand
y^{lev} = Final demand level
y^{mix} = Final demand mix
b = Imports

The IS effect occurs because increased imports can substitute for domestic production. This phenomenon is captured by the "I/O model including imports," in which imports are recorded as negative final demand. The advantage of this framework is that the changes in the intermediate demand can be interpreted as technological changes that are unaffected by import substitution. Note that in this type of I/O model, the production processes of domestic and imported goods are assumed to be the same.

The index approach proposed by Dietzenbacher and Los (1998) is used in these calculations (it is commonly applied to SDA). For a complete discussion of the use of indices in decomposition analysis, see Hoekstra and van den Bergh (2003). The SDA results for the physical commodities are presented in table 4.5.[7] In addition, this table records the FDM/FDL and IOC/FDL ratios.

The FDL effect in table 4.5 reflects the influence of economic growth on physical output and is therefore always positive. It represents the most important driving force of increased physical output for eight out of ten commodities examined in the table. Only for office equipment and computers and for electronics is it surpassed in importance by the FDM effect. For cars and engines, the FDL effect is the largest in relative terms: 416 percent. However, this product group is relatively small in terms of absolute physical output. In absolute terms, the level effect is the most important for primary iron and steel, where it is responsible for a 2.63×10^9 kg increase in output.

The data for the FDM effect show a more mixed set of positive and negative results. They exhibit negative effects for transportation equipment (-192 percent), machines (-77 percent) and primary iron and steel (-15 percent). This indicates that the relative shift in the final demand

Table 4.5
Results of decomposition of output (10^9 kg and percentage)

Product group	FDL	FDM	IOC	IS	Change in output	FDM/FDL	IOC/FDL
Primary plastics	1.31 81%	0.94 58%	0.06 4%	−0.68 −42%	1.63 100%	0.72	0.04
Plastic products	0.48 146%	0.13 39%	0.03 8%	−0.31 −93%	0.33 100%	0.26	0.06
Primary iron and steel	2.63 190%	−0.20 −15%	0.59 43%	−1.63 −118%	1.38 100%	−0.08	0.22
Iron and steel products	1.00 89%	0.24 21%	0.33 29%	−0.45 −40%	1.12 100%	0.24	0.33
Machines	0.40 188%	−0.16 −77%	0.02 8%	−0.04 −19%	0.21 100%	−0.41	0.04
Office equipment and computers	0.04 45%	0.05 64%	0.01 17%	−0.02 −26%	0.08 100%	1.43	0.39
Electrical appliances	0.17 69%	0.11 43%	0.12 50%	−0.15 −62%	0.24 100%	0.62	0.72
Electronics	0.11 34%	0.14 42%	0.10 30%	−0.02 −6%	0.32 100%	1.24	0.89
Cars and engines	0.33 416%	0.01 10%	−0.02 −23%	−0.24 −304%	0.08 100%	0.03	−0.05
Transportation equipment	0.09 80%	−0.23 −192%	−0.02 −13%	0.03 25%	−0.12 −100%	−2.40	−0.17

Note: Percentage values are relative to the change in commodity output.

package of products is leading, through direct and indirect demand, to reductions in the output of these goods. However, the FDM effect contributes toward increasing output of office equipment and computers (64 percent) and primary plastics use (58 percent).

Surprisingly, eight of the IOC effects presented for the ten products in the table are positive; that is, the intermediate input structure has shifted toward production processes that use these physical products, either directly or indirectly. Whereas technology is often assumed to lead to dematerialization, these results show that for this specific time period, country and materials, this is only the case for two commodities. Only cars and engines, and transportation equipment, have (small) negative IOC effects. Shifts in intermediate demand have contributed to 43 percent and 50 percent of the growth in the physical output of, respectively, primary iron and steel, and electrical appliances.

The IS effect shows that increased imports of iron and steel, and plastic products, have led to lower domestic output levels. This indicates that the growth in domestic physical output has been tempered by an increase in imports. For eight out of ten products, it is the most influential driving force that exerts downward pressure on the physical output. An exception is transportation equipment, which saw a decrease in imports, and therefore a positive IS effect.

4.4.2   Comparison with Rose, Chen, and Adams (1996)

The results for the United States (1972–1982) reported in table 4.1 differ from those for the Netherlands (1990–1997) in table 4.5. The results will be compared here based on the ratios reported in tables 4.1 and 4.5. The TC/FDL ratios from table 4.1 can be compared with the IOC/FDL ratios from table 4.5. In table 4.1, the results for plastics show that the ratio for FDM/FDL is negative and the TC/FDL ratio has a small positive value. The results in table 4.5 show that the FDM/FDL and IOC/FDL ratios are positive for primary plastics and plastic products. The divergence in the sign of the FDM/FDL ratio is the most noticeable difference: The change in the mix of the final demand package contributes to increased physical output in table 4.5, whereas in Rose, Chen, and Adams (1996), it contributes to lower use of plastics.

For iron and steel, the FDM/FDL and TC/FDL ratios are both negative in table 4.1. The TC/FDL ratio is less than −1, which is the main driving force for the absolute dematerialization occurring in this period. The IOC/FDL ratio for primary iron and steel and for plastics products in table 4.5 shows the opposite: Technological change caused increased output of iron and steel and of plastics products during the period 1990–1997. Only the FDM/FDL ratio for primary iron and steel is negative.

The overall impression is that the FDM and TC effects help to reduce environmental pressures in the United States for the period 1972 to 1982. On the other hand, the results for the Netherlands (1990–1997) suggest that FDM and IOC effects are generally causing increases in physical throughput. Possible reasons for the differences in the results of the two studies relate to

1. *The period* The driving forces of materials for the 1970s and 1990s might have been different.
2. *Country* The Netherlands and the United States might have experienced different types of structural change.
3. *Decomposition method* The decomposition methods of the two studies differ.

A conclusive comparison of the results is impossible because of the difference in decomposition methods that have been used. Nevertheless, a number of general insights can be derived. The dematerialization patterns differ significantly between the two studies. Table 4.1 shows that for the period 1972 to 1982, the United States was characterized by absolute dematerialization for three materials, including a 30 percent reduction in iron and steel use. The development for the Netherlands over the period 1990 to 1997 shows an increases for nine out of ten physical commodities. These differences suggest that determinant effects differ as well.

The decomposition methods used in the two studies could have influenced the difference in results in a number of ways. The use of distinct indices is likely to affect the size of the decomposition effect, although it is unlikely to affect the sign. Another methodological difference, the use of a closed I/O model for the United States analysis versus the use of an open I/O model for the Netherlands in the current study could have affected both the size and the sign of the results.

4.4.3 Decomposition of Final Demand Categories

The previous decomposition of output has underlined the importance of final demand effects. An additional SDA, shown in equation (4.5), takes a closer look at the influence of changes in separate categories of final demand—exports ($L \cdot \Delta y^{\text{exp}}$), household consumption ($L \cdot \Delta y^{\text{cons}}$), government consumption $(L \cdot \Delta y^{\text{gov}})$, investments $(L \cdot \Delta y^{\text{inv}})$, and stocks $(L \cdot \Delta y^{\text{stock}})$:

$$\Delta q = \Delta L \cdot y + L \cdot \Delta y^{\text{exp}} + L \cdot \Delta y^{\text{cons}} + L \cdot \Delta y^{\text{gov}} + L \cdot \Delta y^{\text{inv}}$$
$$+ L \cdot \Delta y^{\text{stock}} + L \cdot \Delta b, \qquad (4.5)$$

where

y^{exp} = Exports
y^{cons} = Household consumption
y^{gov} = Government consumption
y^{inv} = Investment
y^{stock} = Stocks

In equation (4.5), the IOC effect and IS effects are included, but since these effects have already been discussed in the previous decomposition, they are not included in the results in table 4.6. The results show that exports are the largest driving force of iron and steel output and plastic products output, which is not surprising for a small open economy such as the Netherlands. The exception to this finding is transportation equipment, which exhibits a negative export effect. This contributes to the absolute dematerialization that is observed for transportation equipment in table 4.4. However, the main driving force of the dematerialization of this commodity is the investment effect. (The export effect dominates the other category effects for all other commodity groups.) Significant investment effects are observed for office equipment and computers (53 percent) and machines (38 percent). Note the large effect of investment on primary iron and steel (15 percent), which is an indirect effect. The effects of changes in household consumption, government consumption, and stock changes are rather small.

4.4.4 Detailed Decomposition of Final Demand Categories
The effect of the individual final demand categories can be decomposed further into level and mix effects, as shown in equation (4.6):

$$\Delta q = \Delta L \cdot y + L \cdot \Delta y^{\text{exp, lev}} + L \cdot \Delta y^{\text{exp, mix}} + L \cdot \Delta y^{\text{cons, lev}} + L \cdot \Delta y^{\text{cons, mix}}$$
$$+ L \cdot \Delta y^{\text{gov, lev}} + L \cdot \Delta y^{\text{gov, mix}} + L \cdot \Delta y^{\text{inv, lev}} + L \cdot \Delta y^{\text{inv, mix}}$$
$$+ L \cdot \Delta y^{\text{stock, lev}} + L \cdot \Delta y^{\text{stock, mix}} + L \cdot \Delta b, \qquad (4.6)$$

Table 4.6
Results of decomposition of final demand categories (10^9 kg and percentage)

Product group	Exports	Household consumption	Government consumption	Investments	Stock	Final demand effect
Primary plastics	2.11	0.10	0.00	0.06	-0.02	2.25
	94%	4%	0%	2%	-1%	100%
Plastic products	0.50	0.07	0.00	0.04	-0.01	0.61
	83%	12%	1%	7%	-2%	100%
Primary iron and steel	1.94	0.27	0.02	0.38	-0.18	2.43
	80%	11%	1%	15%	-7%	100%
Iron and steel products	0.85	0.13	0.02	0.29	-0.04	1.24
	68%	10%	1%	23%	-3%	100%
Machines	0.15	0.06	0.00	0.09	-0.06	0.24
	63%	24%	2%	38%	-26%	100%
Office equipment and computers	0.04	0.00	0.00	0.05	0.00	0.09
	41%	4%	0%	53%	0%	100%
Electrical appliances	0.23	0.02	0.00	0.03	-0.01	0.27
	83%	9%	1%	10%	-2%	100%
Electronics	0.19	0.03	0.00	0.03	0.00	0.25
	78%	11%	0%	11%	0%	100%
Cars and engines	0.24	0.06	0.00	0.09	-0.05	0.33
	72%	17%	0%	26%	-15%	100%
Transportation equipment	-0.03	0.01	0.00	-0.11	0.00	-0.13
	-26%	7%	1%	-81%	-1%	-100%

Note: Percentage values are relative to the final demand effect per commodity. Final demand effect is the sum of the FDM and FDL effects of table 4.5.

where

$y^{\text{exp, lev}}$	= Export level
$y^{\text{exp, mix}}$	= Export mix
$y^{\text{cons, lev}}$	= Household consumption level
$y^{\text{cons, mix}}$	= Household consumption mix
$y^{\text{gov, lev}}$	= Government consumption level
$y^{\text{gov, mix}}$	= Government consumption mix
$y^{\text{inv, lev}}$	= Investment level
$y^{\text{inv, mix}}$	= Investment mix
$y^{\text{stock, lev}}$	= Stock level
$y^{\text{stock, mix}}$	= Stock mix

The variables are the same as those of the previous decomposition of final demand but have a "mix" or "lev" superscript. The I/O effect and IS effects are included in the equation, but since these effects have already been discussed in the output decomposition, they are not included in the results in table 4.7.

The table shows that all level effects are positive, except for some of the effects of stock changes. The level effects are generally larger than the corresponding mix effects. The main exceptions concern office equipment and computers, for which the investment mix effect exceeds the investment level effects. This reflects the fact that the mix of investment goods has shifted in such a way that it has led to increases in the output of computers. For electronics, too, the export mix effect is greater than the export level effect.

The household consumption effects are only a small driving force in the overall material throughput. Nevertheless, it is interesting to see the influence of changes in consumption patterns. Nearly all consumption mix effects lead to a small reduction in physical output. However, although the Dutch consumption package is causing lower demand for these physical products, the level of consumption has increased to such an extent that the net consumption effect is an increase for all commodities.

4.4.5 Decomposition of Value Added

Equation (4.7) shows that value added can be decomposed in a similar way to output:

$$\Delta v = \Delta u \cdot L \cdot y + u \cdot \Delta L \cdot y + u \cdot L \cdot \Delta y + u \cdot L \cdot \Delta b, \qquad (4.7)$$

Table 4.7
Results of detailed decomposition of final demand categories (10^9 kg and percentage)

Product group	Exports		Household consumption		Government consumption		Investments		Stocks		Total
	Mix	Level	Mix	Level	Mix	Level	Mix	Level	Mix	Level	
Primary plastics	0.56 / 25%	1.56 / 69%	-0.04 / -2%	0.14 / 6%	0.00 / 0%	0.01 / 0%	-0.05 / -2%	0.10 / 5%	0.03 / 1%	-0.05 / -2%	2.25 / 100%
Plastic products	0.09 / 16%	0.41 / 67%	-0.02 / -4%	0.10 / 16%	0.00 / 0%	0.00 / 1%	-0.04 / -6%	0.08 / 13%	0.01 / 2%	-0.02 / -4%	0.61 / 100%
Primary iron and steel	-0.50 / -21%	2.44 / 101%	-0.03 / -1%	0.29 / 12%	0.00 / 0%	0.03 / 1%	-0.14 / -6%	0.52 / 21%	0.00 / 0%	-0.18 / -7%	2.43 / 100%
Iron and steel products	0.16 / 13%	0.69 / 55%	-0.01 / -1%	0.14 / 11%	0.00 / 0%	0.02 / 1%	0.00 / 0%	0.29 / 23%	-0.01 / -1%	-0.03 / -2%	1.24 / 100%
Machines	-0.11 / -45%	0.25 / 108%	0.00 / -2%	0.06 / 26%	0.00 / -1%	0.01 / 2%	-0.04 / -15%	0.12 / 53%	-0.10 / -44%	0.04 / 17%	0.24 / 100%
Office equipment and computers	0.01 / 12%	0.03 / 29%	0.00 / 1%	0.00 / 4%	0.00 / 0%	0.00 / 1%	0.04 / 40%	0.01 / 13%	0.00 / 1%	0.00 / 0%	0.09 / 100%
Electrical appliances	0.11 / 39%	0.12 / 44%	0.00 / -2%	0.03 / 11%	0.00 / 0%	0.00 / 1%	-0.02 / -5%	0.04 / 15%	0.00 / 0%	-0.01 / -2%	0.27 / 100%
Electronics	0.10 / 41%	0.09 / 37%	0.01 / 3%	0.02 / 8%	0.00 / 0%	0.00 / 1%	0.01 / 3%	0.02 / 8%	0.00 / -1%	0.00 / 1%	0.25 / 100%
Cars and engines	0.05 / 15%	0.19 / 57%	-0.01 / -4%	0.07 / 22%	0.00 / 0%	0.00 / 0%	-0.01 / -2%	0.09 / 28%	-0.03 / -10%	-0.02 / -5%	0.33 / 100%
Transportation equipment	-0.08 / -57%	0.04 / 31%	0.00 / -4%	0.01 / 10%	0.00 / -1%	0.00 / 2%	-0.14 / -109%	0.04 / 28%	0.01 / 7%	-0.01 / -8%	-0.13 / -100%

Note: The percentage values are relative to the final demand effect per commodity. Note that, in this specification, the sum of the mix effects of all categories of final demand is not equal to the aggregate FDM effect in table 4.5. This also holds for the level effects. However, the sum of mix and level effect of each individual final demand category is equal to the results per category in table 4.6.

where

v = Value added
u = Value-added coefficient
y = Final demand
b = Imports

The change in value added (Δv) for each commodity group is decomposed into a value-added coefficient ($\Delta u \cdot L \cdot y$), an I/O coefficient ($u \cdot \Delta L \cdot y$), final demand ($u \cdot L \cdot \Delta y$) and IS effect ($u \cdot L \cdot \Delta b$). The value-added coefficient is equal to the ratio of value added and output. The results of the calculations are presented in table 4.8.

The table includes a row that indicates the total impact of each decomposition effect on the ten commodities. The value-added coefficient effect contributes −60 percent to the total value-added growth of these ten commodities. In other words, the overall earnings per physical output are decreasing and are responsible for a significant downward pressure on the total value added of physical commodities. Nevertheless, the final demand effect leads to sufficient growth in value added to overcome the value-added coefficient effect.

The six commodities that experience rematerialization (see table 4.3) all have a negative value-added coefficient effect, because these commodities are earning less per kilogram over the period 1990–1997. The commodities that exhibit relative and absolute dematerialization have a positive value-added coefficient effect. Overall, the effect of the lower value-added coefficient is responsible for a 60 percent reduction in value added for the ten physical commodities in this study. The IS effect is large and negative for many commodities: This shows that increasing imports have reduced domestic earnings.

What do these SDA results tell us about dematerialization? Table 4.4 shows that six commodities exhibited rematerialization, three experienced relative dematerialization, and only one shows absolute dematerialization. The SDA results can help to explain these developments. Dematerialization has two components: change in physical output and change in value added.

The growth in physical output is caused primarily by the export level effect. This is the most important driving force for increased material use for eight out of the ten physical commodities studied. Furthermore, the export mix effect is positive for seven out of the ten commodities. The investment and IOC effects are an important source of growth for a few

Table 4.8
Value-added decomposition (millions of €1997 and percentages)

Product group	Value-added coefficient	I/O coefficient	Final demand	Import substitution	Change in value added
Primary plastics	271	14	548	−166	667
	41%	2%	82%	−25%	100%
Plastic products	−104	30	658	−332	252
	−41%	12%	261%	−132%	100%
Primary iron and steel	−214	135	556	−374	103
	−208%	131%	541%	−364%	100%
Iron and steel products	−425	210	796	−288	294
	−145%	72%	271%	−98%	100%
Machines	−325	54	761	−132	357
	−91%	15%	213%	−37%	100%
Office equipment and computers	50	20	126	−30	166
	30%	12%	76%	−18%	100%
Electrical appliances	−670	318	717	−400	−35
	−1,917%	910%	2,049%	−1,143%	−100%
Electronics	−418	155	390	−32	95
	−440%	163%	411%	−33%	100%
Cars and engines	266	−35	646	−460	417
	64%	−8%	155%	−110%	100%
Transportation equipment	169	−19	−160	36	26
	640%	−71%	−606%	137%	100%
Total	−1,400	882	5,038	−2,179	2,342
	−60%	38%	215%	−93%	100%

Note: Percentage values are relative to the change in value added per commodity.

of the commodities. The growth of physical output is lowered mainly by the growth in imports. However, this negative impact is always dominated by the export effect. The final demand is the most important driving force of value added. However, the value-added coefficient plays an important role in diminishing this growth for the commodities that experience rematerialization.

4.5 Scenario Analysis

Scenario analysis is common in the I/O literature (see, for example, Duchin and Lange 1994 and Proops, Faber, and Wagenhals 1993). However, scenario analysis in the context of SDA, or in other decomposition analysis methods, is rare. To our knowledge, only two SDA studies and one index decomposition analysis (IDA) study also link to a scenario analysis. In the first environmental SDA study ever performed, Leontief and Ford (1972) projected the increase of air pollutants to 1980. This increase was calculated using estimates of final demand in 1980 and historical emission coefficients. Siegel, Alwang, and Johnson (1995) provide a conceptual framework for SDA policy models, but it is not applied empirically. In the IDA setting, Ang and Lee (1996) produce several projections based on changing one of the determinant effects, while keeping others constant.

The results of SDA can be used as basis for forecasting scenarios. For example, figure 4.1 shows a number of scenarios for primary plastics for the period 1997–2030, using the results reported in table 4.6 as a basis. The figure shows the business-as-usual situation in which the use of primary plastics would develop at the same pace as over the period 1990 to 1997. This would lead, according to the figure, to an almost quadrupling of primary plastic use.

The other scenarios in figure 4.1 are "optimistic" in the sense that it is assumed that over the period 1997 to 2030, the decomposition effects will follow a certain trend that will lead to lower material use. In particular, the positive decomposition effects (which lead to increased material use) in the period 1990–1997 will halve over the period 1997 to 2030, whereas all negative effects (which lead to lower material use) will double over the same period. The figure shows the impact of an optimistic change in one of the decomposition variables, if all the other decomposition effects remain at business-as-usual levels. Clearly the impact of the IOC, household consumption (final demand consumption, or FDC), government consumption (final demand government, or FDG), and

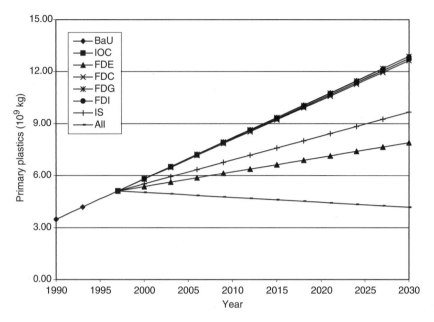

Figure 4.1
Optimistic "adjusted business-as-usual" scenario for primary plastics. BaU = business as usual; IOC = input/output coefficients; FDE = final demand export; FDC = final demand consumption; FDG = final demand government; FDI = final demand import; IS = import substitution.

investment effects have little impact. However, a 50 percent reduction of the export effect (final demand export, or FDE) or the doubling of the IS effect leads to significant slowing of the growth rate of primary plastic use. However, a reduction in the use of primary plastics is obtained only if all optimistic decomposition effects occur simultaneously (All). This suggests that absolute dematerialization for primary plastics can be achieved for the period 1997 to 2030 only if the decomposition effects alter significantly compared to the period 1990 to 1997. Further forecasting as well as backcasting scenarios can be found in chapter 9 of Hoekstra (2003).

4.6   Conclusions

This chapter has analyzed the driving forces of bulk-materials through structural decomposition analysis. The empirical results of a study by Rose, Chen, and Adams (1996), which decomposes plastics, rubber, glass, iron and steel, and nonferrous metals (United States, 1972–1982), were reviewed. The general conclusion is that the technological changes effect was the most important driving force for overall reductions in material use over the period. The final demand level effect was the largest driving force for increased material use. The final demand mix effect was generally small but negative; that is, it contributed to lower material use. Similar studies for the United States show that the TC effect is less important for overall energy use and CO_2 emissions for the same period.

This chapter introduced a new SDA study that examined the driving forces of iron and steel and of plastics. Four SDA calculations were presented for the Netherlands for the period 1990–1997. Overall, this was a period of rematerialization for the country. Only one commodity group, transportation equipment, experienced absolute dematerialization over this period. Three commodity groups experienced relative dematerialization, and the other six commodities exhibited rematerialization.

The first three decomposition calculations analyzed the driving forces of output. The simplest decomposition of output showed that economic growth—the FDL effect—contributes most to physical output growth. Nevertheless, the FDM effect and I/O coefficients effect also contributed to increases in output for, respectively, seven and eight commodities. The import substitution effect contributed to lower physical output for nine out of ten commodities. In the subsequent decompositions of the final demand categories, it becomes clear that the change in the level of exports was the largest driving force for eight of the ten commodities. Although increase in import substitution led to lower physical output, these effects were always outweighed by the export effect. Changes in the other categories of final demand—household consumption, government consumption, investments, and stocks—generally had little impact on output. Nevertheless, the investment effect and TC effect were large for some of the commodities.

The SDA results for value added showed that changes in final demand were the largest driving force behind increases in value added. However, for the six commodities that experienced rematerialization, the decrease in the value-added coefficient led to a lower level of value added.

The results of the basic decomposition of output were different from those of Rose, Chen, and Adams (1996) for the United States. Whereas the latter study suggested that TC and FDM effects are contributing to lower environmental pressures, in our study these effects generally represent the main driving forces behind increases in physical throughput. The differences cannot be explained conclusively, because the results are based on distinct decomposition methods.

This chapter also introduced the use of SDA results as a basis for scenario analysis. As an example, an optimistic scenario was analyzed that shows that absolute dematerialization of primary plastics requires that all decomposition effects assessed for the period 1990 to 1997 undergo large, favorable changes during the period 1997 to 2030.

Notes

1. Moll et al. (1998a, 1998b) will not be reviewed because these studies do not decompose individual bulk materials, but an aggregate material indicator called total material requirement (TMR).

2. In a closed I/O model, households and sometimes investment are included in the intermediate input matrix (Miller and Blair 1985). In the case of Rose, Chen, and Adams (1996), the columns for replacement investment and consumption, which are final demand categories in the open I/O model, are included in the I/O coefficient matrix. The corresponding rows of wage and capital depreciation of the primary inputs are part of the intermediate input matrix.

3. Rose, Chen, and Adams (1996) report that the level effect is not precisely the same for all materials because of rounding errors and problems in obtaining constant price tables.

4. The physical I/O table that has been constructed for the Netherlands, Germany, and Denmark is similar to the physical supply and use tables derived here (Konijn, de Boer, and van Dalen 1997; Stahmer, Kuhn, and Braun 1997; and Gravgård-Pedersen 1999).

5. See Konijn (1994) and Konijn and Steenge (1995) for a detailed description of the construction of I/O tables. These publications propose an activity-by-activity I/O table that takes the production process as the basis for classification.

6. Rose, Chen, and Adams (1996) report a sharp decrease in the demand for iron and steel for the United States over the period 1972 to 1982. Table 4.4 shows that the use of primary iron and steel, and iron and steel products, are increasing for the Netherlands.

7. Because we are interested in developments in the physical economy, the results for the forty-six other commodities have been excluded from the table.

References

Ang, B. W., and P. W. Lee. (1996). Decomposition of industrial energy consumption: The energy coefficient approach. *Energy Economics* 18: 129–143.

Casler, S., and A. Z. Rose. (1998). Carbon dioxide emissions in the U.S. economy. *Environmental and Resource Economics* 11(3–4): 349–363.

CBS (Centraal Bureau voor de Statistiek). (2001). IJzer, staal en kunststoffen in de Nederlandse economie, 1990 en 1997 (Iron, steel and plastics in the Dutch economy, 1990 and 1997). Nota no. 01753-01-MNR, Sector Nationale Rekeningen, Voorburg, the Netherlands.

de Bruijn, S. M., and R. J. Heintz. (1999). The environmental Kuznets curve hypothesis. In J. C. J. M. van den Bergh (ed.), *Handbook of Environmental and Resource Economics*, 656–677. Cheltenham, England: Elgar.

Dietzenbacher, E., and B. Los. (1998). Structural decomposition techniques: Sense and sensitivity. *Economic Systems Research* 10(4): 307–323.

Duchin, F., and G. M. Lange, with K. Thonstad and A. Idenburg. (1994). *The Future of the Environment: Ecological Economics and Technological Change.* New York and Oxford: Oxford University Press.

Gravgård-Pedersen, O. (1999). *Physical input-output tables for Denmark: Products and Materials, 1990; Air Emissions, 1990–92.* Copenhagen: Statistics Denmark.

Hoekstra, R. (2003). Structural Change of the Physical Economy: Decomposition Analysis of Physical and Hybrid-Unit Input-Output Tables. Tinbergen Institute Research Series no. 315. Ph.D. diss., Thela Thesis Academic Publishing Services, Amsterdam.

Hoekstra, R., and J. C. J. M. van den Bergh. (2002). Structural decomposition analysis of physical flows in the economy. *Environmental and Resource Economics* 23: 357–378.

Hoekstra, R., and J. C. J. M. van den Bergh. (2003). Comparing structural and index decomposition analysis. *Energy Economics* 25: 39–64.

Konijn, P. J. A. (1994). The Make and Use of Commodities by Industries: On the Compilation of Input-Output Data from the National Accounts. Ph.D. diss., Universiteit Twente, Enschede, the Netherlands.

Konijn, P. J. A., S. de Boer, and J. van Dalen. (1997). Input-output analysis of material flows with application to iron, steel and zinc. *Structural Change and Economic Dynamics* 8: 129–153.

Konijn, P. J. A., and A. E. Steenge. (1995). Compilation of input-output data from the national accounts. *Economic Systems Research* 7(1): 31–45.

Leontief, W., and D. Ford. (1972). Air pollution and economic structure: Empirical results of input-output computations. In A. Brody and A. P. Carter (eds.), *Input-output techniques*, 9–30. Amsterdam: North-Holland.

Miller, R. E., and P. D. Blair. (1985). *Input-output analysis: Foundations and extensions.* Englewood-Cliffs, NJ: Prentice-Hall.

Moll, S., F. Hinterberger, A. Femia, and S. Bringezu. (1998a). Ein Input-output-Ansatz zur Analyse des stofflichen Ressourcenverbrauchs einer Nationalökonmie: Ein Beitrag zur Methodik der volkswirtschaftlichen Materialintensitätsanalyse (A preliminary input-output analysis of resource use by the national economy: A contribution to the method of economic analysis of material intensity). Paper presented at Sixth Stuttgarter Input-Output Workshop, Stuttgart, February 19–20.

Moll, S., F. Hinterberger, A. Femia, and S. Bringezu. (1998b). An input-output approach to analyse the total material requirement (TMR) of national economies. Paper presented at Conaccount Workshop: Ecologizing Societal Metabolism. Designing Scenarios for Sustainable Materials Management, Amsterdam, November 21.

Proops, J. L. R., M. Faber, and G. Wagenhals. (1993). *Reducing CO_2 Emissions: A Comparative Input-Output Study for Germany and the UK.* Berlin: Springer-Verlag.

Rose, A. (1999). Input-output decomposition analysis of energy and the environment. In J. C. J. M. van den Bergh (ed.), *Handbook of Environmental and Resource Economics*, 1164–1179. Cheltenham, England: Elgar.

Rose, A., and S. Casler. (1996). Input-output structural decomposition analysis: A critical appraisal. *Economic Systems Research* 8(1): 33–62.

Rose, A., and C. Y. Chen. (1991). Sources of change in energy use in the U.S. economy, 1972–1982. *Resources and Energy* 13: 1–21.

Rose, A., C. Y. Chen, and G. Adams. (1996). Structural Decomposition Analysis of Changes in Material Demand. Working paper, Department of Mineral Resources, Pennsylvania State University.

Siegel, P. B., J. Alwang, and T. G. Johnson. (1995). Decomposing sources of regional growth with an I/O model: A framework for policy analysis. *International Regional Science Review* 18(3): 331–353.

Stahmer, C., M. Kuhn, and N. Braun. (1997). *Physical Input-Output Tables for Germany, 1990.* Working paper no. 2/1998/B/1, German Federal Statistical Office, Wiesbaden.

Wier, M., and B. Hasler. (1999). Accounting for nitrogen in Denmark—A structural decomposition analysis. *Ecological Economics* 30: 317–331.

III Projective Analysis of Structural Change

5 Dynamic Industrial Systems Analysis for Policy Assessment

Matthias Ruth, Brynhildur Davidsdottir, and Anthony Amato

5.1 Challenges for Dynamic Industrial Systems Analysis

Assessing the dynamics of industrial systems and their environmental performance can be tricky business, yet is important to help guide today's policy choices. If, for example, constraints on technological and managerial improvements are underestimated, then future material use, energy use, and emissions of an industry may exceed expectations. As a consequence, policies implemented today to reduce material use, energy use, or emissions may be too lax to achieve desired environmental performance goals (see the discussion of dematerialization in chapter 3). Conversely, if assessments are primarily guided, for example, by an analysis of periods in which an industry has changed only slowly, then future reductions in material use, energy use, and emissions are likely to be underestimated if the constraints that were relevant in the past are no longer as binding. Such underestimates may potentially prompt the enactment of policies today that are overly restrictive and may possibly even hamper the future environmental and economic performance of an industry. These policies may include higher than necessary taxes or tougher prescriptions for technology choice.

Dealing with constraints on, and potentials for, improvements in an industry's environmental performance poses formidable methodological challenges as well as challenges for decision support. In creating dynamic industrial systems models for the U.S. Environmental Protection Agency (EPA) to explore likely consequences of climate change policies for energy-intensive industries, we encountered many such challenges. These two types of challenges are discussed, in turn, in sections 5.2 and 5.3, respectively. Subsequently, in section 5.4, we present a modeling framework for dynamic industrial system analysis followed by three case studies in which the framework is implemented for three industrial systems: the U.S. pulp and paper, iron and steel, and ethylene industries. Simulation results from the three industrial models are discussed, and the chapter closes in section 5.5 with a set of lessons for dynamic industrial system analysis and climate change policy.

5.2   Methodological Challenges

5.2.1   System Boundaries

Research that attempts to illuminate the likely causes and potential future trajectories of industrial energy use and emissions faces several formidable methodological and empirical challenges. First, industrial systems are complex hierarchical systems, encompassing a plethora of processes and technologies that are highly interconnected at the firm and industry levels. Adjustments in one process or sector may require, or presume, adjustments in other processes or segments of the same sector or other sectors of the economy. For example, increasing the use of recycled fibers by the U.S. pulp and paper industry has implications for the extent to which self-generation of electricity, heat, and steam can take place. This is because black liquor, a significant energy carrier for self-generation, is derived from the chemical processing of virgin fibers. An increase in the use of recycled fibers may reduce availability of black liquor and thus limit the potential for further expansion of self-generation in the pulp and paper industry. Shifts in material markets can also have implications beyond industrial boundaries, such as in cases in which firms self-generate more electricity from wastepaper incineration or from energy recovery than they themselves consume. These firms may feed excess generation into the electricity grid and thus meet peak load demands for electricity that might have otherwise been met by generation from traditional power plants.

In principle, input-output analysis, and specifically energy input-output analysis (Hannon 1973; Bullard and Herendeen 1975; Bullard, Penner and Pilati 1978) may be used to trace implications that changes in one part of the economy have for the performance of other parts of the economy and to explicitly account for the indirect effects that such changes may have in the economy as a whole. However, input-output analysis is often difficult to apply within a dynamic context at the level of detail relevant for decision makers in industry. Methodological and data problems compound each other to the point that dynamic input-output analyses are often mere academic exercises with little relevance to decision makers in industry and policy (see a discussion and use of input-output models in chapters 4 and 7).

Understanding the dynamics of industrial technology change and its associated impacts on energy and emissions profiles requires multiple perspectives to adequately capture changes at the process, firm, and industry levels. Ideally, industrial systems analysis would capture industrial processes in sufficient detail to be relevant and meaningful to decision makers at the process level and, at the same time, provide opportunities to capture the ramifications of process-specific changes at the industry aggregate level. Since it is often not entirely clear a priori what constitutes a specific industry, it becomes even more important to retain process-specific information for industrial systems analysis. As system boundaries change, new processes may be added, or some that were previously considered, removed.

Furthermore, how an industry may be defined for analytical purposes can vary significantly depending on what those purposes are. For the steel industry, markets of both inputs and outputs differ between integrated mills and electric arc furnaces, and as a consequence, analysis of the markets in which the steel industry operates will differ depending on whether both parts of the industry are treated separately or lumped together into one aggregate. Similarly, because the different parts of the steel industry use fundamentally different technologies, an assessment of the impacts of emissions from steel production on environmental quality will differ as well. For example, a significant share of energy use and emissions by integrated steel mills is associated with the production of oxygen, which is typically not within the system boundaries set, for example, by the system of industrial classifications (SIC) or by the new North American Industry Classification System (NAICS) classifications. In contrast, the bulk of energy conversion and greenhouse gas emissions associated with the use of electric arc furnaces are generated in the electricity sector of the economy. A shift from integrated mills to electric arc furnaces will thus have fundamentally different implications for the industry's energy use and emissions profiles based on whether or not the ancillary processes of oxygen production and electricity generation are accounted for.

5.2.2 Industry Dynamics

A second challenge for research dealing with causes and potential future trajectories of industrial energy use and emissions is associated with the treatment of industrial change through time. Many analyses of the performance of industrial systems for the purpose of climate change policy assessment use general equilibrium models that describe, at one point in time, interactions among demand and supply of marketed inputs and outputs; characteristics of technologies, such as energy efficiencies and emissions coefficients; and policy variables, such as subsidies and tax rates (e.g., Goulder 1995; Goulder and Schneider 1999). A set of equations is solved for prices that achieve equilibrium at one point in time in the various energy, material, labor, and product markets captured in the model. Then a set of new technological or policy conditions is imposed for a subsequent period for which the model is solved. Comparison of model results from period to period can then be used to infer potential industrial change over a series of equilibrium points.

An alternative viewpoint posits that industrial systems are in constant flux, or disequilibrium. Choices of "optimal" investment in new technology are guided by essentially unknown future conditions in input and output markets; thus perfect foresight is not obtainable, and instead behavior may be myopic. In addition, since there is often a significant time lag of several years between investment in new capital and bringing that capital on line in the production process, investment in capital that may be optimal from the perspective of one point in time may be too large or too small by the time that capital can actually be used (Jorgenson 1996). As a consequence of uncertainties and time lags, industry continuously adjusts its investments in efforts to close the gap between desired and actual investment yet is unlikely ever to be in equilibrium (Jorgenson 1996).

5.2.3   Representation of Technology

Closely related to the choice between an equilibrium and a non-equilibrium perspective are the ways in which technology is represented. One convenient representation of production processes is with aggregate production functions that relate quantities of industry output to input quantities. For purposes of convenience and analytical tractability, aggregate production functions are typically assumed to meet certain mathematical criteria, such as substitutability among inputs and homogeneity of the capital stock (Doms 1996). The latter assumption may prove problematic when investments are lumpy and occur at irregular intervals and when technical change leads to marked differences between existing and new capital.

As an alternative to using aggregate production functions, an industry's capital stock may be disaggregated into different age classes, each of which may be characterized by vintage-specific efficiencies, substitution possibilities, and capacity utilization rates. Aggregate industry performance is then a function of the industry's capital vintage structure (Meijers 1994; Mulder, de Groot, and Hofkes 2003).

Since a vintage analysis avoids the assumption of a homogenous capital stock, it enables an explicit analysis of the vintage-related potential of input substitution via putty-putty, putty-clay, and clay-clay models. In putty-putty models, input substitution is equally possible for new and existing capital; in putty-clay models, substitution is possible only for new capital; and in clay-clay models, substitution among inputs is possible neither for new nor for existing capital (Meijers 1994; Mulder, de Groot, and Hofkes 2001; Jacoby and Wing 1999). The choice of an appropriate representation of substitution possibilities depends upon the level of aggregation. In highly aggregate models, such as models that represent the entire manufacturing sector as one aggregate, a putty-putty representation may be appropriate, whereas in models examining a particular technology, a clay-clay representation may be appropriate, since little flexibility is available within individual technologies (Davidsdottir 2002). A model of a particular industry falls somewhere between the two levels of resolution and is likely to be best represented by a putty-semiclay representation.

Technological change, defined as a reduction in the intensity of use of one input while holding the intensity of other inputs constant, may occur either in embodied or disembodied form. Embodied technological change occurs as a change in input efficiencies of the new capital stock and can be brought about through capital investment in new, more efficient technologies (Solow 1957; Berndt, Kolstad, and Lee 1993). Thus, it is through investment in new capital that these more efficient technologies become embodied in the capital stock. Yet the structure of new investments is not immune to the structure of the existing capital stock, because as a result of path dependency and capital inertia, the structure of the existing stock influences the choice of which new technology to purchase (Arthur 1994; Unruh 2000). Disembodied changes imply low-cost changes in the input efficiency of the already installed capital stock, such as through improved housekeeping and learning (Ross 1991a, 1991b).

The two types of technological change influence new and existing capital differently. Furthermore, the potential for input substitution varies among vintages. Consequently a heterogenous representation of the capital stock is important for a realistic representation of the dynamics of technological change and input substitution and thus ultimately for a realistic representation of industrial responses to policies that affect technology choice.

5.2.4   Bottom-Up and Top-Down Modeling

Perhaps the most frequently debated issue in the energy and climate change modeling literature is whether to choose a bottom-up or a top-down approach for the representation of energy use technologies (IEA 1998). The bottom-up perspective relies on detailed information about the cost and performance of individual technologies to choose the appropriate combination of energy-using technologies to minimize the cost of energy use (e.g., Market Allocation, or MARKAL [Fishbone and Abilock 1981]). The top-down approach does not represent individual technologies but rather examines the impact of policy on aggregate energy demand and aggregate energy efficiency (e.g., Dynamic General Equilibrium Model, or DGEM [Jorgenson and Wilcoxen 1990], and Regional Integrated Climate Model, or RICE [Nordhaus and Boyer 2000]). The two opposites usually reach very different conclusions regarding the cost of carbon emission reductions: The bottom-up approach is often considered overly optimistic and the top-down approach too pessimistic. The former approach overwhelmingly suggests that many reductions in carbon emissions can be achieved at low economic cost or even with economic benefits, whereas the latter suggests that reductions in carbon emissions may come at considerable cost to society.

The main reasons for these two divergent conclusions stem from the way the two approaches represent technology and the possibilities for technological change. The top-down approach assumes that the economy is operating efficiently at its production possibility frontier. Consequently, there cannot exist any "no regrets" efficiency improvement. In contrast, the bottom-up approach accounts for a wide range of (potentially) available technologies—whether they are used or not—and thus allows for the possibility that the economy is not operating efficiently, and that there are technologies that can be implemented at low or even negative cost. However, the bottom-up approach does not account for any market imperfections, such as lack of information, transaction costs, or path dependency or inertia (IEA 1998). Ruth et al. (1999) argue that models that occupy the space between the extremes delineated by traditional bottom-up and top-down models may be most appropriate for analysis of industrial dynamics in the context of policy change. Examples of such models include, most notably, National Energy Modeling System (NEMS) and Intra-Sectoral Technology Use Model (ISTUM) (Ruth et al. 1999).

5.2.5   Regional Aggregation

In industrial analysis the choice of regional scale should depend on the structure of the industry and the policy questions asked. To date, climate change modeling studies have largely been carried out at the national or global levels. The results of analyses at these highly aggregate levels are used to infer impacts and responses at local levels, despite the lack of regional specificity (Wilbanks and Kates 1999). If an industry is regionally homogenous, a national analysis of industrial dynamics is sufficient. However, if the industry is regionally heterogenous, a lack of regional detail is likely to result in the omission of important industrial dynamics, which potentially carry relevance for decision making. For example, in the iron and steel industry, energy-intensive integrated production is concentrated in the Midwest region of the United States. Energy and climate policies that have an impact on the industry may affect Midwestern steel firms and their employees differently than they affect firms and employees in regions in which steel is predominantly produced in electric arc furnaces.

5.2.6   Time Frames

A sixth methodological challenge lies in choosing appropriate representations of time frames for analysis and of associated uncertainties about an industry's internal structure, technology, or marketplace. Over long time frames, new technologies may emerge, or unforeseen shocks to market dynamics may occur, rendering deterministic specifications of technology change and production and investment decisions increasingly inadequate. Evolutionary models, in contrast, attempt to capture qualitative changes in boundary conditions and their influence on industry behavior. The purpose of these models is typically to explore alternative long-term trajectories of system change, placing significantly greater emphasis on the mechanisms by which change is achieved and the direction that change is taking under a range of alternatives, rather than on the numerical accuracy of model results under narrowly specified assumptions.

5.3   Challenges for Decision Support

In addition to challenges associated with methodological choice and empirical support for studies on the dynamics of industrial systems and their responses to policy intervention, research attempting to inform investment and policy decision making must be sensitive to, and address, the following issues. First, data collection, data analysis, model development, and model calibration are often costly in terms of the time and money required to carry out each task. In the research process, a host of methodological and empirical choices must be made that may ultimately limit the range of policy and investment decisions that can be meaningfully captured by the resulting model, because the choice of methodological and empirical approach is closely intertwined with the policy space to which the analysis can be applied. For example, a general computable equilibrium analysis that represents production with aggregate production functions will be inadequate for investigating the degree to which accelerated depreciation schedules influence the turnover of existing capital within specific industries.

Second, should the model be predictive or descriptive? There is widespread and growing consensus that models of industrial material use, energy use, and emissions cannot be predictive, because such models involve a multitude of psychological, social, technological, economic, and political factors that can never be known with sufficient certainty to render an analysis useful for forecasting. For instance, the breakdown of the Soviet Union led to a collapse and subsequent major restructuring of several metals markets that could not be readily anticipated or evaluated with respect to their implications for the metals industries' material, energy, and emissions profiles (Ruth 1995). Similarly, the invention of Corex production and direct reduced iron (DRI) may revolutionize iron and steel production. In light of the limited predictability of industrial material use, energy use, and emissions, modelers increasingly employ models for descriptive purposes. One way to do so is to play out model scenarios under alternative assumptions about data, functional relationships, and investment and policy decisions. Emphasis remains on the model as a computational tool with which to derive insights into potential systems trajectories, given certain assumptions, but not as an exact predictor of the future.

Third, industrial systems models are frequently large—often encompassing many hundreds of equations derived from disparate data and solved simultaneously with numerical algorithms—and as a consequence are prone to a multitude of errors, many of which may confound each other. Yet at the same time, such models ideally present a consistent way of dealing with disparate data: the best job that can be done, given various constraints on the assessment project and assuming one wants to do more than simply deal with the already better-understood issues surrounding smaller subsystems within an industry. Consequently, researchers must walk a fine line between emphasizing the various strengths of their analysis and paying tribute to the uncertainties that surround their findings. Walking that fine line is often a tricky balancing act in which scientific credibility and policy relevance may quickly be lost.

Fourth, if it is indeed a purpose of an analysis to inform investment and policy decisions, then it will not be sufficient to simply claim in a study's conclusions that there is "policy relevance." Instead, decision makers in the industry and policy arenas need to be involved as judge and jury (Ruth 2001). Which stakeholders to include, and at what stages of a project to bring them in, is important to the credibility of a study in both the scientific and decision-making arenas. Ideally, in a democratic society, those who have a stake in a decision should be invited to participate in the decision-making process. If the decision is made, in part, on the basis of a formal analysis or model, then stakeholders should contribute to the identification of relevant system boundaries and level of system aggregation, choice of data, methodology, generation of scenarios, and interpretation of results. In the process of being involved, they may provide valuable information typically not reported in the scientific literature, and they may offer challenges to modelers and models that can help either verify or modify basic assumptions. However, stakeholders may not perfectly represent society at large, and thus their input may skew the study toward the interests and opinions of a few. Their influence in system boundary definition, data selection, scenario development, and model interpretation may reduce the scientific neutrality of a study and, as a result, reduce its usefulness in policy discourse. How to maximize the benefits from stakeholder involvement in industrial systems analysis without suffering from its pitfalls is often in part determined by the issues at hand and the leadership styles of the individuals involved in the process.

The goal of involving stakeholders in industry analysis also has implications for the choice of modeling environment. Traditional computer models, with their line-by-line code and lack of user-friendliness, are less likely to engage stakeholders than some of the more recently developed graphical programming languages. Yet the latter are often derided by "hard-core modelers" for their limited availability of solution algorithms and limited ability to interface with other languages or across computational platforms. The debate about modeling languages on occasion deteriorates into comparisons of pros and cons of the pet languages of individual participants at the expense of more substantive debate about the issues at hand. As so often is the case, the balance will need to be found on the basis of the policy questions and the expertise of scientists and stakeholders who are charged to find answers to these questions (Ruth and Hannon 1997; Hannon and Ruth 2001).

5.4 Industrial System Modeling: Three Case Studies

In this section we first describe a general framework for dynamic industrial systems analysis followed by the implementation of that framework in three case studies. The three case studies were performed at the request of the EPA to explore the likely consequences of climate change policies for energy-intensive industries in the United States. The first subsection presents the general modeling framework. The case studies themselves are presented in subsections 5.4.2–5.4.4, followed by simulation results in subsection 5.4.5.

The general modeling framework is based on the same choices among the methodology options laid out in the previous section. System boundaries are defined such that the main processes that matter for material and energy flows are included explicitly within the framework, and the level of aggregation is such that it enables us to capture specific production processes in sufficient detail to be relevant and meaningful to decision makers and, at the same time, provide opportunities for capturing the ramifications of process-specific changes at the industry aggregate level. Thus in the framework we distinguish the energy used to drive the main processes by means of which raw materials are converted to semi-finished or finished products. Technology descriptions are more detailed than can be found in many other industry models designed to explore climate change policies (e.g., DGEM, NEMS, LIEF [Long-term Industrial Energy Forecasting (Ross et al. 1993)]), but less detailed than is typical for engineering analyses (e.g., MARKAL). As a result, since the modeling framework presented in this chapter includes technological descriptions combined with a top-down representation of the economics that stimulate changes in the capital stock, this framework exists at the interface between bottom-up and top-down models.

The framework is based on a nonequilibrium approach to portraying time-varying behavior in choices about capital investment, technology, and outputs. What distinguishes the framework the most, however, from that of other models that exist at the interface between bottom up and top down (e.g., LIEF, NEMS) is the careful accounting of the size and structure of the capital stock, using so-called vintage accounting. This allows us to capture the heterogeneity of the capital stock, enabling us to include both embodied and disembodied technological change as well as path dependency in technology choice. We also explicitly model self-generation of energy as well as electricity generation outside the industrial system.

5.4.1 Methodology
The general modeling framework is built from six interacting modules:

- The production module, which describes annual production levels.
- The capital vintage module, which describes the vintage structure of the capital stock and simulates changes in the size of existing vintages and the addition of new vintages.
- The energy intensity module, which relates energy intensity by vintage based on engineering and economic parameters, giving total aggregate energy use as a function of the size of each vintage, capacity utilization rates, and vintage-specific energy intensity.
- The energy mix module, which quantifies changes in the fractional shares of each energy type, thus enabling the total energy use vector to be broken up into energy use by type.
- The electricity production module, which simulates changes in the type of energy required to produce purchased electricity.
- The carbon dioxide emissions module, which simulates total carbon dioxide emissions as a function of purchased energy use.

Each module consists of several equations that either are econometrically estimated based on twenty-six years of historical data or are based on engineering parameters. Each estimated equation is then entered into dynamic simulation model, and run simultaneously to explore potential energy use and carbon emission futures.

The *production module* describes production levels as an econometrically estimated function of demand and exogenous economic variables, for example, income or GDP and output/input prices as well as the trade-weighted value of the dollar. In addition, desired production levels are limited by the size of the installed productive capital stock.

The *capital vintage module* captures the actual structure of the capital stock in 1996; the capital stock is broken down into vintages dating back to 1960. Then changes in the structure and size of the capital stock are simulated using a perpetual inventory (Jorgenson 1996). A perpetual inventory assumes that the capital stock $K(t)$ of an industry in year t is a function of the remaining stock (K) of the previous year $t-1$ and gross investments (I) that occurred in time period $t - x$. After a lag of x years, gross investment is added to the productive capital stock in year t:

$$K(t) = I(t - x) + (1 - \mu(t)) \cdot K(t - 1).$$

Consequently, the actual size of the capital stock is never equal to the desired size of the capital stock because of a system-specific time lag. The deterioration rate $\mu(t)$ is a physical measure of the extent to which capital becomes unavailable. The magnitude of the deterioration rate is a function of the capital stock's maximum lifetime—its service life—and gross new capital investment. Gross capital investment is the sum of expansion and replacement investment, where replacement investment directly replaces retired and deteriorated capital. We assume that deterioration, and thus retrofits, equal a fixed percentage of gross investment, distributed equally across all vintages. The maximum lifetime of the capital stock is fixed, exogenously defined, and derived from data reported by government agencies (e.g., EIA 2000) and industry organizations (e.g., AF&PA, *Paper, Paperboard and Woodpulp*, various years; *Iron and Steelmaker*, various years, *Oil & Gas Journal*, various years). Capital utilization rates are assumed to be uniform across vintages and equal to the average capital utilization rate. Average capacity utilization is estimated as the ratio of installed capacity to total production volume and thus varies from year to year and is endogenous to the model.

The *energy intensity module* captures total use of bundled process energy measured in British thermal units (BTUs) as a function of vintage-specific energy intensity and vintage-specific production levels. Energy intensity is defined for each new vintage class (embodied change) as a function of the average efficiency of the existing capital stock and the relative energy intensity (REI) of new versus old capital as described by technology possibility curves by production process as estimated by EIA (2000). The REI is thus an exogenous parameter that describes the average economically feasible reduction in energy intensity of new capital as a function of the average energy intensity of existing capital. As an example, if the REI equals 0.9, then new capital is 10% more efficient than the average energy intensity of the existing capital stock. Consequently, technological change is simulated as embodied in new vintages, and new investment gradually reduces the energy intensity of the aggregate capital stock based on the assumption that the efficiency of new capital is slightly higher than the average efficiency of the existing capital stock. Changes in energy intensity of the existing capital stock are influenced by disembodied change and are either driven by autonomous change or endogenously by learning and changes in energy prices. Substitution between energy and capital is thus represented as putty, semiclay.

The *energy mix module* captures econometrically estimated changes in the fractional shares of each fuel type as a function of own and cross-price elasticity and output mix. Switching between fuel types is represented as putty-putty and thus possible and equally easy for all vintages. The fractional share of each fuel type multiplied by total energy use from the energy intensity module gives total energy use by type.

The *electricity production module* captures the fuel mix used to generate electricity, capturing both generation and transmission losses to depict accurately the quantity and type of fuel needed to produce the electricity purchased and consumed in each industrial system.

Then finally, the *carbon dioxide emissions module* captures total carbon emissions by summing the product of fuel-specific carbon emissions coefficients and fuel use by fuel type. Great care is taken not to double-count carbon emissions from the use of the various by-fuels. Carbon emissions from the combustion of biomass-based fuels are traced by the model framework, but not reported as a part of total carbon emissions, because carbon emissions from biomass are assumed to be carbon neutral. Furthermore, no carbon emissions are traced for that portion of fossil fuels that is used as feedstock, for example, in ethylene production as raw material, instead of being combusted for its energy content. However, it must be kept in mind that some of the fossil fuels that get embodied in the product are ultimately released into the environment because a share of the materials produced from, for example, ethylene ultimately get incinerated. Consequently, the framework underestimates total carbon emissions due to nonenergy use of fossil fuels to the extent that the carbon captured in fossil-based products is ultimately released into the environment.

After the generic modeling framework is developed, it is implemented for three industrial systems, pulp and paper (see section 5.4.2), iron and steel (see section 5.4.3), and ethylene production (see section 5.4.4) by establishing the initial vintage structure and by estimating each industry-specific equation within each individual module. Fixed parameters are compared across the three industrial systems in table 5.1, and main significant drivers of change in key endogenous variables are compared in table 5.2. Time series analyses quantify changes in industrial fuel mix, energy intensity, production levels, gross investment, and other key variables. Changes in fuel mix are captured using seemingly unrelated

Table 5.1
Comparison of fixed parameters in the three industry models

	Pulp and paper	Iron and steel	Ethylene
Maximum lifetime of capital (years)	35	20 (electric arc furnaces)	25
REI	0.85	.96	.96 (ethane, naphtha) .976 (butane, propane)
Time lag between investment and actual use of capital (years)	2	1	4
Number of fuel types represented in model	7 (including electricity)	6 (including electricity)	5 (4 purchased and electricity)

regressions (Zellner 1962) to capture simultaneity in fuel choice. Simultaneity in fuel choice occurs if an industry expands use of natural gas as a fuel source. Then that expansion has an impact on the choice of all other fuels. Furthermore, all regression equations are subjected to diagnostic tests for heteroscedasticity and serial correlation (Breusch 1978; Breusch and Pagan 1979).

The six estimated modules based on capital vintage, engineering, and time series analyses are then combined into a dynamic simulation model (figure 5.1), one for each industrial system, which then is used to infer possible energy and carbon emissions futures, either without any change in the "business-as-usual" external economic environment or alternatively in the presence of policy interventions. Intervention in the models' dynamics can occur through a variety of investment and policy decisions, many of which can be chosen interactively by model users. To facilitate interaction with the model, for each model a user interface is made available, through which the user can choose alternative data and functional specifications. User interfaces include background information on the particular industry, various model components, and instructions for model use.[1] A small set of possible interventions is discussed in section 5.4.5.

5.4.2 Pulp and Paper

The pulp and paper industry is the second-most energy-intensive industry in the United States. It accounts for over 12 percent of total manufacturing energy use and contributes 9 percent of total U.S. manufacturing carbon emissions (Martin et al. 2000). Over 55 percent of the energy used within the industry is self-generated as a by-product of chemical recovery,

Table 5.2
Main drivers of change in key endogenous variables

Endogenous variable	Drivers		
	Pulp and paper	Iron and steel	Ethylene
Demand	Output prices, GDP, GDP per capita	Output prices, GDP, GDP per capita	assumed = supply
Supply	Demand, input prices	Demand, trade-weighted value of dollar	GDP
Gross investment	Input prices, production levels	Output prices, production levels	Capacity utilization, input prices, production
Capital turnover	Fixed fraction of gross investment; Lifetime of capital	Electric arc furnaces: lifetime of capital; Blast furnace: variable retirement rate	Lifetime of capital
Fuel shares	Relative and absolute fuel prices, output mix	Electric arc furnace capacity share, relative and absolute fuel prices	Feedstock shares

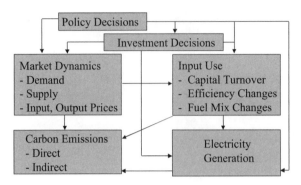

Figure 5.1
Structure of the dynamic industry models.

with the remainder divided among natural gas, coal, electricity, and residual fuel oil. The expansion of self-generated energy is limited by the
use of chemical pulping processes of virgin fibers, which indicates that
increased use of recycled fibers may slow the rate of increase in energy
self-generation of heat, steam, and electricity from waste-to-energy conversion by the industry. Conversely, an increase in the use of recycled
papers reduces total energy requirements in the industry but increases
purchased energy requirements if waste pulping replaces chemical pulping, which immediately has implications for carbon emissions (Ruth and
Harrington 1997; Davidsdottir 2002).

In the past two decades, new capital expenditures in the paper industry
as a proportion of sales have averaged 9 percent, which is twice the average for all manufacturing industries and is matched only by that for the
chemical industry (Slinn 1992). Yet the expansion in output has been only
modest and not significantly higher than in other manufacturing industries. These figures highlight the immense capital intensity of the paper
industry. A comparison of replacement and expansion investment in the
industry demonstrates that expansion investment is consistently larger
than replacement investment and has resulted in a constant increase in
capacity throughout the years without a concomitant turnover of older
vintages (AF&PA, *Paper, Paperboard and Woodpulp,* various years).
Equipment as old as one hundred years or more is still in use within the
industry (Miller Freeman 1997).

The high capital intensity of the industry, coupled with the relatively
low retirement rates (on average 1.8 percent), may have provided a formidable barrier to changes in material and in particular in energy inten-

sity, since the vintage structure of the capital stock tends to be long and rigid, and any change expensive (AF&PA, *Paper, Paperboard and Wood-pulp*, various years). This may translate into a slow response to any policy since the short-run effects of policies will be only small changes in the industry's performance, yet at the same time these small changes are very expensive.

Average material intensity has not changed markedly for the last thirty years, yet over the same time frame, energy intensity has decreased from 39.3 million BTUs per ton of output in 1972 to 29.32 million BTUs per ton of output in 1998. This represents an annual decline in energy intensity of 0.9 percent. Despite those gains in energy efficiency, energy use in the industry has continued to increase as a result of increases in total paper production, which have averaged more than 2.9 percent annually for the last thirty years, but the rate of increase has declined in recent years and was 2.2% in 1999 (AF&PA, *Statistics of Paper, Paperboard and Woodpulp*, various years).

System boundaries are drawn to enclose both purchased and self-generated energy used to drive all the major processes in the production of paper and paperboard, from the pulping stage to the refinement of paper products (figure 5.2). We exclude the energies used in the forestry sector to produce capital and to collect wastepaper. Generation of purchased electricity is included as a separate model component. The calculation of carbon emissions includes emissions from the use of all purchased fuels and electricity. Emissions from self-generation are excluded, because such emissions result from the use of biomass-based fuels, which may be considered carbon neutral under the assumption that no net deforestation of tree species used for pulp and paper production takes place.

5.4.3 Iron and Steel

The iron and steel industry is the fourth-most energy-intensive industry in the United States and accounts for 9 percent of manufacturing energy use or 2.3 percent of all energy use (Office of Industrial Technologies 2002). The industry relies heavily on coal, as approximately 60 percent of energy consumed is directly derived from coal, and indirectly a considerable share of the electricity purchased is produced in coal-fired power plants. Energy represents 15–20 percent of production costs in the industry (AISI 2002), which in 1994 translated into a $6.5 billion energy expenditure, roughly 9 percent of manufacturing energy expenditures in the United States (EIA 1997).

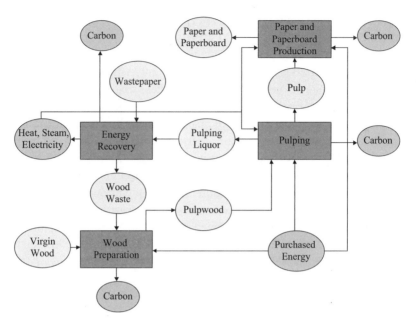

Figure 5.2
System boundaries for pulp and paper production.

U.S. iron and steel is a $57 billion industry with capital expenditures of $2.7 billion or 7.5 percent as a proportion of sales (AISI 1999). Today, the industry is the third-largest producer of steel in the world, with 54 percent of its output from integrated mills and 46 percent from electric arc furnaces. Pressures from other steel-producing nations have led to a consolidation of domestic producers in a series of plant closures and mergers and have prompted increased protection against foreign imports. No new capacity investments in basic oxygen furnaces have occurred in the United States since the 1960s, and future expansions of electric-arc-based production significantly depend on the industry's competitiveness with foreign producers.

Integrated mills predominantly use virgin iron ore to produce high-quality products commonly used in automobile manufacturing and appliances. The integrated process, however, is highly energy intensive, requiring blast furnace temperatures in excess of 3,000 degrees Fahrenheit in order to melt iron ore, reduce iron oxides to iron, and produce steel. In contrast, electric arc furnaces use recycled scrap steel, are relatively less energy intensive, and use electricity to produce their products. During the last twenty years the industry has drastically reduced capacity and increasingly shifted production to electric arc furnaces, which in turn has increased capacity utilization rates and decreased energy consumption. In fact, since the mid-1970s the industry's energy consumption per ton of steel produced has been reduced by 45 percent (AISI 2002). Increased energy efficiency and decreased production rates combined have resulted in significant decreases in energy use.

The iron and steel model distinguishes integrated and electric arc furnace production processes. Changes in their respective output shares are, in part, determined by changes in relative energy costs. Although products from the two routes of steel production have historically served separate markets, with output from electric arc furnaces somewhat confined to bars and beams and integrated mills more able to provide high-quality speciality products, those distinctions have eroded primarily as scrap preparation and removal of product impurities improved in both types of steel production.

The iron and steel model distinguishes integrated and electric arc furnace production processes, and changes in their respective output shares are, in part, determined by changes in relative energy costs. Integrated production (figure 5.3) includes preparation of coke from coal and associated by-products, such as coke oven gas (COG in the figure) used elsewhere in the industry as an energy carrier. Also included are production of pig iron from ore and scrap, generation and use of blast furnace gas (BFG), as well as conversion of several purchased fuels, most notably electricity, natural gas, and coal, and the calcination of limestone in basic oxygen furnaces. Carbon emissions result from energy conversions and

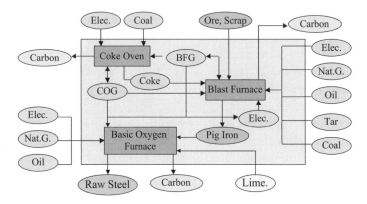

Figure 5.3
Coke oven, blast furnace, and basic oxygen production.

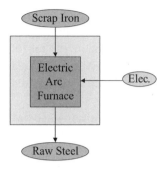

Figure 5.4
Electric arc furnace production.

limestone calcination and are calculated for each stage of the aggregate coke oven–blast furnace–basic oxygen route, including carbon emissions from electricity generated off site. In contrast, all carbon emissions from electric arc furnace production (figure 5.4) are from electricity generated outside the industry. Energy and emissions from the mining of iron ore, the recycling of scrap metal, and the conversion of raw steel to finished products remain outside the system boundaries of the model.

5.4.4   Ethylene

The U.S. chemicals industry accounts for 25 percent of manufacturing energy use, 7 percent of total energy consumption, and 2.6 percent of carbon emissions (Office of Industrial Technologies 2001). New capital expenditures in the industry as a proportion of sales are over 9 percent, and research and development expenditures are 5 percent. One of the largest chemical industry segments is devoted to the production of ethylene, which is the principal building block for the production of plastics and resins. The United States is the world's largest ethylene producer and currently accounts for 28 percent of world ethylene production capacity (*Oil & Gas Journal* 1998). Since 1980, U.S. ethylene production has grown by nearly 5 percent annually.

In the production of ethylene, hydrocarbon feedstocks—ethane, propane, butane, and naphtha—are heated or "cracked" in a pyrolysis furnace, separated into gaseous products, and then cooled and compressed into final products (figure 5.5). Most of the energy requirements are to carry out the pyrolysis process. Ethylene yields and process energy requirements are determined by the type of feedstock used and, to a lesser extent, processing conditions such as pressure, temperature, and residence

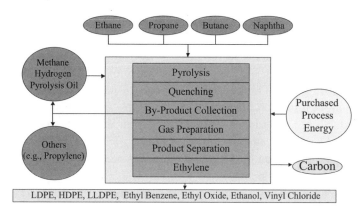

Figure 5.5
Ethylene production. LDPE: low-density polyethylene; HDPE: high-density polyethylene;
LLDPE: linear low-density polyethylene.

time (Worrell et al. 2000). When ethane is used as the feedstock, ethylene yields are highest, but additional energy needs to be imported for pyrolysis. In contrast, the other feedstocks yield relatively less ethylene, yet are self-sufficient in running the cracking process.

The U.S. ethylene industry is based on ethane (45 percent), propane/butane (27 percent) and naphtha (27 percent), which are by-products of natural gas and oil production. Because the refining industry is the major supplier of raw materials for ethylene production, more than 50 percent of all ethylene plants are located at refineries (Office of Industrial Technologies 2000). Ethylene is produced by cracking hydrocarbons under intense heat and then rapidly cooling the product. The majority of hydrocarbons in the production of ethylene are not combusted as a fuel but rather are used as a feedstock and contained in the product itself.

5.4.5 Results
For the results described in this section, each industrial model is run first under the assumption that no new external influences, such as external shocks to input or output markets or policies that otherwise affect an industry's choice of input and output, are exerted on the industry. All observed changes are the consequences of the models' underlying assumptions, which are derived on the basis of past observations. The resulting scenarios constitute a base case, against which we compare different alternatives for various policy interventions. One of these interventions occurs in the form of a policy that raises the cost of

carbon-containing fuels by $75 per ton of carbon for fuels purchased by the industry and used as an energy carrier. A $75 increase per ton of carbon translates into an increase in the price of coal of $1.92 per million BTU, a $1.61-per-million-BTU increase in the price of residual fuel oil, and a $1.08-per-million-BTU increase in the cost of natural gas. Those fuels used as input into a production process, such as some feedstocks in ethylene production, are assumed to be exempt from the policy because they do not lead to carbon emissions by the industry. The increase in cost of carbon may be prompted by a carbon tax or the need for industry to purchase carbon emissions permits.

A second policy scenario assumes a 10 percent decline in the energy intensity of new technology relative to the intensity of the aggregate existing capital stock in the industry. This improvement is over and above the efficiency gap that historically has existed between new and existing capital in each industry, thus essentially decreasing REI to 10 percent below its historical value. That does not mean that each year capital becomes 10 percent more efficient than in the previous year, because the efficiency gap is defined on the basis of the aggregate existing capital stock, which, for example, in the paper industry contains the past thirty-five years' worth of investments in capacity. What it does mean is that new capital becomes 10 percent more efficient than in the base scenario, essentially widening the gap between the efficiency of new capital and the average efficiency of the capital stock as the result of an improvement in the efficiency of new cost-effective technologies added to the capital stock. An increase in REI of 10 percent is entirely plausible. For instance, in the paper industry, a substitution of the more efficient new black liquor gasification technology for the traditional Tomlinson recovery boilers as the technology of choice for energy recovery would succeed in generating an increase at that level (Larson et al. 1998).

Since capital investments and retirements continuously change an industry's aggregate efficiency, widening the efficiency gap (for example, by stepping up research and development and promoting pilot projects), aggregate efficiencies improve faster than they otherwise would, and as a consequence, carbon emissions may be reduced. We refer to the corresponding policy scenarios as REI policy scenarios.

A third set of policy scenarios combines an increase in the cost of carbon with improvements in relative energy intensities. Here, we assume an added cost of $25 per ton of carbon in purchased fuels used as energy carriers and a 5 percent REI improvement.

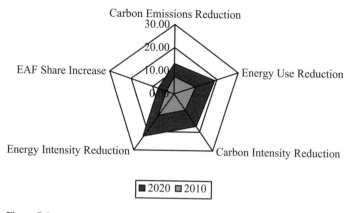

Figure 5.6
Steel industry scenarios with base case.

All policies are assumed to have been implemented in the year 2002. Each model is run for an annual time step to calculate the resulting market dynamics, changes in input use, and corresponding carbon emissions in each year and across time. To make it easier to compare the impacts that alternative policies have for each industry and across the various industries, we plot for key industry characteristics the observed percentage changes over the last year before policy implementation, that is, 2001. To at least partially reflect the dynamic element of industrial adjustments, we report the percentage changes for two years: 2010 and 2020. For example, figure 5.6 shows, for the years 2010 and 2020, respectively, a 7 percent and 13 percent decrease in carbon emissions by the iron and steel industry over the 2001 levels without any policy intervention. With the increase of $75 per ton of carbon in purchased fuels, carbon emissions reductions are 12 percent and 16 percent, respectively, in 2010 and 2020 (figure 5.7). The $75 cost-of-carbon increase policy also leads to a more rapid expansion in the share of electric arc furnaces, a slightly larger reduction in energy intensities (measured in BTU per ton of raw steel), a significant lowering of carbon intensities (measured in tons of carbon per ton of raw steel), and reduction in total energy use.

Comparison of the three policies investigated here (figures 5.7–5.9) shows that a 10 percent REI improvement results in less-aggressive expansion of electric arc furnace output (figure 5.8). To illustrate, in 2010 under the 10 percent REI improvement policy, the electric arc furnace share of output increases by just over 3 percent, whereas the $75 cost-of-

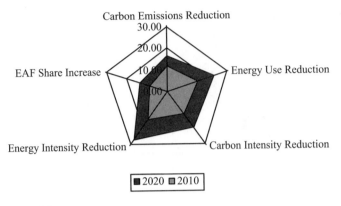

Figure 5.7
Steel industry scenarios with $75 increase in cost of carbon.

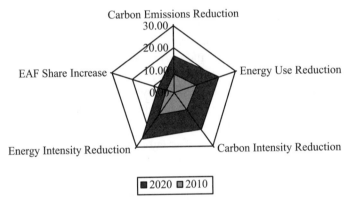

Figure 5.8
Steel industry scenarios with 10 percent REI improvement.

carbon increase policy induces a 9 percent increase in electric arc furnace share. As expected, the mixed policy of a $25 increase in the cost of carbon and a 5 percent REI improvement (figure 5.9) leads to electric arc furnace share expansion that lie between the results from the two "pure" policies. However, changes in energy intensity, carbon intensity, energy use, and carbon emissions are only marginally different from those in the "pure" policy scenarios.

A $75 increase in the cost of carbon stimulates expanded use of self-generation in the pulp and paper industry (figure 5.11) by more than is observed in the absence of new policies (figure 5.10) or any other policy

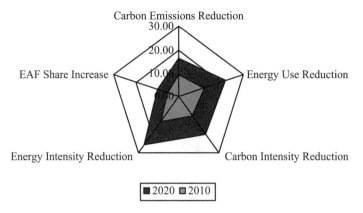

Figure 5.9
Steel industry scenarios with $25 increase in cost of carbon and 5 percent REI improvement.

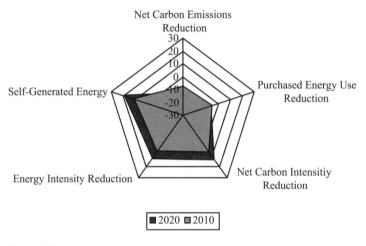

Figure 5.10
Paper industry scenarios with base case.

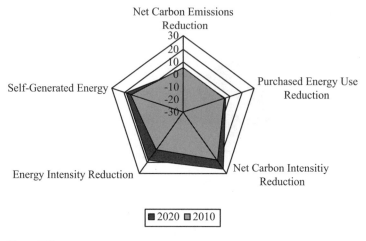

Figure 5.11
Paper industry scenarios with $75 increase in cost of carbon.

(figures 5.12 and 5.13), because of potentials still available in the industry to convert biomass to heat, steam, and electricity, and thus to cut the cost of purchased energy and the increased use of chemical processing as fuel prices increase. This result complements the results of a companion study (Davidsdottir 2003) that shows that an increase in the cost of carbon will reduce expansion toward increased use of recycled fibers, whereas in the pure REI scenario, such expansion still occurs. However, all scenarios show that expansion of self-generation is more rapid for the years 2002–2010 than from 2010 onward, indicating decreasing returns to fuel switching and technology change.

Under all scenarios (figures 5.10–5.13), only small changes in net carbon emissions by the pulp and paper industry occur, because the effect of fuel switching and technology change on emissions is nearly counteracted by the industry's output expansion. A $75 cost-of-carbon increase lowers net carbon emissions because of a more rapid switch to (carbon-neutral) self-generation. In contrast, policies that stimulate REI improvements provide less of an incentive for that switch to occur and consequently result in a slight increase in net carbon emissions while industry output continues to rise. Net carbon intensity and energy intensity decline more in the REI improvement scenario than in the base scenario and decline the most in the $75 cost-of-carbon increase scenario. Of course having system boundaries is crucial in obtaining this favorable view of a policy that is based on an increase in the cost of carbon. If self-generated energy

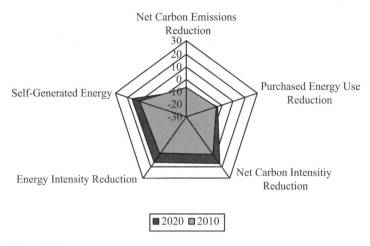

Figure 5.12
Paper industry scenarios with 10 percent REI improvement.

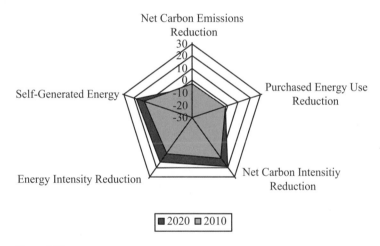

Figure 5.13
Paper industry scenarios with $25 increase in cost of carbon and 5 percent REI improvement.

was not considered carbon neutral, an increase in the cost of carbon would in fact increase carbon intensity. Furthermore Davidsdottir (2002, 2003) demonstrates that if system boundaries are expanded even further to include carbon emitted as methane as a result of paper decay—which then is corrected for its global warming potential—carbon emitted as methane accounts for over 50 percent of total carbon emissions. Thus since an increase in the cost of carbon will discourage paper recycling and effectively increase landfilling of paper, such expansion of system boundaries will result in an even less favorable view of an increase in the cost of carbon with respect to its ability to reduce the industry's impact on global warming.

Some of the largest percentage changes can be observed for the ethylene industry. Even in the absence of new policies, purchased energy consumption and carbon emissions from purchased energy increase, respectively, by 65 percent and 194 percent in 2010 and by 41 percent and 118 percent in 2020, compared with the year 2001 (figure 5.14). Raising the cost of carbon of purchased fuels used as energy carriers by $75 per ton leads to a lower increase of purchased energy than in the base case (figure 5.15). This increase is accompanied by diversion of intermediate products within the industry, leaving many of the other key characteristics of the industry relatively unaffected.

A 10 percent REI increase policy is by far the most effective approach to cutting consumption of, and emissions of carbon from, purchased en-

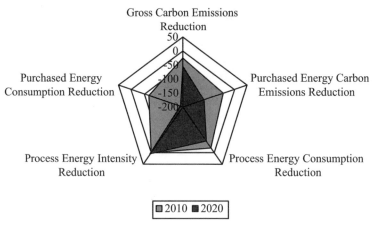

Figure 5.14
Ethylene industry scenarios with base case.

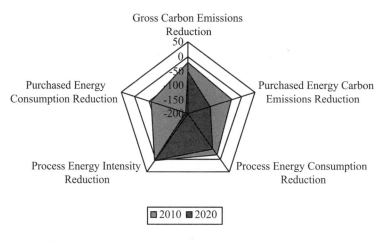

Figure 5.15
Ethylene industry scenarios with $75 increase in cost of carbon.

ergy (figure 5.16). The 10 percent REI increase policy has more uniform impacts on key industry characteristics than a pure cost-of-carbon increase policy. Combining a 5 percent REI increase with a $25 cost-of-carbon increase (figure 5.17) leads to results that reflect more the 10 percent REI increase policy than the $75 cost-of-carbon increase policy, providing a strong indication of the overwhelming influence that the existing capital stock has on energy use and emissions profiles. It is worth mentioning that any combination of different REI levels and an increase in the cost of carbon could have been implemented. Different combinations would change absolute numerical results, but not the relative differences or the interpretation of the results. However, the results presented here are roughly comparable to the results of other studies that use different methodologies and data sets (e.g., Interlaboratory Working Group 1998; EERE 2000).

5.4.6 Discussion
Comparisons across the three industries considered reveal that cost-of-carbon and capital-oriented policies have vastly different implications for energy use and carbon emissions profiles. These differences are largely attributable to each industry's production and capital vintage structure. For example, the steel industry basically consists of two different production sectors using very different energy sources and different investment dynamics. An increase in the cost of carbon makes the blast furnace

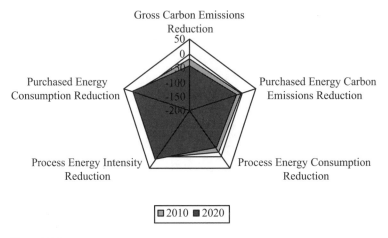

Figure 5.16
Ethylene industry scenarios with 10 percent REI improvement.

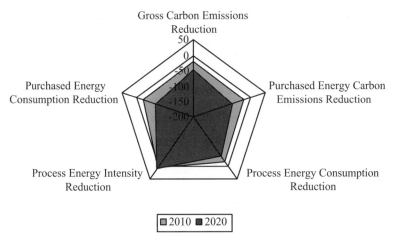

Figure 5.17
Ethylene industry scenarios with $25 increase in cost of carbon and 5 percent REI improvement.

route of steel production significantly more costly because of its high reliance on carbon-rich coal. As electric arc furnaces take over more market share, efficiencies of the aggregate capital stock in that segment of the industry increase because of investments in new, more efficient equipment and learning by doing. At the same time, the efficiency of aging blast furnaces declines because of deterioration, lower capacity utilization, and lack of positive net investment in new capacity. In contrast, policies targeting relative energy efficiencies lead to less of an increase in the share of electric arc furnaces, and consequently a lower reduction in carbon emissions.

The pulp and paper industry's inclination to self-generate heat, steam, and electricity from biomass and waste products increases with higher costs of carbon for purchased fuels. As a result of expanded self-generation, net carbon emissions (total emissions minus emissions from self-generated energy) increase less than they otherwise would. However, in contrast to the scenario in the iron and steel industry, improvements in the energy intensity of new capital relative to existing capital *do* have noticeable positive effects on the industry's total energy and carbon emissions, since a simple increase in REI does not facilitate any fuel switching, and thus technological change is unable to outweigh the increase in production level. Energy and net carbon intensities, on the other hand, decline.

A comparison between the pulp and paper industry and the other two industries investigated here in regard to REI (see table 5.2) also suggests the potential for comparatively large energy and carbon savings in pulp and paper production. Yet that potential is not fully realized, because of longer capital life and fast growth of demand.

Our choice of system boundaries, and in particular the assumption that biomass-based fuels are carbon neutral, obviously does influence the results obtained. If carbon emissions from self-generated energy were included, an increase in the cost of carbon would in fact increase carbon intensity as the result of an increase in the use of self-generated energy, which has higher carbon content than that generated from other fuels. Furthermore, if emissions from landfilled wastepaper were included, the carbon intensity would increase even further (Davidsdottir 2003).

These findings are markedly different for the ethylene industry, in which process energy is a small share of total energy use, and in which, as a consequence, a rise in the cost of process energy would not lead to a noticeable decrease in the industry's carbon emissions. In contrast, studies that increase the cost of carbon for feedstock energy as well as process energy find large emission reductions as biomass-based feedstocks replace fossil-fuel-based feedstocks (Groenendaal and Gielen 1999). Improvement in the energy intensity of new capital relative to existing capital would help replace old capital with more-efficient equipment, thus lowering carbon emissions.

A reduction in the carbon and energy intensity of all three industries is dependent on increased investment in new capital. An increase in capital investment may indeed increase energy use elsewhere in the economy, since energy is required to produce capital. Thus, because we do not include the energy required to produce new capital, or the energy required to produce, extract, or transport the raw material input, we are unable to draw any conclusions on the economy-wide impact of the policies we chose for this study.

5.5 Lessons for Industrial System Analysis and Climate Change Policy

The results of our analysis of energy use and carbon emissions profiles of three capital-intensive industries in the United States under alternative policy scenarios point to a set of broader issues relevant for the debate of such policies and raises some important issues for industrial systems analysis. One such issue raised by the results above—albeit implicitly— lies in the role that modeling can play in exploring the implications of alternative policies for future energy use and carbon emissions profiles. The track record of energy models over the last thirty years in accurately predicting industrial energy use and energy prices has been rather dismal (Laitner et al. 2003). There is little hope that current modeling can be significantly more successful in predicting the future: "All forecasts are wrong in some respect" (Koomey 2002, 516). But unlike in past (and many present) modeling efforts, emphasis is increasingly placed on models as exploratory tools rather than as forecasting devices. By playing out the implications of alternative assumptions about market dynamics, technology change, and industrial behavior through carefully designed scenario analysis, modelers hope to identify policy strategies that are robust under a wide range of assumptions. In addition, testing the sensitivity of industrial behavior to key parameters reveals where to concentrate modeling efforts in the future. The modeling framework presented in the foregoing easily enables such scenario analysis.

The models and results previously discussed help identify qualitative differences in the impacts of different policies and indicate that each policy instrument may trigger different kinds of responses by different industries—from shifting production among segments in the industry, as is the case in iron and steel, to changing the fuel mix, as is done by pulp and paper, or changing the use of intermediate products, as is happening in ethylene. Conversely, various instruments may have different abilities to leverage opportunities in industry to achieve desired policy goals, such as shifts toward renewable energy sources, reductions in total energy use, or reductions in carbon emissions. A mix of policy instruments may be useful for simultaneously improving different industry features. Yet to be able to assess the industrial impact of different policy responses, the models applied must include enough industry-specific features to enable the analysis of industry-specific responses to price-based policy instruments as well as to alternative approaches, such as accelerated depreciation schedules. Clearly, price-based policy instruments are potentially an important policy tool, but the effectiveness of other tools needs to be assessed, particularly in light of the current policy atmosphere both in Europe and in the United States, where the emphasis has shifted toward policy initiatives based more on voluntary measures than on legislation. It seems that the lack of studies that analyze the impact of non-price-based policies is mostly due to the small pool of models that are able to capture successfully industrial response to such alternative policies.

A second issue raised by the foregoing results is that the system boundaries for climate change policy may be defined in various ways. Policymakers may determine an economy-wide or industry-wide emissions target and set in place mechanisms that lead, for example, to a change in the price of carbon in order to meet the target within a set time frame, irrespective of different industries' propensities to respond to cost-of-carbon changes. Changes in the cost of carbon may be prompted, for example, by levying energy or carbon taxes, by removing subsidies for the extraction, distribution, or use of fossil fuels, or by requiring the purchase of carbon emissions permits. This is, in essence, the approach chosen for the Kyoto protocol, which—now that it has gone doolally in the United States—prompted debate about adjusting boundary constraints, such as emissions targets, time frames, and the inclusion of sinks, instead of implementing policies that could trigger quick changes in energy use and carbon emissions and at the same time provide incentives to fundamentally change, over the long haul, the structure of fossil-fuel-using industries.

Alternatively, fixed emissions targets may be set for each industry, leaving industry to figure out how to meet these targets. However, setting

such targets individually for each industry will require models that actually are able to analyze the impact on individual industries in addition
to sophisticated accounting methods to ensure that reductions actually
take place and are not simply the effect of "system boundary effects"—
as could be the case with an industry that, in response to emissions constraints, purchases more of its inputs from elsewhere in the economy or
shifts emissions toward life cycle stages that are considered outside of
conventional system boundaries. Similarly, industry-specific targets may
limit opportunities for cost-effective emissions reductions by combining
forces across sectors: For example, increases in emissions by a supplier
may allow for production changes downstream that lead to overall
reductions in emissions.

Yet another alternative lies in policy instruments (or policy mixes)—
as implied by our analysis in the foregoing—that have industry-specific
features in mind. Although such an approach may be more effective
in leveraging the different potentials of different industries to meet environmental goals, it requires considerable information on each industry
and may be perceived as unfair if the policies create different (uncertain) economic incidence for the various industries. Collaborations between industry and policymakers may provide unique opportunities and
challenges in designing these types of policies. On the one hand, collaborations do follow the democratic principle of including in the decision-
making process those parties who are potentially affected by a decision.
Collaborations may also reveal information that may otherwise not be
available and build mutual understanding and trust. On the other hand,
there is the challenge of policy capture by specific interest groups, which
needs to be addressed in the policymaking process.

A third issue raised by the results reported earlier in the chapter—
and in the analysis of capital-intensive industries in general—lies in the
relationship among the modeling approach, the choice of policy instruments, and the time frames over which policies are evaluated. Past modeling efforts and past debates about climate change policies have paid
notoriously little attention to the capital vintage structures of different
industries, often assuming, for mathematical convenience, (near-perfect)
malleability of capital. In contrast, the analyses presented in this chapter
point toward considerable capital stock inertia in the three industries
studied, highlighting the importance of capturing the structure of capital
vintage. Existing capital stock inertia can lead to large costs of changing energy use and emissions profiles (Lecocq, Hourcade, and Ha Duong
1998) and extend the time frame over which the impact of policies is

actually seen in the energy and carbon intensity profiles. Capital stock inertia, a common characteristic of energy-intensive industries, makes energy and emission trajectories rigid, makes it unlikely that ambitious short-term policy objectives can be met or highly expensive to do so (Jacques, La France, and Doucet 2001), and points to the need for far-sighted policymaking. The modeling exercise presented in this chapter indicates that simply raising the cost of carbon of purchased fuels may be insufficient to overcome existing capital vintage effects, again raising the necessity to be able to capture the impact of alternative policy instruments. Alternatively, stimulating research and development in the hope of maintaining a sizable gap between new capital equipment and the capital in use in an industry may make sense only if adequate incentives are present to actually replace existing with new capital. Without incentives to replace existing capital, rapid improvements in technology may provide an incentive for industry decision makers to wait to make their investments, in fear of locking in equipment that will soon be outdated. Policy incentives that have the potential to increase the rate of capital turnover are, for example, tax credits for investing in new, more efficient technology; research and development subsidies; and demonstration projects that reduce the risk and uncertainty of the benefit of new technologies (EERE 2000).

Since increased capital turnover requires energy expenditures and carbon emissions elsewhere in the economy, it is not obvious a priori what the net effects of an investment-led policy are (though in most cases energy use associated with capital stock turnover is likely to be dwarfed by energy use during the lifetime of the capital itself). However, since both the change in an industry's capital stock and the associated changes in upstream capital suppliers are likely to play themselves out over the course of many years, even more credence in the modeling arena must be given to models that are able to capture the impact of policies on the turnover of the capital stock and its associated impact on energy use and carbon emissions. Without such modeling frameworks, policy analysts are not able to assess the importance and impact of policies that are able to accelerate capital turnover, which will limit the policy options considered, as it seems that the available analysis tools somewhat dictate the policy route taken.

Paying more attention to longer time frames both in policy and in modeling, however, does not mean that short-term goals for emissions reductions can or should be abandoned. On the contrary, meeting long-term emissions reduction goals requires that changes in the existing capital stock be undertaken *now* so that aggregate industry efficiencies can improve and so that a basis for learning by using new capital equipment can be generated. Policy analysis must therefore identify how to make short-term and long-term emissions goals consistent with one another and how to design and implement policy instruments that leverage the unique potential of industry to meet these short- and long-term goals.

Acknowledgment

The project reported on in this chapter was made possible by support from the U.S. Environmental Protection Agency under grant number X 826822-01-0 and benefited from many discussions with John "Skip" Laitner, members of the U.S. Department of Energy, and representatives from the U.S. iron and steel industry, pulp and paper industry, and ethylene industry. However, the chapter does not necessarily reflect the views of EPA or the Department of Energy or those of the individuals who provided input into, or commented on, the study on which this chapter is based.

Note

1. A copy of the models and software is available by request to the first author at mruth1@umd.edu.

References

AF&PA (American Forest and Paper Association). (various years). *Statistics of Paper, Paperboard and Woodpulp*. Washington, DC: AF&PA.

AF&PA (American Forest and Paper Association). (various years). *Paper, Paperboard and Woodpulp: Fiber Consumption*. Washington, DC: AF&PA.

AISI (American Iron and Steel Institute). (1999). *Annual statistical report*. Washington, DC: AISI.

AISI (American Iron and Steel Institute). (2002). *Energy Efficiency*. Available at http://www.steel.org/facts/power/energy.htm.

Arthur, B. (1994). *Increasing returns and path dependence in the economy*. Ann Arbor: University of Michigan Press.

Berndt, E., C. Kolstad, and J. K. Lee. (1993). Measuring energy efficiency and productivity impacts of embodied technological change. *Energy Journal* 14(1): 33–55.

Breusch, T. (1978). Testing for autocorrelation in dynamic linear models. *Australian Economic Papers* 17: 334–355.

Breusch, T., and A. Pagan. (1979). A simple test for heteroscedasticity and random coefficient variation. *Econometrica* 47: 1287–1294.

Bullard, C. W., and R. Herendeen. (1975). Energy impacts of consumption decisions. *Proceedings of the Institute of Electrical and Electronic Engineers* 63: 484–493.

Bullard, C. W., P. Penner, and D. Pilati. (1978). Energy analysis handbook. *Resources and Energy* 1: 267–313.

Davidsdottir, B. (2002). A Vintage Analysis of Regional Energy and Fiber Use, Technology Change and Greenhouse Gas Emissions by the US Pulp and Paper Industry. Ph.D. diss., Boston University.

Davidsdottir, B. (2003). Modeling Industrial Behavior and Feedback between Energy and Material Flows and Capital Vintage: Implications for Material, Energy and Climate Change Policy Design. In *Proceedings from the 2003 ACEEE Summer Study on Sustainability and Industry*, vol. 1, 25–36. Washington, DC: ACEEE (American Council for Energy Efficient Economy).

Doms, M. E. (1996). Estimating capital efficiency schedules within production functions. *Economic Inquiry* 34: 78–92.

EERE (Office of Energy Efficiency and Renewable Energy). (2000). *Scenarios for a Clean Energy Future*. Prepared by the Interlaboratory Working Group on Energy-Efficient and Clean-Energy Technologies. U.S. Department of Energy, Office of Energy Efficiency and Renewable Energy, Washington, D.C.

EIA (Energy Information Administration). (1997). *Manufacturing Energy Consumption Survey, 1994*. Washington, DC: EIA.

EIA (Energy Information Administration). (2000). *Industrial Sector Demand Module of the National Modeling System*. Washington, DC: Office of Integrated Analysis and Forecasting, U.S. Department of Energy.

Fishbone, L. G., and H. Abilock. (1981). Markal—A linear programming model for energy systems analysis: Technical description of the BNL version. *International Journal of Energy Research* 5: 353–375.

Goulder, L. H. (1995). Effects of carbon taxes in an economy with prior tax distortions: An intertemporal general equilibrium analysis. *Journal of Environmental Economics and Management* 29: 271–287.

Goulder, L. H., and S. H. Schneider. (1999). Induced technological change and the attractiveness of CO_2 abatement policies. *Resource and Energy Economics* 21: 211–253.

Groenendaal, B. J., and D. J. Gielen. (1999). The Future of the Petrochemical Industry: A MARKAL-MATTER Analysis. ECN Report C-99-052, Energy Research Centre of the Netherlands, Petten.

Hannon, B. (1973). *System Energy and Recycling: A Study of the Beverage Industry*. Energy Research Group Document no. 264, Office of the Vice-Chancellor for Research, University of Illinois, Urbana-Champaign.

Hannon, B., and M. Ruth. (2001). *Dynamic Modeling*. 2nd ed., New York: Springer-Verlag.

IEA (International Energy Agency). (1998). *Mapping the Energy Future: Energy Modeling and Climate Change Policy*. Paris: OECD/IEA.

Interlaboratory Working Group. (1998). *Scenarios of US Carbon Reductions: Potential Impacts of Energy Technologies by 2010 and Beyond*. U.S. Department of Energy, Office of Energy Efficiency and Renewable Energy, Washington, D.C.

Iron and Steelmaker. (various years). Warrendale, PA.

Jacoby, H. D., and I. S. Wing. (1999). Adjustment time, capital malleability and policy cost. *Energy Journal* (Kyoto special issue): 73–92.

Jacques, C., G. LaFrance, and J. A. Doucet. (2001). Inertia in the North American Electricity Industry: Is It Realistic to Think That the Kyoto Protocol Objectives Can Be Met? *Energy Policy* 29: 453–463.

Jorgenson, D. W. (1996). *Investment*. Cambridge, MA: MIT Press.

Jorgenson, D. W., and P. J. Wilcoxen. (1990). Intertemporal general equilibrium modeling of U.S. environmental regulation. *Journal of Policy Modeling* 12: 715–744.

Koomey, J. G. (2002). From my perspective: Avoiding the "big mistake" in forecasting technology adoption. *Technological Forecasting and Social Change* 69: 511–518.

Laitner, J. A., S. J. DeCanio, J. G. Koomey, and A. H. Sanstad. (2003). Room for improvement: Increasing the value of energy modeling for policy analysis. *Utilities Policy* 11: 87–94.

Larson, E. D., W. Yang, K. Iisa, E. W. Malcolm, G. W. McDonald, W. J. Frederick, T. G. Kreuz, and C. A. Brown. (1998). A Cost-Benefit Assessment of Black Liquor Gasification/ Combined Cycle Technology Integrated into a Kraft Pulp Mill. Paper presented at the TAPPI International Chemical Recovery Conference, Tampa, FL, June 1–4.

Lecocq, F., J. Hourcade, and M. Ha Duong. (1998). Decision making under uncertainty and inertia constraints: Sectoral implications of the when flexibility. *Energy Economics* 20: 539–555.

Martin, N., N. Angliani, D. Einstein, M. Khrushch, E. Worrell, and L. K. Price. (2000). *Opportunities to Improve Energy Efficiency and Reduce Greenhouse Gas Emissions in the U.S. Pulp and Paper Industry*. Report Number LBNL-46141. Berkeley, CA: Lawrence Berkeley National Laboratory. Environmental Energy Technology Division.

Meijers, H. (1994). On the Diffusion of Technologies in a Vintage Framework: Theoretical Consideration and Empirical Results. Ph.D. diss., Maastricht University.

Miller Freeman. (1997). *Lockwood-Post's Directory of the Pulp, Paper and Allied Trades*. San Francisco: Miller Freeman.

Mulder, P., H. L. F. L. de Groot, and M. W. Hofkes. (2001). Economic growth and technological change: A comparison of insights from a neo-classical and an evolutionary perspective. *Technological Forecasting and Social Change* 68: 151–171.

Mulder, P., H. L. F. L. de Groot, and M. W. Hofkes. (2003). Explaining the energy efficiency paradox: A vintage model with returns to diversity and learning-by-using. *Resource and Energy Economics* 25(1): 105–126.

Nordhaus, W., and J. Boyer. (2000). *Warming the World: Economics Models of Global Warming*. Cambridge, MA: MIT Press.

Office of Industrial Technologies. (2000). *Energy and Environmental Profile of the U.S. Chemical Industry*. U.S. Department of Energy, Washington, D.C.

Office of Industrial Technologies. (2001). Chemical Industry Profile. Available at http://www.oit.doe.gov/chemicals/page11.shtml.

Office of Industrial Technologies. (2002). Steel industry profile. Available at http://www.oit.doe.gov/steel/profile.shtml.

Oil & Gas Journal. (1998). Worldwide ethylene capacity grows in spite of warnings. *Oil & Gas Journal* 13(96): 41–47.

Ross, M. (1991a). The potential for reducing the energy intensity and carbon dioxide emissions in US Manufacturing. In R. Howes and A. Fainberg (eds.), *The Energy Sourcebook*, 415–440. New York: American Institute of Physics.

Ross, M. (1991b). Efficient energy use in manufacturing. *Proceedings of the National Academy of Science* 89: 827–831.

Ross, M., P. Thimmapuram, R. Fisher, and W. Maciowski. (1993). *Long-term Industrial Energy Forecasting (LIEF) Model* (18-sector version). Argonne National Laboratory, Argonne, IL.

Ruth, M. (1995). Technology change in U.S. iron and steel production: Implications for material and energy use, and CO_2 emissions. *Resources Policy* 21(3): 199–214.

Ruth, M. (2001). Dynamic modeling for consensus building in complex environmental and investment decision making. In M. Mathies, H. Malchow, and J. Kriz (eds.), *Integrative Systems Approaches to Natural and Social Dynamics*, 379–399. Berlin: Springer-Verlag.

Ruth, M., S. Bernow, G. Boyd, R. N. Elliot, and J. M. Roop. (1999). Analytical Approaches to Measuring the Potential for Carbon Emission Reductions in the Industrial

Sectors of the United States and Canada. Paper presented at the IEA International Workshop on the Technologies to Reduce Greenhouse Gas Emissions, Washington, DC, May 5–7.

Ruth, M., and B. Hannon. (1997). *Modeling Dynamic Economic Systems.* New York: Springer-Verlag.

Ruth, M., and T. Harrington. (1997). Dynamics of material and energy use in US pulp and paper manufacturing. *Journal of Industrial Ecology* 1: 147–168.

Slinn, R. J. (1992). The paper industry and capital. *Tappi Journal* (September): 16–17.

Solow, R. W. (1957). Technical change and the aggregate production function. *Review of Economics and Statistics* 39: 312–320.

Unruh, G. C. (2000). Understanding carbon lock-in. *Energy Policy* 28: 817–830.

Wilbanks, T. J., and R. W. Kates. (1999). Global change in local places: How scale matters. *Climatic Change* 43: 601–628.

Worrell, E., D. Phylipsen, D. Einstein, and N. Martin. (2000). *Energy Use and Energy Intensity of the U.S. Chemical Industry.* Report LBNL-44314, Energy Analysis Department, University of California, Berkeley.

Zellner, A. (1962). An efficient method of estimating seemingly unrelated regressions and tests for aggregation bias. *Journal of the American Statistical Association* 57: 348–368.

6 Modeling Physical Realities at the Whole Economy Scale

Barney Foran and Franzi Poldy

6.1 Introduction

6.1.1 The Physical Economy

Although this chapter is primarily a description of physical economy modeling in Australia, it also acts as a bridge between the more focused ideals of industrial ecology and broader concepts such as the dynamics and resilience of socioeconomic and ecological systems. Without such a bridge, broad implementation of industrial ecology could languish between scales of space and time. Regional applications of industrial ecology need to pervade and then saturate a whole national economy before material and energy flows moderate and then begin to decline. Models of the physical economy can test the physical feasibility of industrial ecology concepts and whether they have sufficient strength and importance to stimulate structural change in the economic system and the material transactions that underpin it. The main issue is that the key elements of a nation's stocks of infrastructure have to change before the flows driven by those stocks change substantially. National stocks, such as houses, vehicles, and electricity generators, are typified by their size, age, and location. Implementing industrial ecology nationally requires that the stocks of infrastructure and the linkages among them are turned over in unison, in sympathy with some grand plan over time scales that span human generations. Physical economy models can help design and test the grand plan, but implementation is another matter.

The complexity of socioeconomic systems makes it unlikely that technological progress alone will force an economy-wide implementation of industrial ecology. To understand this complexity, Holling (2001) suggests that an analytical framework be as simple as possible (but no simpler), be dynamic and prescriptive (rather than static and descriptive), and embrace uncertainty and unpredictability. He further suggests that human systems are distinguished from natural systems by having foresight and intentionality, communicating and storing experience and using technology to amplify the effect of management actions. If the development of resilience in socioeconomic systems is seen as a necessary attribute and one of the driving rationales to implement industrial ecology, for example, then Carpenter et al. (2001) stress the importance of time scales and of separating and understanding slow- and fast-moving variables. In their context, the slow-moving variables define the underlying structure of the system (e.g., stocks of infrastructure in a modern economy), whereas the fast-moving variables, such as production values or pollutant flows, depend in turn on the dynamics of the underlying structure. These broader philosophies of complexity science and resilience science help define some characteristics of physical economy models that are useful in designing the implementation of industrial ecology. Such models should be reasonably simple, dynamic, and prescriptive, track information flows, and focus on slow-moving variables that act as controlling variables of the whole socioeconomic system.

In implementing these concepts at a more practical level, Ayres (1998a) makes the case for more measurement and modeling of the physical economy for two reasons. The first is that national decisions are generally guided by computable generalized equilibrium (CGE) models, which seldom recognize the material and energy flows on which the function of the socioeconomic system is based. The second is that national accounting procedures generally ignore hidden flows such as the removal of mine overburden, since little economic value is derived from them. The volume of visible and hidden material flows is increasing in line with economic growth and development. Prior to the Industrial Revolution, the volume of material flow was inconsequential relative to the assimilation capacity of natural ecosystems. However, von Weizsäcker (1998) notes that material flows induced by humans now compare with large-scale global processes and that it is unlikely that they can continue to grow exponentially forever. This realization has led to the concept of "ecological rucksacks" intended to increase awareness of hidden material flows behind products, and to the concept of a Factor Four and a Factor Ten economy, in which economy-wide reductions in material and energy flows are implemented. The ability to foresee the requirements for dematerialization at an economy-wide level, as well as the design and implementation of pathways for individual processes, creates the niche where models of the physical economy become useful to the concepts of industrial ecology.

This chapter describes the design and implementation of two physical economy models in Australia.

6.1.2 Physical Economy Models in Australia

The development of physical economy modeling in Australia was stimulated by the national population debate within the context of long-term sustainability issues. The concepts of population targets and carrying capacity have a long history in Australia, starting in the 1920s, when a Sydney university geographer, Thomas Griffith Taylor, set Australia's estimated carrying capacity at sixty-five million people and later reduced this estimate to twenty million people (Cocks 1996). During the 1980s and 1990s there were several national inquiries on population, the most recent of which was the Jones Inquiry (Long-Term Strategies Committee 1994), which stopped short of recommending a national population policy (Cocks 1996). By default, Australia's population seems to be moving toward a more or less stable population of twenty-three to twenty-five million people in one to two human generations' time. During the 1990s, the national population debate evolved to include a wide range of linked issues, such as resilience of ecological systems, material consumption levels, and changes to the structure and function of the economic system.

It was against this background that the Commonwealth Scientific & Industrial Research Organization (CSIRO), a national science agency, initiated a strategic project to underpin the population debate, and its linkages to resource use and environmental quality, with scientific analysis. The project's initial aim was to focus on the environmental aspects of population impact with particular emphasis on the quality and quantity aspects of water, soils, biodiversity, atmosphere, and natural amenity. Initially, the work proceeded along a traditional scientific route, in which plans were made to examine the effect of population on water resources, land resources, and so on. However, because of the complex linkages among all sectors of society and the economy, the traditional approach of defining tight boundaries around a well-defined problem prior to analysis was judged difficult to implement. In addition, the project faced the challenge of tackling a future-oriented and long-term topic that required integrated advice and a range of possible solutions. At this point in the project, the project members became aware of two important methodological approaches. The first involved Godet's (1991) work on *strategic prospectives* and thence the use of foresighting and scenario development by multinational companies such as Royal Dutch Shell. The second was the implementation of population-development-environment simulators (particularly the work by the International Institute for Applied Systems Analysis [IIASA] in Mauritius) (Lutz 1994), the physical analysis paradigm using the design approach (Gault et al. 1987), and the embodied energy approach of Slesser (1992; Slesser, King, and Crane 1997) and colleagues.

The project design then evolved to the development of new designs for the socioeconomic system, in order to lessen the effect of humans on resource depletion and environmental quality. The first theme of work was the development of a number of robust and well-documented national scenarios to lead and inform debate on national development and sustainability issues. Three scenarios, *economic growth, conservative development*, and *postmaterialism*, were published as part of a book that arose out of the project (Cocks 1999). The second theme of work developed to underpin the scenario work with physical economy analyses. Within this theme, two system simulators were developed based on different paradigms of physical analysis. One of these, OzEcco (Foran and Crane 1998), used the embodied energy approach of Slesser (1992; Slesser, King, and Crane 1997) to construct a top-down and aggregated simulator of Australia's physical economy. This analytical approach assumed that the delivery of goods and services to a domestic economy is a function of the extraction, delivery, and efficiency of use of energy resources, most of which are derived from fossil sources.

The second simulator, the Australian Stocks and Flows Framework (ASFF), was a disaggregated set of linked models that access a database describing the last fifty years of Australia's physical function or physical metabolism. The *design* approach used by the ASFF is philosophically attractive for two reasons. First, it treats the complete range of physical function as separate entities (crops, animals, people, cars, steel production, chemical production) and allows a detailed treatment of vintaging or age for most big-ticket items of physical infrastructure. Secondly, the physical functioning is retained within the modeling code and termed "machine space." The management and policy decisions that guide this physical functioning are retained as part of a scenario under development and testing by the user or policy analyst and are termed "control space." Gault et al. (1987) describe the design approach as follows:

The design approach is a philosophy for building computer-based simulation frameworks, which represent socio-economic systems, and for using the simulation framework to design alternative futures through repeated simulation. It is the exploration of alternative futures by the user, who forms part of the system, which distinguishes this approach from that of macro-economics with its emphasis on prediction. The exploration and the involvement of the user result from the absence of optimisation or equilibrating mechanisms in the physical representation of the socio-economic system. This ensures that the user, working alone or with the aid of a model of decision processes, controls the system. The policy decisions necessary to exercising this control are required to be explicitly stated, and they form a record of how the future, resulting from the simulation, was arrived at. (23–24)

In section 6.2, the physical economy simulators will be described in more detail. Section 6.3 will give some current examples of model use within policy and science processes. Section 6.4 will describe some challenges for analytical approaches in achieving their goal of influencing national policy. The chapter will conclude with some insights in section 6.5 into the many conundrums that face integrative modeling of the physical economy in implementing concepts such as industrial ecology.

6.2 Model Descriptions

6.2.1 The OzEcco Embodied Energy Model
The OzEcco model is designed to integrate the driving forces of population, lifestyle, organization, and technology and to explore their possible impacts on environmental loadings. It is a systems dynamics representation of Australia's metabolism, based on the philosophy of embodied energy analysis. This analysis evolved from the integration of financial input-output tables (Leontief 1970; Rose and Miernyk 1989), with national physical accounts, such as those for energy (Herendeen 1998). Slow-moving variables such as capital stocks are expressed as a physical measure in petajoules of embodied energy rather than in monetary terms. Fast-moving variables have been expressed as energy flows, again in petajoules per year rather than dollars per year. In this way economic activity has been transformed into physical activity, which is consistent with the first and second law of thermodynamics. All economic transactions are thus represented by the physical transformations that underpin them. This simplification is consistent with the slow-moving variables that determine the rates of growth and development of any modern economy.

Conceptually the OzEcco model has five broad components: natural resource stocks, the energy transformation sectors, consumption activities, pollution generation, and whole system indicators. The core modeling concept is that access to and transformation of energy (typically

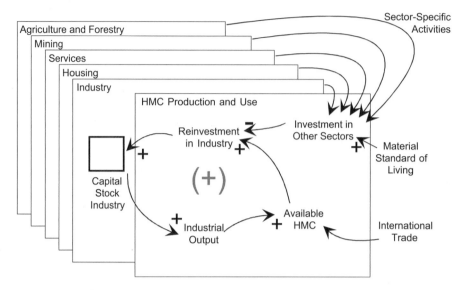

Figure 6.1
A diagram of the central growth-determining loop in the OzEcco model, with the aggregated
industrial sector depicted here as the core resource on which growth depends. The processes
of fixed and human-made capital (HMC in the diagram) are depicted as an influence diagram, illustrating the main causative features represented in the model. The total human-
made capital available is the sum of imports and domestic production.

stocks of fossil fuel) are the determinants of physical growth in a modern
industrial economy. Thus all goods and services are expressed in terms of
the chain of energy processes that eventually become included (embodied)
in a final good (a motor car) or a service (banking or education). Some
sectors, such as domestic housing, act as long-term accumulators of fixed
energy capital (embodied energy), whereas personal consumption dis-
sipates embodied energy relatively quickly. The concept is shown in figure
6.1. The capital stock of industry (stock of embodied energy) is the primary focus through which human-made capital is created. Industry contributes to other sectors such as agriculture (fertilizer, machines) and
domestic housing (bricks, carpets, stoves).

The rate at which the aggregated industrial sector can grow in any one
year is limited by its contribution to other sectors of the physical economy and the consumption activities of the population at large. Both of
these activities (industrial growth and personal consumption) act as negative feedbacks to restrict the rate at which the physical economy may
grow. The effects of international trade and financial flows can be either
positive or negative. Exports are classified as a negative drain on the

amount of embodied energy available nationally. Physical imports and monetary inflows are positive additions, because they increase the capability to provide physical transactions and services. All of these factors are linked in a systems dynamics framework (Richardson and Pugh 1981). The simulated economy has an endogenous growth mechanism constrained by the availability of renewable and nonrenewable energy and by the need to maintain national infrastructure with personal consumption activities. Global financial issues, such as the balance of payments and international debt, are regarded as flows and stocks of virtual embodied energy that, in the short term, help overcome resource and infrastructure issues limiting the expansion of the physical economy.

To date, acceptance of the OzEcco approach by both the science and the policy community is restricted. The use of integrating concepts such as "embodied energy" is limited by the background of policy analysts, although embodied energy is similar to money as a numeriare for economic analysis. However, two recent developments in energy and greenhouse policy have increased the potential acceptability of this approach. Using input-output tables to compute energy embodiment for different economic sectors and to highlight energy use and greenhouse gas emissions (Lenzen 1998) is gaining general acceptance. There is also interest in simulations of a methanol economy based on biomass (Foran and Mardon 1999).

6.2.2   The Australian Stocks and Flows Framework

6.2.2.1   General Description   ASFF is a highly disaggregated simulation framework that keeps track of all physically significant stocks and flows in the Australian socioeconomic system. In this context, stocks include people, livestock, trees, buildings, vehicles, capital machinery, infrastructure, land, air, water, energy, and mineral resources—disaggregated, as appropriate, according to their physical characteristics and, importantly, age or vintage. Flows, resulting from physical processes of many kinds, represent the rates of change of stocks and constitute the development of the system in more or less desirable directions. In the context of this chapter, ASFF is a nationally scaled framework that provides both a database and a simulation model in which industrial ecology concepts can be tested.

The simulation model consists of thirty-two hierarchically connected modules or calculators, which account for the physical processes of demography, consumption, buildings, transport, construction, manufacturing, energy supply, agriculture, forestry, fishing, mining, land, water and air resources, and international trade. Each calculator deals with the stocks and flows relevant to a sector and with the physical processes through which they interact.

Calculator assumptions are based on the technical and scientific understanding of the processes involved and are intended to provide a plausible representation in physical terms of the workings of the sector concerned. Indeed, it is a criterion of validity for the calculator that a professionally informed person should be able to follow the structure of the representation and conclude that it and the values of the parameters are plausible and appropriate to the level of aggregation of the treatment.

An overview of the whole framework is given in figure 6.2, in which the arrows link calculators arranged in functionally similar and hierarchically related groups (note that the arrows do not represent sector linkages or information flows; these are shown in figure 6.3).

6.2.2.2 The Model Calculators In figure 6.2, the unshaded boxes with heavy borders represent hierarchical groupings and the shaded boxes represent calculators. At the highest level, the Australian socioeconomic system is conceived of in terms of people (Demography) and the physical needs of their way of life (Materials and Energy). Population is an important driver in the framework, and other things being equal, more people require more materials and energy. However, other things are not necessarily equal, and one of the goals of the ASFF approach is to explore the interplay and trade-offs among issues such as population, affluence, institutional organization, and technological innovation.

The five Demography calculators account for various population issues, including overseas and internal migration. These calculators also transfer information to other calculators that depend directly on population and its distribution over age, sex, and location, such as education, morbidity and health requirements, internal travel, household formation, labor force participation, demand for personal services, and inbound tourism. Population and inbound-tourist numbers are independent drivers in the framework. The parameters that determine their level and growth are specified exogenously outside the model in the control space. Information from these demography calculators is passed to later calculators and used to determine the requirements for infrastructure, goods, and services of all types.

The Consumables calculator determines the need for food and other consumable items directly from population (including overseas visitors)

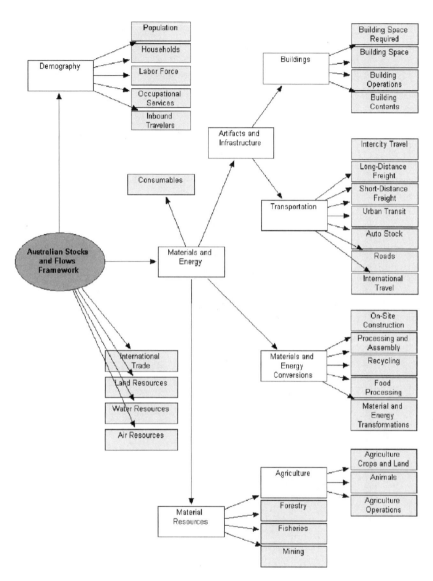

Figure 6.2
Hierarchy of calculators in the Australian Stocks and Flows Framework. (See figure 6.3 for information flow among calculators.)

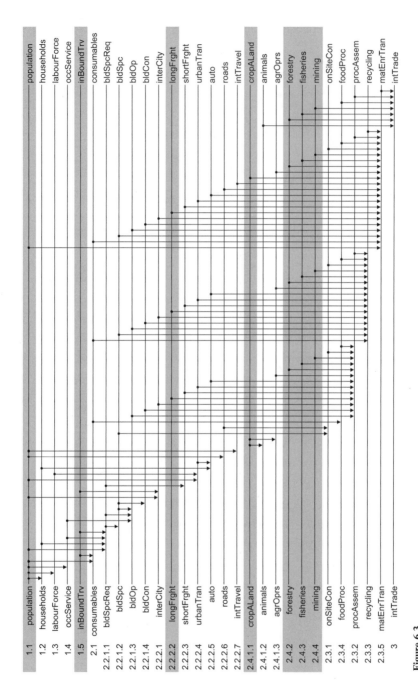

Figure 6.3
One-way information flow (vertical arrows) among calculators (horizontal lines) of the Australian Stocks and Flows Framework. Shaded calculators receive only exogenous input (no arrowheads on shaded lines).

on a per capita basis. The four Buildings calculators use information from demography to determine the needs of the population for residential, commercial, educational, health care, and institutional buildings.

Seven calculators deal with various aspects of Transportation. Broadly, these cover domestic passenger and freight transport in urban and rural areas. Separate calculators deal with the car fleet, roads and their maintenance, and fuel for international travel. In most cases, a transport task is determined in relation to demographic parameters, and with the help of load factors and average yearly distance traveled, the task is translated into a need for vehicles. The Material Resources calculators describe production processes in the country's primary industries: agriculture, forestry, fisheries, and mining. Like population, tourism, and long-distance freight, these industries are independent drivers in the framework and receive no information from earlier calculators. Their planned levels of production are specified exogenously, because much of the produce from Australian primary industries is destined for export.

Agriculture is covered by three calculators, which deal with crops and land, livestock, and agricultural operations in each statistical division. Cropping deals with the areas of land devoted to each of ten different crops (or land may remain fallow or idle), the impact of cropping activity on four indicators of soil quality (acidity, dryland salinity, irrigation salinity, and soil structure), and the effect on yield of genetic improvements to crop varieties, of the application of fertilizer and irrigation, and of declining soil quality due to the cumulative effects of previous cropping. The Animals calculator deals, in each statistical division, with the stocks of animals of different types, the quantities of animal products they yield, and their feed requirements in terms of crops and area of grazing land.

The Forestry calculator deals with fifteen different types of forest, managed under regimes that vary from full protection to clear cutting and managed plantations. Fire frequency and tree growth and survival rates are taken into account. Inventories are kept of land areas and of tree numbers and wood volumes by age. The Fisheries calculator deals with both wild fishing and fish farming. Wild fish stocks vary in response to their own natural rates of reproduction and mortality—and to the level of fishing. Each fishery can sustain some moderate level of fishing, but if overfished, the stock collapses to levels at which catch per unit effort no longer warrants fishing. Fishing effort is allocated among fisheries in an attempt to meet planned production levels with minimum effort.

The Mining calculator covers exploration for mineral and energy resources, evaluation and classification of resources as reserves, and extraction of minerals and energy materials to meet planned production. "Resources ever found" are the current estimate of the nation's total endowment of a material. Unless augmented by new discoveries, cumulative production will never exceed this quantity. The Materials and Energy Conversions group of calculators covers construction, manufacturing, and energy supply. These calculators deal with the need for materials, energy, goods, and infrastructure identified in earlier calculators. The Processing and Assembly calculator consolidates the requirements for vehicles, machinery, building contents, and operating goods of all types from previous calculators and, allowing for imports and exports, determines the level of domestic production of these goods. The Recycling calculator consolidates all discarded goods, vehicles, and machinery and determines the proportions to be recycled or disposed of to landfill. The material content of the recycled fraction is determined from a knowledge of the material composition and vintage of the goods and vehicles. The Material and Energy Transformations calculator ensures that the needs of the whole economy for materials and energy are met.

The International Trade Balance calculator consolidates domestic production and domestic requirements for primary materials, secondary materials, vehicles and machinery, and intermediate and final demand goods and determines import and export quantities. These are combined with a set of import and export prices and an interest rate to determine the value of the trade flows, the current merchandise trade balance in nominal dollar terms, and its contribution to the international debt (or surplus), again in nominal dollar terms. Finally, the Land Resources, Water Resources, and Air Resources calculators consolidate information from the whole framework into accounts which provide an overview of the state of these important resources.

The framework is grounded in a database for a fifty-year historical period which is complete (all data gaps are filled) and in which variables are consistent with one another and with the assumptions in the calculators. These assumptions are based on technical and scientific understanding of all the processes required to describe the physical stocks and flows underlying the Australian socioeconomic system. At the basic level, this ensures that fundamental requirements, such as the conservation of matter and energy and the laws of thermodynamics, are observed. For particular calculators, the assumptions must be consistent with a relevant specialist's understanding of the processes involved.

The process of model validation is a variant of normal model calibration procedures. Essentially, validation is based on reconciling stocks
(people, houses, cars) with the flows (births and deaths, electricity use,
petrol use) that are derived from those stocks. It can be seen as the application of accounting principles across many dimensions in which a
number of more robust national data series over fifty-year time frames
(total population, total primary energy use) set constraints and reality
checks for the operation of all thirty-two submodels. Any important
irregularities in sets of nested output variables (people, per capita car
ownership, car numbers, per car gasoline use, total gasoline use) is traced
back and adjusted until the fifty-year historical foundation of the model
has real data and modeled data in close agreement. For future scenarios,
the control variables from history are used as reasonable guides for
assumptions about future trajectories out to 2050 or 2100.

6.2.2.3 Calculator Linkage, Feedback, and Tensions The calculation
linkages are shown in figure 6.3, in which arrows flow downward only,
indicating that feedbacks caused by demand and supply imbalances are
controlled by the user, who separates control space from design space. In
order to calculate the quantities demanded within the physical economy,
the population calculator (1.1 in figure 6.3) passes down

- the requirements for households (1.2) through an age- and sex-
determined household formation rate);
- the availability of a labor force (1.3) through an age- and sex-
determined participation rate;
- the demand for employment in nonphysical sectors of the economy,
such as services (1.4), as a proportion of the total population;
- consumables, such as food, plastics, paper, pharmaceuticals and chemicals (2.1), on a per capita per year basis;
- the demands for building space (2.2.1.1), intercity travel (2.2.2.1), urban
transit (2.2.2.4), roads (2.2.2.6), international travel (2.2.2.7), and material transformations (2.3.5).

This process is continued down the hierarchy of calculation procedures, which provides a complete set of quantities demanded by the
population driver and the subsequent flow on effects. In order to
supply the quantities demanded, production or control variables are set
in the primary material sectors (agriculture, forestry, fishing, mining) or
the international trade sector, so that the quantities demanded by the

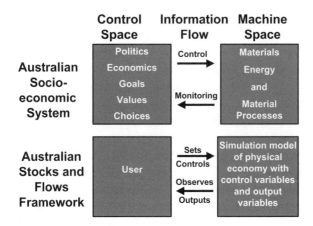

Figure 6.4
Content and information flow between control space and machine space in the reality of the Australian socioeconomic system and in the control and machine space of the Australian Stocks and Flows Framework.

population may equal the quantities supplied over the period of the simulation.

The *design* approach that lies behind the implementation of the ASFF model distinguishes control space from machine space (figure 6.4). Control space is occupied by the user or analyst, who makes assumptions on the basis of current knowledge and future expectations and then alters control variables in the ASFF model. Machine space is occupied by the modeling code and the equations that describe the processes that drive the physical economy. This is the domain of the materials, energy, and physical processes, which are central to the implementation of industrial ecology. What happens in machine space depends on physical laws, but it also depends on choices made in control space according to people's values. However, people's control of the physical world is imperfect, both because the physical world is very complex and because their goals and values conflict with other people's. From control space, the analyst can monitor what happens in machine space during model simulation and evaluate the outcomes according to goals and values set by a policy analyst or a research group. In practice, the iterative nature of design and testing can be slow and spasmodic, because simulation outcomes are delivered to clients as documents with scenario graphs and written interpretations. In theory, a policy client and a simulation analyst could sit together at the computer screen and facilitate the process of learning and design.

In the design approach used in ASFF, generally the physical processes in machine space are modeled. The analyst occupies control space, observes the situation in machine space, and makes decisions about the settings of the control variables. The analyst is therefore an integral part of the feedback loop, using a wide variety of information sources from society and its political and economic agents. As such, the analyst is in a position to learn a great deal about the system-wide effects of new industrial ecology designs being tested.

Resolving tensions (imbalances between quantities demanded and quantities supplied) may be obligatory or optional. If a tension indicates a physical or accounting inconsistency, it must be resolved. If insufficient primary energy is supplied to meet electricity and transport requirements, then its supply and delivery must be increased. Another form of tension might indicate the failure to meet some nonphysical goal or desirable criterion. In this case, its resolution is judged to be optional, as illustrated by an imbalance between the labor demanded and the labor supplied. If there is more labor supplied than is demanded, then this is called unemployment, and the scenario is still physically feasible. If there is more labor demanded than supplied, then the production goals might be regarded as infeasible. Production goals might have to be decreased, or the labor force increased.

Currently, the model operation focuses on the larger material flows and the slow-moving dynamics of important stock variables. To deal with small important flows, such as the liberation to the environment of heavy metals or toxic organic materials, the material and energy input-output table at the heart of the model is being redeveloped to account for small, potentially harmful flows as well as the major volumetric flows. Where short-run dynamics are important, particular model calculators are redeveloped with shorter time steps. This was particularly important in the fisheries calculator, in which five-year time steps were able to deal effectively with long-lived species such as bluefin tuna. However, for short-lived species, such as squid and prawns, a yearly time step was needed. There is a constant challenge in model redevelopment to constrain the types of issues that are analyzed and to stay with the underlying theory that the stock vintages driving slow-run dynamics are important and must remain the prime focus of the modeling activity. There are many other modeling frameworks and research groups that focus only on the quick short-term variables. In terms of resilience theory, healthy, well-balanced stocks of resources and built infrastructure will probably confer more resilience than stocks that are characterized by a limited range of ages and sizes.

6.3   Applications and Results

6.3.1   The OzEcco Design for a Methanol Economy

For a policy client interested in alternative land use scenarios that might help restore landscapes suffering from dryland salinity, scenarios were implemented within OzEcco to produce alcohol fuels from woody biomass (Foran and Mardon 1999). A number of assumptions underpinned this methanol production scenario: (1) The scenario would aim to supply 90 percent of Australia's total oil requirements specifically to meet 100 percent of the requirements for transportation fuels. (2) The feedstock share would be 100 percent woody material from plantation biomass resources managed as forests with a twenty-year rotation and an average mean annual increment of twenty cubic meters per year. (3) Approximately 60 percent of the woody biomass would be derived as logs and the remainder as branches and waste wood. (4) The rate of plantation biomass establishment would be four hundred thousand hectares per annum. (5) The capital cost in constant dollar terms of the methanol plant was fifty million dollars per petajoule of production capacity, and the lifetime of plant was twenty years.

The high-level indicators simulated by the OzEcco model with these scenario assumptions are shown in figure 6.5. The simulated growth rate in GDP for the methanol scenario dips below the base case scenario for the first twenty years of the transition and then tracks with it. The first dip in the base case curve due to domestic oil depletion is avoided, and the second drop due to the depletion of natural gas stocks is not as large. The per capita affluence measure (gigajoules per capita of energy embodied in personal consumption) tracks slightly below the base case for most of the simulation. The energy intensity of GDP (megajoules of fossil energy per constant dollar of GDP) is decreased by a factor of four (from 8 MJ per dollar to 2 MJ per dollar) by 2050. The emissions of carbon dioxide from the energy sector diverge from the base case after 2005 and rise gradually to 800 million tonnes per annum by 2050, a reduction of 400 million tonnes per year compared with the base case.

Successful simulations of the modeled economy do reveal a number of perverse outcomes, as revealed in figure 6.5. One of these is the intersectoral rebound effect, in which CO_2 emissions rebound after the investment period has made the transition to a methanol economy. This results from the release of capital, both from older, now-defunct sectors (domestic oil mining and refining) and new, more efficient sectors (as gas turbines replace coal-fired electricity generators). The capital release stimulates

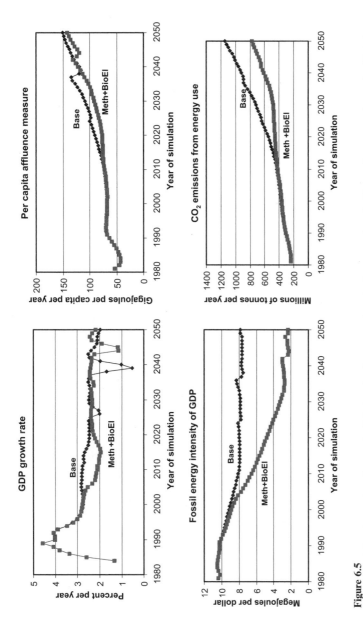

Figure 6.5

Report card for the methanol scenario (Meth+BioEl) compared to the base case (Base). Figure shows growth rate in GDP (top left), per capita embodied energy in personal consumption index (top right), energy intensity of GDP (bottom left), and carbon dioxide emissions from energy use (bottom right).

personal consumption in a general sense. This consumption in turn restimulates the requirement for oil, gas, and coal, some of which is imported if domestic supplies become constrained. Perverse outcomes are difficult to manage in a modeling sense, as well as in a real market-based economy operating in a globalized world. In order to constrain carbon emissions, it may in the end be necessary to restrict the physical amount of carbon used as inputs to the physical economy. A twenty-year process is now underway in Australia's Murray Darling Basin to restrict, in an absolute sense, the use of water for irrigation. That such a restriction will ever be imposed on fossil energy inputs seems unlikely under current political ideologies.

Analyses such as these are not predictions in a traditional sense. Rather, they test the likely behavior of the simulated physical economy in response to the implementation of large-scale industrial ecology. A measure of scenario success is the degree to which indicators for a scenario under test diverge from, or remain with, the base case scenario. Although the OzEcco model is driven by physical processes, it is possible to derive a number of economic indicators, such as nominal GDP, because of the strong relationship, in the current structure of the economy, between flows of dollars and the flows of fossil energy that underpin them.

The evaluation of what constitutes a successful scenario is a difficult one in a policy or industry context. Compared with the indicator sets commonly used in state-of-environment reporting, the advantage of the physical modeling approach (compared with series of reporting indicators obtained from a wide variety of partially linked national statistics) is that modeling indicators are structurally linked to one another within the operations of the physical economy. Provided that the modeling has a sound philosophical and biophysical basis, a simulated output provides a thorough basis for interpretation and understanding, as well as a cogent and robust look-ahead capability.

6.3.2   The Australian Stocks and Flows Framework

The intended use of the ASFF analytical approach is now shown through two complementary analyses. They are extracted from a population study undertaken for a government policy client who sought to compare high, medium, and low rates of population growth, driven in turn by three different rates of net immigration. Material flow analysis and greenhouse gas emissions from the energy sector are selected as higher-level issues that integrate industrial ecology concepts at a whole-of-economy level.

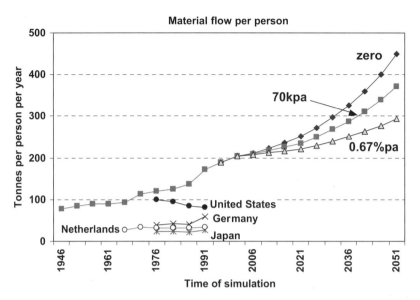

Figure 6.6
Total material flow in tonnes per person per year for three population scenarios: the medium growth rate driven by 70,000 net immigration per year (70 kpa), the lower growth rate driven by zero net immigration per year (zero), and the higher growth rate driven by 0.67 percent of current population as net immigration per year (0.67% pa). Data for four industrialized countries are also displayed (Adriaanse et al. 1997).

6.3.2.1 Material Flow Analysis Australia has maintained a materially intensive economic system for many reasons, and the standard assumption of this study is a continued expansion for many primary exports. This results in a material flow account that continues to expand beyond a contemporary level of 200 tonnes per capita per year to 300 tonnes for the high population growth rate, 370 tonnes for the medium growth rate, and 450 tonnes for the low growth rate (figure 6.6). The higher population scenarios give lower indices of per capita material flow because of a dilution effect of higher population numbers within a material flow driven primarily by export trade decisions. For comparison purposes, the analyses of Adriaanse et al. (1997) are presented in the figure for the material requirements of the United States, Germany, the Netherlands, and Japan. For the period 1970 to 1990, the structural and trade arrangements of those countries allowed much lower material flows on a per capita basis, although higher populations in Germany, the United States, and Japan would give comparable or larger material flows on an aggregated whole nation basis.

There are many ways in which these data may be examined. For the medium population growth rate, the direct and hidden flows for domestic requirements are maintained at below 100 tonnes per capita for the duration of the simulation, as the result of a stabilizing population. Most of the effect is due to hidden flows of material tied to the nation's exports and specifically refers to items such as overburden for open cut mines, material removed in ore concentration activities, and effects of crop and animal agriculture. In general, the mining industry for both metals and energy materials accounts for most of the increase from current levels of the per capita material flows.

In comparison with other developed economies, a variety of material flow indicators for the Australian economy will be higher and will also trend upward. This is because of a mixture of historical antecedents, contemporary policy directions, and future strategic directions already well underway within a variety of major commodity groups with production bases in Australia. None of these are preordained, and global trade and political forces may cause major changes to the base case scenario and the analyses derived therefrom. A carbon tax on coal usage in countries such as Japan and South Korea and those of the European Union would cause a significant reduction in per capita material flow but have large implications for the level of export income. What energy source might replace coal in those countries also presents a large imponderable for both industry and policy. The transition to a Factor Four or Factor Ten economy in countries such as Japan, South Korea and China and those of the European Union, which currently take the majority of Australia's minerals exports, would also have large repercussions. However, many Factor Four transitions rely on advanced composite materials for lightness and strength, and many of these materials rely on large hidden material flows themselves.

In terms of national policy issues, there are three important areas that determine the implications of Australia's future material flow account. The first and most immediate link is to energy use and greenhouse gas emissions. The more material that is moved, the more energy that is required. In physical law terms, these are the realities of thermodynamics and mass balance that lie behind all modern economic systems. Thus, if more material is moved, even allowing for changes in a wide range of efficiencies, then total energy use may increase. Depending on the source of the energy, greenhouse gas emissions do not necessarily increase, but for all practical purposes they must.

The second important issue for material flows would arise if there were to be negotiations between countries on how to account for, and apportion, the responsibility for such flows. In the analysis presented here, the material flows are apportioned directly to the nation and each person classified as a citizen. The rationale for this is that all citizens reap the reward of the material transactions, whether it be a direct effect (employment, food, and housing) or an indirect effect (export income to purchase a video recorder or an overseas holiday). However, there are equally valid arguments that the material flows should be apportioned to the countries that use the material that Australia exports to them. Thus, in both material and energy terms, Australia's major trading partners would take on the direct and hidden flows of the material exported to them.

However, the chain of attribution would not stop there if a full life cycle analysis were implemented and full system boundary applied. Logically, the next step would be to attribute the hidden material flows embodied in the array of goods that each nation imports. Thus, the copper, aluminium, steel, and magnesium in each imported car would be finally attributed to the country in which the consumption finally takes place. Most OECD countries would be disadvantaged in this system of accounting, which would see the accounting responsibility for 80 percent of the world's material and energy flows attributed to 20 percent of the world's citizens in the richer countries currently. However, it is possible that the balance might change over the next fifty years as populous less-developed countries become more developed. Wernick and Ausubel (1999) make the policy assertion that without the data collection and the development of GDP-like metrics that describe material flows, an economy and its political system are navigating blind on the course that leads inexorably upward to higher and higher levels of material consumption.

The third important issue in the material flow dilemma concerns the type of economy (materially heavy or materially light) that the nation's citizens wish to maintain. In commenting on the structure and performance of the U.S. economy, Greenspan (1998) questioned "whether over the past five to seven years, what has been without question, one of the best economic performances in our history, is a harbinger of a new economy, or just a hyped-up version of the old, will be answered only in the inexorable passage of time" (85). In examining the progress of transition to the knowledge-based economy for Canada, Gera and Mang (1998) concluded that "Canadian industrial structure is becoming increasingly knowledge-based and technology-intensive, with competitive advantage being rooted in innovation and ideas, the foundations of the new economy" (177). However, despite the undeniable growth in employment and economic activity in the services portion of the U.S. economy over the last three decades, Salzman (1999) notes that manufacturing in America has not declined. In a comprehensive analysis of the U.S. economy, Ayres, Ayres, and Warr (2003; chapter 3 of this volume) note that total consumption mass and per capita measures of mass and exergy have continued to increase over the one-hundred-year time frame in spite of considerable technological innovation and awareness of environmental and resource issues. The dilemma is that the service economy exists to service the old economy and to make it more efficient in terms of finance, labor, quality, and delivery schedule. What has been saved materially through efficiencies in production processes has been consumed in increasing the diversity of products and opportunities, few of which have zero material and energy contents. The dilemma of material flows could be that in order to halve material flows, the total material consumption of each citizen would also have to be halved, while properly accounting for direct and indirect flows as well as the exported and imported components of globalized trade (see chapter 3).

6.3.2.2 Greenhouse Gas Emissions from the Energy Sector The continuing emissions of carbon dioxide and other greenhouse gases from the energy sector are viewed by many political and scientific groups as an issue of global concern. The possible effects within the span of two to eight human generations include the increased frequency and intensity of weather events, the displacement of agricultural systems, the loss of amenity and infrastructure close to regions of possible sea level rise, and the loss of process diversity in natural systems as key elements of function are lost as a result of the interaction of many factors. Although Australia is a small emitter of greenhouse gas in total terms, an affluent lifestyle and a lower population base in relation to the sum total of its physical transactions makes it a high per capita emitter, among the top five in the world. As a relatively advanced country in technological terms, Australia might be expected to have the capacity to reduce greenhouse emissions through a mixture of technological innovations and changes in the volume and composition of personal consumption. Alternatively, in the future, new institutional arrangements at an international level might implement greenhouse accounting measures that allocate the responsibility for greenhouse emissions to the consumer of the final product or service in the country where the consumption happens. Again, Australia would

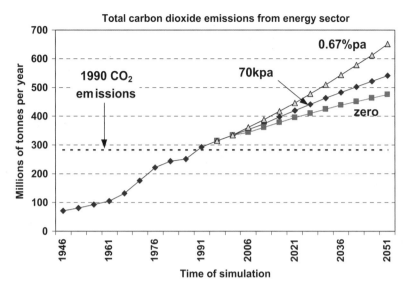

Figure 6.7
Carbon dioxide emissions in million tonnes per year from the energy sector to 2050, for three population scenarios: the medium growth rate driven by 70,000 net immigration per year (70 kpa), the low growth rate driven by zero net immigration per year (zero), and the high growth rate driven by 0.67 percent of current population as net immigration per year (0.67% pa).

not necessarily be advantaged in these new arrangements, as its imports contain more embodied energy, and therefore carbon emissions, than its exports.

The standard assumptions used as a starting position underlying all population scenarios suggest that carbon dioxide emissions will continue to expand up to 2050 for all population scenarios (figure 6.7). In spite of a wide range of technologically optimistic assumptions, throughout most sectors of the physical economy, the carbon dioxide emissions continue to rise. Even under the lower assumed rate of population growth, they continue to grow strongly because of four factors. The first is the inbuilt demographic driver, which, over the next 50 years, meets the requirement for additional houses and cars for younger people who continually enter the consumption economy. Three other important factors are the growth in per capita affluence, large increases in international inbound tourism, and steady expansion of a wide range of export commodities and goods that have energy-intensive production systems.

When a series of technological innovations are combined, then the emission trajectories do change markedly. The band of population-driven emissions shown in figure 6.7 could be reduced by 200 million tonnes per year by 2050 if best practice technologies were implemented throughout all physical sectors. Alternatively, the emissions band could occupy a zone 200 million tonnes higher if current technological options are retained and if lifestyles become more energy intensive overall. Implementing individual technologies, such as better car engines, in the main produces only marginal effects on the overall emissions trajectories. This emphasizes that for industrial concepts to have a major influence on national aggregate indicators, they must be fully implemented throughout all corners of the physical economy.

6.4 Strategy for National Policy Influence

The strategic plan for the physical economy project in which the OzEcco and ASFF models are used has three linked goals. The first goal is to pin the debate about transforming the physical economy to more sustainable modes of operation, such as the dematerialized Factor Four or Factor Ten economy, as detailed by von Weizsäcker, Lovins, and Lovins (1997), Ayres (1998b), and others. The second goal is to have accepted, at national policy levels, the concepts of physical analysis of the national economic function. The third goal is to contribute to changing national policy on a number of key aspects that relate to the physical economy.

The route to achieving these goals is a complex and difficult one, with two important considerations. The first is the dominance of economic theory and debate in assessments of national population development issues, combined with the belief that market mechanisms will deal with environmental problems if they are sufficiently important to require a solution. Allied with these economic arguments is the view that technological innovation is a driver of progress in its own right that will implement industrial ecology if the market sees an economic advantage. The second consideration is that the integration and modeling within these physical economy models is a challenging one in which scientific proof in a traditional sense is difficult. In addition, a modeling framework is always open to improvement. In a project management sense, this can result in an imbalance among investment in modeling, the outputs from scenario simulations, and subsequent contributions to the long-term analysis of national policy issues.

Given these constraints, the route chosen to the project's strategic goals is a partial and iterative one, with an overall integration phase in the final two years of the process. Although the insights into strategic policy require an analysis of the whole physical economy, the way forward requires that twenty important sectors, such as agriculture, building, manufacturing, and energy, each be investigated in a partial sense for an identified client who will underwrite the task. This allows deeper scrutiny and appropriate model development for each main sector for a client with which we might investigate and learn about an important physical sector. The base case scenario must also be further developed in an iterative manner, with additional insights from the client and the particular analysis undertaken. Using this client-focused partial approach, it is possible that important insights that might help accelerate an industrial ecology transition might be lost in a welter of detail or simply not recognized.

6.5 Discussion

6.5.1 The Approach

This chapter has described a framework of analysis around future design issues for Australia's physical economy. The approach aims to create new whole economy designs, as well as revealing physical flaws in national policies. It relies on three key criteria to influence national policy directions. The first is that policy designers, with model analysts, should become active learners within the iterative simulation process. Central to the analytical approaches is that the analyst or user is seen as the human dimension within the modeling procedure, rather than being a value-free controller outside the simulation process. The second criterion focuses on the relationship between the physical economy and the monetary economy. The physical economy should conform to the physical laws of thermodynamics and mass balance. Because of its behavioral basis, the monetary economy is open to a wider array of innovations and beliefs than the physical economy. Creating harmony between the two analytical disciplines will remain a challenge.

The third criterion concerns the nature of predictive analysis versus the nature of scenario analysis. The concept of scenario analysis used in this approach relies on a wide array of expert opinion and data analysis. These help set the control variables that drive simulation outcomes in a transparent and explicit manner. A simulation of a scenario may seek to test the physical feasibility of a particular national policy. Alternatively, it may seek to design the pathways along which a policy must progress, if it is to reach an explicit goal by a future point in time. Whether the policy analyst thinks of himself or herself as an observer or as an architect in national affairs could be seen as an important distinction. An observer might anticipate incremental policy changes at the margin. An architect might seek to redesign and foster the entirely new structures that could force the transition toward concepts of economy-wide industrial ecology.

It is true that many of the issues on which ASFF and OzEcco focus are dealt with in part by many specialist modeling groups in both academia and consulting in Australia. Thus, national statistical agencies undertake national population projections, national resource agencies model the development of petroleum, minerals, agriculture, and water use, and so on. National econometric models also analyze the effect of policy innovation and shocks on a wide range of whole economy variables, usually, but not always, described by financial-reporting variables. Many of the econometric models are focused on equilibrium concepts and contain only rudimentary dynamics. Assumptions about technological change and learning are seldom constrained by natural resource realities (land, water, oil, and gas availability) or process realities in mass balance and thermodynamic terms. The issues that are missed by these other approaches are well described in chapter 5's analysis of the capacity for change in the steel, paper, and ethylene industries in the United States. Their analysis of the dynamics of key infrastructure stocks in these industries, and the degree to which stock inertia confers resistance to change in the face of aggressive financial penalties for carbon emissions, provides a salutary lesson for policy designers who ignore the importance of stock dynamics in setting real limits to the pace of change and innovation.

6.5.2 Advantages and Disadvantages of the Physical Modeling Approach

A design and testing approach that combines the process of understanding how a physical economy functions with a complete and consistent database that underpins the process model is seen as the main advantage for physical economy models. Such an approach relies on complex calibration and validation procedures that construct a foundation for the physical economy in the historical period before the future scenario is simulated. These "grounding" procedures enable a proof of concept to be displayed and an acceptance gained that the underlying modeling procedures compute appropriately in historical time. The treatment of slow-moving variables such as stocks of people, cars, houses, and agricultural fields is central to the concepts of momentum and inertia, central to complexity and resilience in the physical economy. The description and vintaging of stocks is seldom implemented fully in economic models, yet it is a key limiting variable to the sustainability transition. The associated concept of physical realism that drives most production processes is also vital and also not included in many economic models.

The modular and stepwise nature of model design and computation procedure allows partial simulations to be undertaken relatively easily and further model development to be undertaken on a component without disturbing the integrity of the whole. The level of detail is reasonably flexible and in the ASFF model ranges from fifty-eight regions for agricultural productivity to sixteen regions for human population dynamics to eight city air sheds for vehicle emissions and one national account for balance of trade computation. One advantage in national and international terms is that a limited amount of simulation modeling of physical economies has been undertaken in a policy context, when compared with the dominant force of econometric modeling. This may provide an advantage in the policy marketplace for concepts and analyses describing physical sustainability. However, there is little policy experience in the promotion and refutation of design theories dealing with the physical economy and its underlying dynamics.

The size and complexity of the analytical undertaking present an immediate disadvantage to scientific management, funding agencies, national policy analysts, and scientific colleagues. The gulf between the constrained boundaries and reputable sureness of traditional reductionist research approaches, on the one hand, and a nationally scaled modeling approach that uses scenarios, on the other hand, has never been greater. Lutz (1994) noted that the challenge of physical economy modeling lies in its being able to combine a "hard-wired model which only includes unambiguous relationships on which scientific consensus can be expected" with a "soft model which can quantify all kinds of feedbacks and interactions that the user wants to define" (362).

This approach mentioned by Lutz in design and implementation appears to be meeting the project goals. But the absence of price mechanisms in both the OzEcco and the ASFF models poses a significant barrier to acceptance by national policy analysts. Some suggest that the physical and the economic approaches should be hybridized and blended. Others are satisfied to keep them as distinct and separate analytical approaches, each of which contributes insights to the policy process. The philosophical approach behind the development of the physical economy simulators argues that prices and market mechanisms are critical to balancing the economic concepts of supply and demand in the short term. However, the strategic intent of the long-term physical modeling approach is to provide information flows from longer-term horizons to

current market, policy, and business agendas. For these long-term horizons, price and market mechanisms become diffuse and indeterminate,
while the workings of the physical economy still depend on people and
the flows of energy and materials.

6.5.3   Economic and Physical Limits

The complexity of model building and validation sometimes hides obvious tensions between viewpoints that are mostly economic or mostly
physical. For the Australian economy at least, each extra dollar of GDP
requires an extra ten megajoules of primary energy, thirty-seven liters of
water, and three square meters of permanently disturbed land. Under
conditions of limited technological progress and economic growth rates of
3 percent per annum, this will mean that physical requirements approximately double every twenty-five years or so. The modeling approaches
implemented for Australia and the long-run analyses of the U.S. economy
by Ayres, Ayres, and Warr (2003; also chapter 3) provide a compelling
picture of these physical realities at the macroeconomy level. That technological progress is constrained by the lock-in of important capital
stocks and industrial processes over long periods is well described in
chapter 5 and their conclusion that "even more credence in the modeling
arena must be given to models that are able to capture the impact of
policies on the turnover of the capital stock and its associated impact on
energy use and carbon emissions" (225–226).

Whereas physical approaches are driven in a longer-term sense by
technological parameters and the dynamics of turnover in capital stocks,
economic approaches are driven by price and policy changes within
shorter-term horizons. Even with a reasonably upbeat interpretation, the
hybrid economic-physical models STREAM (chapter 7) and DIMITRI
(chapter 8) give little suggestion that the physical approaches are substantially in error. The conclusions of the STREAM modeling are that
there is no absolute decline of material use in OECD countries and that
tax measures to reduce the use of materials are largely ineffective. The
DIMITRI modeling emphasizes that policies constraining energy and
material use have to be applied to all trading partners of a nation-state,
because of the complex interdependencies of international trade and the
degree to which the market realities of loopholes in national policies
make it easy to shift production to lower-cost locations where energy and
material use are often not constrained.

This leads to the vexing questions of physical limits within an economic
world that is seemingly limitless. There is no doubt that the processes of

material and process substitution will continue, allowing some measure of affluence growth for the next century. However, the emphasis in "limits" identification has shifted from the sources of material progress to the sinks, where the by-products of energy use and material transformation are eventually deposited. Whether it is the atmospheric commons of the global change debate or the local problems of water quality and toxified agricultural soils, only the most myopic of policymakers are unable to recognize that real physical limits are now more than just a blip on this century's optimistic horizons. However, what most citizens and policymakes do not appreciate is the huge challenge posed by the inability of advanced measures of technological progress to penetrate the capital stocks of most modern economies and thus to moderate or reduce material use. The challenge is even greater when an economy is maintaining high rates of growth fueled by the expanding personal consumption that underpins it.

References

Adriaanse, A., S. Bringezu, A. Hammond, Y. Moriguchi, E. Rodenburg, D. Rogich, and H. Schütz. (1997). *Resource Flows: The Material Basis of Industrial Economies*. Washington, DC: World Resources Institute.

Ayres, R. U. (1998a). Rationale for a physical account of economic activities? In P. Vellinga, F. Berkhout, and J. Gupta (eds.), *Managing a Material World: Perspectives in Industrial Ecology*, 1–20. Dordrecht, the Netherlands: Kluwer Academic. Boston and London.

Ayres, R. U. (1998b). *Turning Point: An End to the Growth Paradigm*. London: Earthscan.

Ayres, R. U., L. W. Ayres, and B. Warr. (2003). Exergy, power and work in the US economy, 1900–1998. *Energy* 28: 219–273.

Carpenter, S., B. Walker, J. M. Andries, and N. Abel. (2001). From metaphor to measurement: Resilience of what to what? *Ecosystems* 4: 765–781.

Cocks, D. (1996). *People Policy: Australia's Population Choices*. Sydney: University of New South Wales Press.

Cocks, D. (1999). *Future Makers, Future Takers: Life in Australia 2050*. Sydney: University of New South Wales Press.

Foran, B. D., and D. Crane. (1998). The OzECCO embodied energy model of Australia's physical economy. In Sergio Ulgiati et al. (eds.), *Advances in Energy Studies: Energy Flows in Ecology and Economy* (conference held at Porto Venere, Italy, May 26–30), 579–596. Siena: Department of Chemistry, University of Siena.

Foran, B. D., and C. Mardon. (1999). Beyond 2025: Transitions to the Biomass-Alcohol Economy using Ethanol and Methanol. CSIRO Resource Futures Working Document no.99/07, available at http://www.cse.csiro.au/research/Program5/RF/publications/Biomassfuels.pdf (accessed June 14, 2002).

Gault, F. D., K. E. Hamilton, R. B. Hoffman, and B. C. McInnis. (1987). The design approach to socio-economic modelling. *Futures* 19: 3–25.

Gera, S., and K. Mang. (1998). The knowledge based economy: Shifts in industrial output. *Canadian Public Policy: Analyse de Politiques* 24(2): 149–184.

Godet, M. (1991). *From Anticipation to Action: A Handbook of Strategic Prospective*. Paris: United Nations Educational, Scientific and Cultural Organization Publishing.

Greenspan, A. (1998). Is there a new economy? *California Management Review* 41(1): 74–85.

Herendeen, R. (1998). *Ecological Numeracy: Quantitative Analysis of Environmental Issues*. New York: Wiley.

Holling, C. S. (2001). Understanding the complexity of economic, ecological, and social systems. *Ecosystems* 4: 390–405.

Lenzen, M. (1998). Primary energy and greenhouse gases embodied in Australian final consumption: An input-output analysis. *Energy Policy* 26(6): 495–506.

Leontief, W. (1970). Environmental repercussions and the economic system. *Review of Economics and Statistics* 52: 262–272.

Long Term Strategies Committee (House of Representatives Standing Committee for Long Term Strategies). (1994). *Australia's Population Carrying Capacity: One Nation—Two Ecologies*. Canberra: Australian Government Publishing Service.

Lutz, W. (ed.). (1994). *Population-Development-Environment: Understanding Their Interactions in Mauritius*. Berlin: Springer-Verlag.

Richardson, G. P., and A. L. Pugh. (1981). *Introduction to System Dynamics Modeling with DYNAMO*. Portland, OR: Productivity Press.

Rose, A., and W. Miernyk. (1989). Input-output analysis: The first fifty years. *Economic Systems Research* 1(2): 229–271.

Salzman, J. (1999). Beyond the smokestack: Environmental protection in the service economy. *UCLA Law Review* 47(2): 411–489.

Slesser, M. (1992). *ECCO User Manual Part 1*. 3rd ed. Edinburgh: Resource Use Institute.

Slesser, M., J. King, and D. C. Crane. (1997). *The Management of Greed: A BioPhysical Appraisal of Environmental and Economic Potential*. Edinburgh: Resource Use Institute.

von Weizsäcker, E. U. (1998). Dematerialisation: Why and how? In P. Vellinga, F. Berkhout, and J. Gupta (eds.), *Managing a Material World: Perspectives in Industrial Ecology*, 45–54. Dordrecht, the Netherlands: Kluwer Academic.

von Weizsäcker, E. U., A. B. Lovins, and L. H. Lovins. (1997). *Factor Four: Doubling Wealth-Halving Resource Use*. St. Leonards, Australia: Allen and Unwin.

Wernick, I. K., and J. H. Ausubel. (1999). National material metrics for industrial ecology. In P. Schulze (ed.), *Measures of Environmental Performance and Ecosystem Condition*. Washington, DC: National Academic Press. Available at http://phe.rockerfeller.edu/NatMatMetIndusEcol (accessed July 26, 2000).

7 Environmental Policy Analysis with STREAM: A Partial Equilibrium Model for Material Flows in the Economy

Hein Mannaerts

7.1 Introduction

Policymakers in Europe have an urgent need for efficient, effective, and equitable environmental policy instruments that stimulate a delinking of economic growth and environmental deterioration. In recent years, the gap between greenhouse gas emissions and the Kyoto emission target widened for most European countries (EEA 2003a). The reason for this undesirable development stems from the relative high growth of energy-intensive production and consumption. Moreover, the use of carbon-intensive fuels tends to increase again after a long period of substantial decline.

Material production is one of the main sources of CO_2 emissions. It increased substantially at the end of the 1990s. The EU Sustainable Development Strategy and the sixth environmental action plan both mention dematerialization of the economy as an important objective for sustainable growth (EEA 2003b). Most EU countries have placed dematerialization of the economy on their research and policy agenda.

However, because of the openness of the European economies, unilateral emission policy measures with respect to material producers are generally offset by increasing emissions in neighboring countries. Increased production costs resulting from the policy measures, together with strong international competition, encourage these industries to reallocate their production capacity outside the region in which the measures apply. In particular, energy-intensive basic industries that produce steel, aluminum, paper, ethylene, and ammonia in large quantities face fierce international competition and may lose market share to equally or even more pollutive foreign competitors.

The main questions addressed in this chapter are, How sensitive are the capacity investments of the basic industries to environmental costs and material taxation? and Which policy instruments are the most cost-effective and induce the smallest displacement effects as well? These questions are examined with the help of the Substance Throughput Related to Economic Activity Model (STREAM), a partial equilibrium model for steel, aluminum, petrochemicals, paper, nitrogen, phosphate, and potassium flows in Europe and the Netherlands. The coefficients of the model are determined by estimation and by calibration of the variables over the period 1960 to 1995. Simulations over the past show to what degree the model can explain the observed material flows and prices. Future developments on the material markets can be explored through scenario analysis.

The first part of this chapter describes the economic mechanisms of STREAM that play a central role in the relation between material flows and economic activities. Next, various environmental policy instruments that affect national and international material flows are analyzed.

The organization of the chapter is as follows: Section 7.2 discusses the general structure of the model and explores in depth some crucial relations in view of the main topics in materials analysis. The model specifications of these relations are summarized in the appendix. The section also draws a comparison with other material-related models. Section 7.3 illustrates the evaluation of the cost-effectiveness of various environmental policy instruments with respect to CO_2 and NO_x abatement. Section 7.4 gives the main conclusions.

7.2   The Model

This section presents the methodological and empirical foundations of STREAM first, paying due attention to system boundaries, industrial adjustment, representation of technology, sources of information, regional specialization, and time frame (chapter 5). Next, it discusses the material flow relations in the economy within the context of five dominant themes in the field of materials analysis: dematerialization, recycling, input substitution, resource scarcity, and international allocation of production capacity.

7.2.1   General Structure

7.2.1.1   System Boundaries   STREAM is an international material-product Chain (IMPC) model. It describes material flows in the successive stages of economic production and consumption in the Netherlands, Western Europe, and the world, as well as the flows among these regions.

The model consists of seven independent submodels, one each for steel, aluminum, paper, plastic, ammonia, phosphate, and potassium. Each submodel distinguishes the following activities in the subsequent stages of the economic process: the mining industry, basic industry, product industry, final users, recycling industry, and waste disposal industry. The steel, aluminum, paper, and plastic submodels also distinguish between primary materials (produced from raw material) and secondary materials (produced from scrap).

STREAM has the structure of a partial equilibrium model. It describes three markets: raw materials, materials, and scrap. These markets determine the material prices through the interaction of supply and demand forces in an economic environment of free competition. Economic choices are based on the cost-minimizing behavior of the economic actors that take the market prices as given. The outcome of this optimizing process can be depicted by demand and supply relations. Demand and supply determine the market prices in the model.

The chain approach may be used to trace the indirect effects of environmental policy measures. The market equilibrium approach provides the dynamics of adaptation of producers and consumers to changes in economic conditions and policies.

Figure 7.1 shows the material flows in the material-product chain that are described by the model. Materials enter the chain as minerals from the mining industries or as imported material and leave the chain as exported material or as disposed waste. The quantity of material flow is determined by prices and the income earned by the final users. Prices depend on the scarcity of natural resources, labor, and capital. Institutions

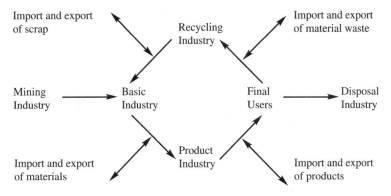

Figure 7.1
International material product chain in STREAM.

and policymakers influence these scarcities by imposing rules, restrictions, and taxation.

7.2.1.2 Technology Restrictions and Dynamics Long-run technology restrictions are represented in the model by aggregate cost functions. These functions describe the producers' substitution possibilities. In the short run, producers have no substitution possibilities at all. The model applies a putty-clay vintage structure in which the installed vintages have fixed input-output ratios. The parameter values of the aggregate cost functions are derived both from historical time series and from bottom-up information. Only new vintages are assumed to operate efficiently at the production frontier. The fixed technology of the old vintages may create a substantial "efficiency gap" if technological progress is great or input prices have changed.

The model describes technological progress by means of an exogenous trend. New production techniques enter the production process through investments in new vintages. Expected input prices determine the optimal production technique. Expected absolute and relative profits on new vintages determine the level of investments in new capacity as well as the scarp rate of the old capacity. Producers form their expectations by means of past experiences.

7.2.1.3 Exploring the Future STREAM is a consistent framework for policy analysis, providing a convenient tool for the construction of material scenarios. These scenarios are useful for developing a good understanding of potential problems and their solutions as well as their predictability. The limited predictability of the future stems from the multitude of factors involved. Only the factors that have been proved to be important in the past thirty years are included in the model. Unexpected changes in market conditions and fundamental changes in technology may reduce the adequacy of the model specification. Such events can be simulated by making alternative assumptions about the exogenous variables and the parameter values. The model computations of these scenarios of changes in market conditions and in technology may generate robust insights into future risks and possibilities.

As an illustration, this chapter presents a "business-as-usual" scenario. So that it will remain realistic, the time span of the scenario is no longer than twenty-five years. Different policy options are analyzed in the context of this scenario.

7.2.2   Dematerialization of the Economy

Material consumption depends on the level of economic development, material scarcity, energy prices, and investment cycles. This subsection presents the empirical relations among these economic conditions and material consumption.

7.2.2.1   Economic Growth and the Use of Materials   The model describes material demand as a material intensity relation at the macroeconomic level for Europe and at sector level for the Netherlands. This relation describes the ratio between material consumption and economic production on the macro or sector level. Empirical studies reveal that material intensity exhibits a bell-shaped relation, sometimes referred to as the "inverted U-curve." This relation reflects different phases in the relationship between materials and economic activity. The upward slope of the inverted U is the phase of innovation, characterized by many new applications of the materials accompanied by strong cost-price reductions. This development levels off when no new applications of the material emerge and consumer saturation is reached. The downward slope of the inverted U is the maturity phase, characterized by substitution of other materials, accompanied by continuous improvement of material efficiency of the product industry.

As an example, figure 7.2 shows that West European intensity of use for plastics is still on the upward slope. Model extrapolation according to a business-as-usual scenario[1] suggests that for the next quarter century, saturation of plastic demand will be approached. Plastic demand will then expand in line with GDP. Other empirical studies (Simonis 1989; Jänicke et al. 1989; Jänicke 2001) have revealed that the intensity of use of the other materials in the model have been declining in the OECD countries from the 1970s on. The transition from a manufacturing economy toward a service economy has largely contributed to this decline. Indeed, in the 1970s the leading role of the manufacturing industry in economic growth has been taken over by the service industry. However, the decline of the intensity of use of tin originated in the 1920s and that of steel in the 1950s, whereas the intensity of use of aluminum, ammonia, and plastic was still rising at the end of the twentieth century (see also chapter 3). Thus, each material has its own development in time with respect to the intensity of use, irrespective of the deindustrialization of the economy.

The bell shape of the material intensity can be specified with the use of a decreasing income elasticity of demand. Table 7.1 shows that the income

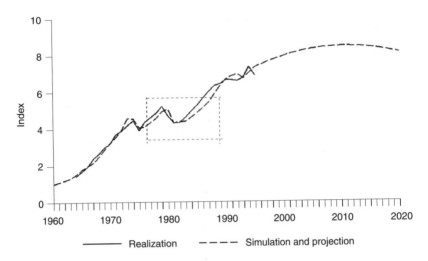

Figure 7.2
Development of the intensity of use of plastics in Western Europe.

Table 7.1
Income and trend elasticities of material demand for OECD countries

Material type	Income elasticity, 1960	Income elasticity, 1995	Trend elasticity
Steel	0.9	0.6	−0.75
Aluminum	2.0	1.5	−1.75
Plastic	3.8	2.7	−0.30
Paper	1.1	1.0	−1.00
Nitrogen	1.8	1.2	−1.50
Phosphate	1.1	0.6	−0.75
Potassium	0.8	0.4	−0.50

elasticities of the materials in 1995 are substantially lower than in 1960. However, most of them are still well above one in 1995. Consequently, increased material demand due to economic growth would have exceeded GDP growth if there were no autonomous dematerialization effect, as also shown in table 7.1. For most materials, the two effects add up to a decreasing intensity of use, as long as economic growth is less than 3 percent per year.

The empirical evidence for the process delinking material use or emissions from economic activity has been questioned. De Bruyn and Opschoor (1994) argue that the inverted U-curve is invalid and that instead an N-shaped relationship can be observed over the period 1970–

Table 7.2
Price elasticities of material demand for OECD countries

Material type	Own price elasticity	Energy cross-price elasticity
Steel	−0.2	−0.1
Aluminum	−0.2	−0.05
Plastic	−0.9	—
Paper	−0.4	−0.025
Nitrogen	−0.5	—
Phosphate	−0.5	—
Potassium	−0.5	—

1990. Our data confirm this finding for this period but also indicate that this is just a deviation from the inverted U-curve that roughly spans a century. The N-curve is caused by fluctuations of the energy price and its economic consequences in that particular period. Figure 7.2 shows the N-curve in the shaded area. The positive trend of the intensity of use is simply interrupted by the second oil crisis in 1979.

7.2.2.2   The Effects of Material Prices on the Use of Materials   Material prices are inversely related to material demand. The expansion of plastic and aluminum demand in the early 1960s largely depended on decreasing plastic and aluminum prices. The first column in table 7.2 shows that most materials' own price elasticities are between −0.2 and −0.5. This suggests that dematerialization can be stimulated through taxation. However, the estimated material demand relations do not yet include the prices of material substitutes. Cross-price elasticities of other materials are not included in the model. Explicit introduction of these prices would alter the picture. For instance, if two materials are taxed at the same rate, the overall demand effect is smaller than their own price elasticities suggest.

7.2.2.3   The Effect of Energy Prices on the Use of Materials   Energy price affects material demand in two ways. Energy is used in material production, therefore increasing energy prices lead to increasing production costs. If the energy price increases worldwide or for large economic regions, the increased production costs can be passed on to the material prices. These higher material prices lower material demand.

Energy is also used complementarily to materials, especially in transportation activities and energy production. Higher energy prices induce less transport activities and more energy savings. Consequently, material demand falls as a result of reduced investments in transport equipment and energy production. The second column of table 7.2 shows that this indirect energy price effect is important for steel, aluminum, and paper demand.

7.2.2.4 The Effects of Investments on the Use of Materials   Besides being influenced by the level of GDP, material demand is affected by the composition of GDP as well. The steel requirement for transport equipment is 110 tons per million euros of investment goods, whereas it is only 7.5 tons per million euros of services (Konijn, de Boer, and van Dalen 1995). Therefore, changes of composition from goods to services will affect the material intensity of GDP. Long-run structural changes are captured in the model by decreasing income elasticities in the material demand specification. Changes in the short-term investment cycle also affect material demand, because of the high material content of investment goods. From empirical data, an investment ratio–to–GDP elasticity can be found of 0.75 for steel demand and 0.4 for paper. For other materials, the investment elasticity is small.

7.2.2.5 Material Policy   Material policy is still in its infancy. Only a few policy measures have been implemented with respect to the use of materials. Moreover, the focus of these measures is on waste reduction and recycling.

Two such measures are worth mentioning. First, the second Dutch packing material agreement[2] in 1997 was quite successful. It accounted for an additional reduction in intensity of use of packing materials by roughly 15 percent. It also increased the recycling of packing materials from 50 percent to almost 65 percent. A successful European measure is the restriction of nitrogen consumption in agriculture. As a result of this restriction, nitrogen consumption dropped by 20 percent between 1985 and 1990.

7.2.3   Recycling of Materials

The recycling of materials reduces emissions substantially. It diminishes the physical production of the mining industry at the beginning of the material-product chain as well as the waste flow to deposits at the end of the chain. Moreover, secondary material production from waste is generally much cleaner than primary production and requires less energy. However, waste collection and recovery activities of the recycling indus-

Table 7.3
Elasticity of substitution between primary and secondary production

Material type	Elasticity of substitution
Steel	−3
Aluminum	−2
Plastic	−2
Paper	−4

try require energy and generate emissions as well. This in turn reduces the overall environmental gain from recycling.

Product industries can choose between primary and secondary materials as input to their production processes. The choice depends on the required quality of the materials produced and on the relative prices of the primary and secondary input materials. In general, secondary materials are of lower quality than primary materials. Therefore the two types of materials are usually not considered pure substitutes. The lower price of secondary materials reflects not only their lower quality, but also their lower production costs. Table 7.3 shows high substitution elasticities between primary and secondary materials, in the range between −2 and −4.

Up to now, substitution between primary and secondary materials has been largely determined by changing scrap and energy prices. Figure 7.3 presents the historical development of secondary steel production and the scrap price, together with the model simulations of both variables.[3] The model extrapolation according to a business-as-usual scenario suggests that the share of secondary steel production in Western Europe will increase from 35 percent in 1995 to 45 percent in 2020. The main driving force in this development is the ongoing decline of the scrap price as a result of waste policy and technical progress in scrap collection and recycling.

7.2.4 Input Substitution and Technical Progress

Input substitution and technical progress are both objects of environmental policy by means of input taxation and subsidies for new technologies. The impact of these measures can be estimated from bottom-up technical information or from a more traditional econometric top-down perspective. Most bottom-up studies suggest that the substitution possibilities for natural resources are considerable and that technical progress can reduce demand for natural resources substantially. According to the Factor Four discussion (von Weizsäcker, Lovins, and Lovins 1997),

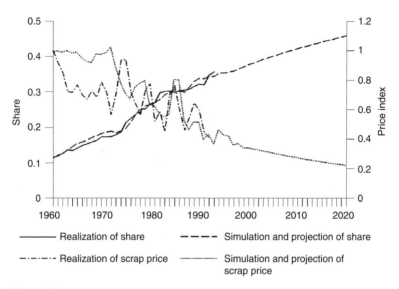

Figure 7.3
Development of the secondary steel share in production and the scrap price in Western Europe.

sustainable economic growth can be achieved even as demand for natural resources is reduced. Bottom-up studies (Blok et al. 1994) also suggest that the energy efficiency can be improved substantially at little (net) investment cost.

For the model discussed here, a conventional neoclassical top-down approach (Slade 1981) is chosen, because of its generality and simplicity, to estimate substitution possibilities and technical progress. First, the chosen approach accounts for substitution among all inputs, not just between energy and capital. Second, the approach makes it possible to derive the marginal production costs, which determine material supply. Third, specific technical information can easily be taken into account in the approach by assigning specific values to elasticities (Koopmans et al. 1999). Finally, the compact representation of the economic system enables a dynamic vintage approach without exceeding the limits of convenience.

Following the conventional practice of neoclassical substitution analysis, the model applies a translog cost function (Varian 1984) with six inputs: labor, capital, coal, oil, natural gas, and electricity. This function is able to represent fixed and switching technologies, as well as all the variants in between. Furthermore, it is based on three assumptions: pre-

Table 7.4
Price elasticities of energy demand of the basic industries of Western Europe used in STREAM and of the Netherlands used in NEMO

Material type	STREAM	NEMO
Steel	−0.15	−0.25
Aluminum	−0.25	−0.16
Plastic	−0.12	−0.08
Paper	−0.20	−0.62
Nitrogen	−0.10	−0.05
Phosphate	−0.20	n.a.

n.a. = not applicable

determined input prices, constant returns to scale, and cost-minimizing entrepreneurs. Cost-minimizing behavior can be represented by a cost function (Varian 1984). The shape of the function summarizes all relevant information about the available technology to the sector. In fact, the cost function represents the minimum production costs that can technically be attained by an entrepreneur at different values of the input prices and production level.

Factor demand functions can be directly derived from the cost function (Shephard 1970), providing equations for labor, capital, energy, and materials. The constant-returns-to-scale assumption implies that if production doubles, factor demand will double too.

The derived energy demand equations consist of not only a price-dependent part for substitutable energy, but also a price-independent part representing the "thermodynamic minimum" in the production process. The latter covers a substantial part of the overall energy use for materials in question. For instance, 45 percent of the current energy use in ammonia production and 60 percent of the energy use in methanol production can never be reduced by energy savings.

Table 7.4 presents the price elasticities of the model for Western Europe. They all range between −.1 and −.25. The values for petrochemicals and nitrogen are low because of the high energy content of the product itself, which cannot be substituted for.

Table 7.4 also compares the STREAM results with those of the Netherlands Energy Demand Model (NEMO). The values of NEMO are based on bottom-up information instead of the historical development of input demand. The differences between STREAM and NEMO figures in table 7.4 are not systematic, but they are substantial, especially for the paper industry. They stem from the different sources of information about energy savings in the two models. The STREAM elasticities are based on the realized energy savings in the past, whereas the NEMO elasticities are based on bottom-up information about the current available specific energy-saving options.

Comparison of observed material prices and calculated production costs on the basis of observed factor productivity and prices indicates a rate of autonomous technological progress in material production of between 1 percent and 2 percent per year. This trend contributes to the continuous decline in most real material prices, which has encouraged demand for materials.

7.2.5 Resource Scarcity

Primary and secondary material production use, respectively, minerals and scrap as raw material input. The development of both processes largely depends on the prices of these raw materials and on the price for energy. This subsection describes the main determining factors of mineral and scrap prices.

7.2.5.1 Mineral Prices The economy extracts an increasing amount of minerals from natural deposits. Those that are easy to exploit get exhausted, and the demand for minerals can then be met only through the exploitation of resources with higher extraction costs. This long-run development of increasing marginal exploitation costs is specified in the model by an exponential function of cumulative mineral production. The identified world reserves of raw materials analyzed by the model are abundant. The marginal extraction costs therefore increase only gradually. This hardly affects the prices of the minerals in question. In fact, the price development is dominated by cost-reducing technological progress in the mining industries, by around 1.5 percent per year.

7.2.5.2 Scrap Prices Increasing marginal costs also arise in the recycling industry. Here, an increasing rate of recovery of secondary materials from waste can be achieved only at rising costs. Waste supply depends on historical material demand and the economic life span of the final products that incorporate these materials. If the waste flow is curbed by taxes or restrictions, material recovery will decrease. Consequently, scrap prices will rise, reducing demand for secondary materials.

Figure 7.4 shows the historical recovery rate of Western Europe. The upsurge in the recovery rate after 1980 stems from increased scrap demand resulting from higher energy and lower scrap prices. These lower scrap prices are caused by cost-reducing technological progress and indi-

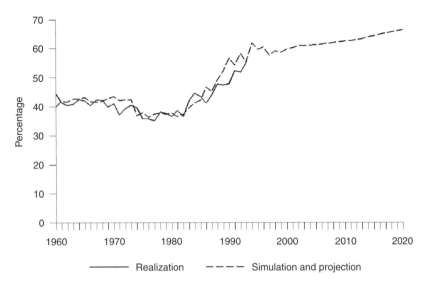

Figure 7.4
Development of the steel scrap recovery rate in Western Europe.

rect subsidies for waste recovery. The price effects of these developments have exceeded that of higher marginal cost because of the increased recovery rate.

The coefficients of the marginal cost equation of the model are based on bottom-up information about various recycling techniques and the composition of the waste flow. Model extrapolation according to a business-as-usual scenario suggests that, from the year 2000 on, the rate of increase in the Western European recovery rate of steel will slow because of the higher marginal recovery costs.

7.2.6 International Allocation

Comparative advantages determine the international allocation of production capacity. The location of the mining industry is determined by the presence of natural resources. Basic industries are generally located near coasts and along rivers to reduce transport and production costs. Indeed, the Western European oil and petrochemical industry makes its presence felt in the Netherlands because of the country's favorable location between the oil production fields and the concentration of final users in the Rhine basin. This advantageous economic position has drawbacks for national energy consumption and emissions. For example, the energy use and the related emissions involved in the transformation of crude oil into oil products destined for a large part of Europe takes place in the Netherlands.[4]

Geographical advantages are not the only factors that determine the location of industrial activities. Other factors, such as qualified labor, efficient markets, and government policy, also affect international allocation of such activities. The strong competition in international material markets makes the allocation highly sensitive to costs, availability of production factors, and policy changes. The material flows in an open economy shift critically if one these economic conditions changes. Historic changes in exchange rates indicate the extent of these shifts. For instance, U.S. steel exports dropped by 75 percent in response to the high dollar exchange rate in the mid-1980s. The firm grip of the U.S. producers on their large domestic market and the limited period of the high dollar exchange rate kept domestic steel production on its feet.

Another example is the booming nitrogen fertilizer production in the Netherlands during the 1960s and 1970s as a result of the low gas prices in the Netherlands. As a result, the Dutch fertilizer industry has a dominant position in the Western European market of fertilizers.

The import and export flows of materials depend on the international demand and the market share of domestic producers on the international material markets. Domestic producers' market share is determined by technological, geological, and geographical factors as well as their cost-price compared with that of foreign producers. Because the prices of most materials are set on the world market, there is only one material price for all countries and world regions. Empirical evidence confirms the notion that material prices will not diverge for long across different parts of the world. Therefore, comparative advantages of countries and regions are not so much reflected in market prices as in cost-price differences, or equivalently, the profit margins.

The price elasticities of trade relations play a crucial role in the debate on displacements effects, such as those caused by unilateral environmental policy measures. The Western European price elasticities in table 7.5 are based mainly on changes in material trade flows due to exchange rate fluctuations in the period 1960–1990. In particular, the consequences of the high dollar exchange rate in the mid-1980s gives a clear picture of what might happen to the Western European material flows if the region's competitive position improves or deteriorates.

Dutch material trade is dominated by trade with Germany. The constant exchange rate that existed between the Dutch guilder and the German deutsche mark makes the values of the import and export

Table 7.5
Price elasticities of import and export of materials and scrap by Western Europe and the Netherlands

	Western Europe		The Netherlands	
	Export	Import	Export	Import
Material type				
Steel	−4	2	−7.5	4
Aluminum	−3	2	−8	6
Plastic	−5	2	−8	4
Paper	−2	2	−8	5
Nitrogen	−4	4	−10	8
Phosphate	−4	4	−8	10
Potassium	−4	4		
Scrap type				
Steel	−5	2	−6	2
Aluminum	−4	2	−6	4
Plastic	n.a.	n.a.	n.a.	n.a.
Paper	−1.5	1	−4	4

n.a. = not applicable

price elasticities for the Netherlands more difficult to identify. However, changes in gas prices and wages give some insight into the sensitivity to changes in competitiveness between the Dutch and German basic industries.

Table 7.5 shows the import and export price elasticities of Western Europe and the Netherlands. The following conclusions can be drawn:

• The Western European price elasticity of most material exports is about −4. The paper industry appears to be a relatively sheltered sector within Europe and therefore has a smaller price elasticity (−2).
• The import price elasticities are half as high as those for exports for steel, aluminum, and plastic. This reflects the fact that Western European material producers are more vulnerable on the export markets than on the domestic markets. This may be the result of import protection measures and a better integration into the local cluster of related economic activities.
• Western European price elasticities for the scrap trade are about the same as those for other materials.
• Netherlands price elasticities for material imports and exports are roughly twice as high as for Western Europe as a whole. This reflects the vulnerability of the competitive position of the Dutch basic industries on the international market.

• In contrast to Western Europe, in the Netherlands, price elasticities of scrap imports and exports appear to be somewhat lower than those of other materials. This is in line with the expectation that the scrap market is relatively sheltered because of the low value of scrap relative to transport costs. In general, it is not profitable to transport scrap over long distances, even if the scrap prices differ substantially between regions. Low-cost overseas transport is a clear exception in this respect. It allows for substantial scrap flows between the continents.

The trade relations in the model follow the well-known Armington (1969) trade specification. In a dynamic simulation, this specification implies that only relative cost-price changes will lead to a change in the market share on the international and domestic markets. However, the ongoing penetration of newly industrialized countries on markets for materials has not been accompanied by continuous reductions in relative prices. To address this issue, cost-price differences are taken into account, as in the traditional Heckscher-Ohlin trade specifications. (Gielen and van Leeuwen 1998).

Figure 7.5 shows the aluminum export ratio for Western Europe. The export ratio expanded rapidly between 1960 and 1973. After 1973, it remained more or less constant as a result of rising energy prices. In the

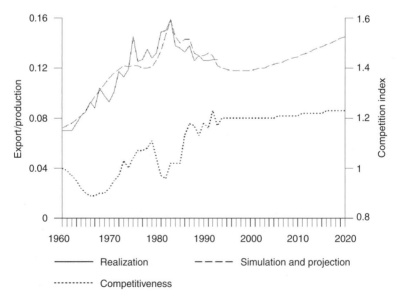

Figure 7.5
Development of the export ratio for aluminum in Western Europe.

first half of the 1980s, the export rate was pushed up temporarily because of the high dollar exchange rate. Model extrapolation, according to a business-as-usual scenario, suggests that for the next quarter century, the aluminum export ratio for Western Europe will increase only a little, because Western European aluminum production costs will remain in line with the world market price.

7.2.7 Empirical Validation

The parameter values of the identified mechanisms of the model are based on historical data between 1960 and 1995 on national and international material flows, material production costs, and material market prices. However, many time series are not completely time consistent as a result of numerous changes in definitions and classifications. Information about the development of the production costs of material producers is generally not available. Nevertheless, the model simulation of the period 1960–1995 shows that the model is able to reproduce the development of the material flows and prices during this period quite well. This underlines the robustness of the model, especially since this period is characterized by subperiods of high and low economic growth, as well as energy and economic crises accompanied by large price fluctuations.

7.2.8 A Comparison with Other Models: Matter-Markal, DIMITRI, FMS-3

The search for sustainability has stimulated the development of a number of models to analyze the relations between material flows and economic activity. This subsection compares STREAM with two other Dutch policy models—Matter-Markal (Energy Resource Center of the Netherlands, or ECN) and DIMITRI (National Institute for Public Health and the Environment, or RIVM)—and a Finnish model—FMS-3 (Thule Institute).

The Matter-Markal model is a representation of the Western European energy and material system (Gielen, Gerlach, and Bos 1998). This system includes material production, product assembly, product use, and waste handling. The model calculates the least-cost system configuration which that a certain energy, materials, and products demand. In other words, it calculates the optimal social outcome for the production system as a whole. The underlying marginal cost relations are based on bottom-up information about existing production technologies. Financial and restrictive policies with respect to the supply side of the economy can be easily analyzed. The main difference between Matter-Markal and STREAM is the former's centralized decision-making process on the supply side and its lack of markets for materials. Moreover, the Matter-Markal model ignores the effects of international competition. On the other hand, the Matter-Markal model is far more detailed than STREAM. It covers 25 energy carriers, 125 materials, 50 products in which these materials are used, and 30 waste categories. Therefore, the model is useful in analyzing the first-order impacts of specific material policies on the supply side of the economy.

DIMITRI (Dynamic Input-Output Model to Study the Impacts of Technology-Related Innovations) is based on an input-output structure (Wilting et al. 2001; also chapter 8). The final demand categories are exogenous, with the exception of the investments. Investments are determined by depreciation, growth of sales experienced in past years, and the gross profit of the previous year. These investments not only influence final demand but also determine the introduction of new technologies. The model aims to study the impact of technology-related innovations. Market-related feedback mechanisms and substitution mechanisms are ignored.

Another material model is the FMS-3 (Finnish long-term model system) of the Thule Institute in Finland (Mäenpää and Juutinen 1996). The core model is a thirty-sector econometric simulation model for the Finnish economy. The model assumes constant returns to scale production and full competition. Consequently, the supply of goods and services meets demand at full-cost prices plus normal profit margin. The core model is extended with satellite models for energy, forest products, and metal products. These satellite models are based on fixed input-output relations in physical terms. Final physical demand depends on relative prices. Intermediate demand depends on production level and technological coefficients, and prices are derived according to the full-cost principle from technological parameters and income distribution parameters. The characteristics of this model show much similarity to those of STREAM. The main difference between the two is the highly detailed structure of the FMS-3 model, as well as its focus on forest and metal products.

Table 7.6 gives an overview of the differences and similarities among the models.

7.3   Policy Modeling Illustrations

7.3.1   Introduction
The following subsections present the model outcomes of some policy measures for energy and emission reduction. They deal with the quanti-

Table 7.6
Comparison of the characteristics of different models

Characteristic	STREAM	Matter-Markal	DIMITRI	FMS satellite
Material flow	Physical	Physical	Monetary	Physical
Optimization	Decentralized	Centralized	None	Decentralized
Aggregation level	Macro level	Process level	Input-output	Input-output
Demand	Price dependent	Exogenous	Exogenous	Price dependent
Supply	Marginal costs and flexible technology	Marginal costs and flexible technology	Fixed costs and fixed technology	Marginal costs and flexible technology
Price	Dynamic market	Static cost	Static cost	Dynamic market
Scale	World, Western Europe, and the Netherlands	Western Europe	The Netherlands, World	Finland
Information base	Time series	Bottom-up	Input-output table	Time series

Table 7.7
Displacement effects in the basic industries in the Netherlands as a result of an excess tax leading to a 100 percent increase of primary energy costs (percentage change)

Material type	Production capacity		Production costs	
	Primary	Secondary	Primary	Secondary
Steel	−75	−22	10	6
Aluminum	−55	−20	10	3
Petrochemicals	−80	−24	15	4
Paper	−20	−10	5	2
Nitrogen	−65		10	

tative effects of taxation, restrictions, and tradable permit systems on the international allocation of the basic industries and on energy or emission reductions. Most models only describe energy savings or emission reduction per unit output. STREAM also describes the changes in the material-product chain.

7.3.2   Taxation versus Restrictions

The conclusion of the Netherlands Steering Committee on Regulatory Taxation of Energy—that a unilateral energy tax charged to large energy users generates substantial displacement effects—is confirmed by the outcomes of the STREAM model. Table 7.7 presents the decline of primary and secondary production in the Netherlands resulting from a 100

Table 7.8
Displacement effects in the basic industries in the Netherlands as result of an imposed
energy-efficient production equal to the taxation example of table 7.7 (percentage change)

Material type	Production capacity		Production costs	
	Primary	Secondary	Primary	Secondary
Steel	−15	−12	2	1
Aluminum	−10	−6	2	1
Plastics	−30	−1	3	1
Paper	−1	−1	0	0
Nitrogen	−20		2	

percent increase in energy costs through taxation, without refunding the
tax revenues.[5] In particular, the energy-intensive primary producers lose
up to 80 percent of their market share and have to reduce their production capacity by 10 to 15 percent as a result of their increased production
costs. The existence of only one world price for bulk products prevents
basic industries from passing on their environmental costs to consumers.
The energy intensity of the remaining production capacity decreases by
about 15 to 20 percent. For the less energy-intensive secondary production, the displacement effects are much more limited.

Alternative measures with the same reduction in energy intensity cause
milder displacement effects. Table 7.8 shows the effect on production of
an energy restriction (or equivalently, a taxation with 100 percent recycling of tax proceeds), with the same reduction in the energy intensity
as in the case of plain taxation. The results in table 7.8 show that in this
case, the increase of the production costs is only one-fifth of that of the
uniform tax variant. Consequently, these measures substantially mitigate
the displacement effects and are therefore more effective than plain taxation. The loss of market share by primary producers is 10 to 30 percent,
and the effect on secondary production, except for secondary steel production, is small. The decrease in the energy intensity is 15 to 20 percent,
the same as in the case of plain taxation. The main disadvantage of restrictive policies is the high transaction costs of setting up, monitoring,
and upholding a system of standards. These costs, not included in the
model, reduce the advantage of such policies.

7.3.3   Performance Standard Policies

The use of stringent performance standard rates (PSRs) for emission reduction in the basic industries is broadly accepted as a policy instrument

Table 7.9
Effects of restrictive and market related NO_x performance standard policies, with and without refunding of the revenues to the Dutch petrochemical industry (percentage change)

| | NO_x restriction to PSR level[a] | NO_x emission trade | | |
		Permit price: 3.7 euro/kg	Permit price: 3.7 euro/kg 100% refunding of the PSR[b]	Permit price: 7.4 euro/kg 100% refunding of the PSR[b]
Shadow price	3.7			
Production	−4	−14	−4	15
Production costs	0.25	1.25	.25	−1.25
Energy costs per gJ	7	7	7	11
Energy efficiency	−2	−2	−2	−4
Energy use	−6	−16	−6	11
NO_x	−53	−58	−53	−65

[a] 50 percent reduction compared with the current level.
[b] By granting tradable permits to the PSR level.

in the Netherlands. Long-term agreements (LTAs) referring to benchmark technologies are the backbone of the current emission policy with respect to these industries. However, these LTA measures are inefficient. A good alternative is an emission trade system with emission permits granted based on PSRs. This subsection presents the model outcomes of various policy variants on NO_x reduction.

The model calculates not only the penetration of NO_x scrubbers in the petrochemical industries, but also the effect on energy demand, production costs, and production level as well. For this particular NO_x analysis, a marginal cost function of end-of-pipe solutions, derived from bottom-up information, is added to the model.

Table 7.9 shows the effects on the Dutch petrochemical industry of various performance-standard-based policies for NO_x emission reduction. The first column shows the effects of an imposed 50 percent reduction of NO_x per gigajoule (GJ) of energy used. The shadow price of this restriction turns out to be 3.7 euros/kg NO_x. Because NO_x emissions are related to fossil fuels, the use of these fuels becomes more expensive. Consequently, it is profitable to save on fossil fuels through energy-saving investments. The overall energy efficiency increases by 2 percent. The NO_x restriction raises production costs by a tiny 0.25 percent. This leads to only a slight deterioration of the competitive position on the international markets and consequently to a decrease of petrochemical production by no more than 4 percent. The total NO_x reduction of the performance standard is 53 percent. The secondary effects on production and energy savings appear to be rather small. This indicates that end-of-pipe solutions are much cheaper than energy-saving investments or re-allocation of capacity. The effect on production is also small because the NO_x emissions below the imposed performance standard are not charged for.

Instead of an imposed NO_x restriction per unit output for every individual firm, policymakers can choose to introduce a system of tradable permits in which NO_x emissions are linked to emission permits that firms are obliged to hold. If the total NO_x quantity covered by permits is 50 percent of the current emissions per gigajoule, emissions per gigajoule will automatically be reduced by 50 percent. The initial permits can be put into circulation by auction. Firms must buy, at auction prices, the permits necessary to cover their NO_x emissions. The second column of table 7.9 shows the effects on the Dutch petrochemical industry of an assumed auction price for permits of 3.7 euro/kg NO_x. This case is equivalent to the introduction of a plain tax of 3.7 euro/kg NO_x. The effect on energy efficiency is the same as in the restriction case in the first column, because the auction price for NO_x is exactly same as the shadow price in the restrictive case. However, the auction price has to be paid for every kg NO_x, not only for the marginal unit. This leads subsequently to higher production costs, a less favorable competitive position, diminishing exports, and increasing imports. Ultimately, petrochemical production falls by 14 percent, which significantly contributes to the decline of energy use by 16 percent. Fifty percentage points of the 58 percent reduction in NO_x emissions stem from end-of-pipe measures.

If NO_x permits are granted free of charge to all firms up to 50 percent of the average current emissions per gigajoule in petrochemical installations, firms need not pay for all the permits they require. This mitigates the increase in production cost considerably. The third column of table 7.9 shows the effects of an NO_x market price of 3.7 euro/kg combined with 50 percent recycling of emission permits. The effects on production, energy efficiency, and NO_x emissions are exactly the same as in the restrictive case as a result of the initial permits granted free of charge.

If an economy-wide NO_x reduction of 50 percent leads to a market price twice as high as the shadow price in the first column, this indicates that other industries have relatively high marginal NO_x reduction costs. It is profitable for the petrochemical industry to sell a part of its granted permits while reducing emissions more than the standard rate. The last column of table 7.9 shows the effects of emission trade on petrochemical production. The high price for NO_x leads to an 11 percent increase in energy costs. However, these additional production costs are exceeded by the revenues from the permits sold on the market. On balance, production costs decrease by 1.25 percent, so that production expands by 15 percent. Although energy use increases as a result of production growth, the NO_x emissions in the petrochemical industry fall far below the emission standard. However, at a national level the picture may be somewhat different. Indeed, the exchange of NO_x permits is emission neutral, but the NO_x effect of the expansion of the petrochemical industry must be compensated for by a decrease of production in other industries that have higher marginal NO_x reduction costs. This can take place only if these other industries have relative high NO_x intensities and are more exposed to national and international competition than the petrochemical industry. In general, these conditions are not met. Therefore, a PSR system with granted permits runs the risk of attracting pollutive industries with low marginal reduction costs, which undermines the aimed-at reduction. On the other hand, if the permits are granted only to incumbent petrochemical producers and not to new entrants, none of the incumbents will pass on the advantage of the granted permits to the new entrants. The price of permits is determined by the marginal producers, which, in this case, are the new entrants, which are not required to have granted permits and have to buy such permits.

7.4   Conclusions

STREAM describes the key dynamics, factors, and parameters that determine material extraction, production, use, recycling, and disposal in the world, Western Europe, and the Netherlands. The model depicts the international material-product chain of steel, aluminum, plastics, paper, ammonia, phosphate, and potassium. It provides a consistent framework for long-term scenario analysis and for understanding future demand for, and supply of, natural resources. It is based on empirical knowledge about dematerialization, recycling, international competition on material and scrap markets, factor substitution, long-run material scarcity, and the price-making process for raw materials, materials, and scrap. The model can also be used for analyzing the effects of policy instruments, such as taxes and subsidies, regulations, and tradable emissions permits, on material flows. Model simulations from 1960 to 1995 show that the model is able to reproduce the development of material flows and prices, as well as the input demand in the various production processes. The conclusions of this chapter are based on empirical data and show that there are powerful driving forces that can be used for a corresponding strategy for the sustainable use of materials.

This chapter confirms the findings of other studies that no absolute decline in material use in OECD countries can be detected, but that a relative decoupling of material use and GDP can be observed. Indeed, the existence of an inverted U-curve with respect to the intensity of material use is still valid.

It is too early to formulate a strategy for material flow management and eco-restructuring on the basis of dematerialization, as the knowledge base about the interaction of the mechanisms underlying dematerialization is still insufficient. Nevertheless, this chapter presents some leads: Higher prices of material intensive products reduce material demand; unilateral measures with respect to material production increase the production costs but do not affect the material prices that are set on the world markets; higher energy prices for end users enhance dematerialization for steel, aluminum, and paper; and higher prices for primary materials stimulate the use of secondary materials.

However, there are two serious obstacles for dematerialization policy in the Netherlands. First, in an open economy, tax measures to reduce the use of materials are largely ineffective because of the substantial induced displacement effects. Second, the de facto industry policy in the country has, up till now, been in favor of the basic industries. Although restrictions on energy and material use generate smaller displacement effects than tax measures, they are applied only on a voluntary basis. This undermines the stringency of the measures and obstructs structural change favorable to the environment. International coordinated environmental and industrial policy allows for stringent measures without inducing large displacement effects, thus providing the effective policy measures that are needed to promote the "green industrial restructuring" of the economy.

Appendix: Basic Equations in STREAM

1. Material demand (see section 7.2.2)

$$D = D_0 \times Y^{\beta - \alpha \times t} \times \left[\frac{I}{Y}\right]^{\gamma} \times \left[\frac{Pe}{Py}\right]^{\eta} \times \left[\frac{P}{Py}\right]^{\varphi}$$

D	material demand (ton)	Py	GDP deflator
D_0	material demand 1960	$\beta - \alpha \times t$	income elasticity
Y	real national income	γ	investment elasticity
I	real investment	η	energy price elasticity
Pe	energy product price	φ	material price elasticity
P	material price		

For parameter values, see tables 7.1 and 7.2.

2. Primary and secondary material demand (see section 7.2.3)

$$\frac{Dp}{Ds} = Sps \times \left[\frac{Pqp}{Pqs}\right]^{\beta} \times \left[\frac{I}{Y}\right]^{\gamma}$$

Dp	demand for primary material	Pqs	cost-price of secondary material
Ds	demand secondary material	β	substitution elasticity
		γ	investment elasticity
Sps	ratio long-run market shares	Sps	long-run market ratio (see equation (7))
Pqp	cost-price of primary material		

$$Dp + Ds = D$$

For parameter values, see table 7.3.

3. Cost-price of materials (see section 7.2.4)

New vintage:

$$\ln(Pq) = \alpha \times \lambda^t + \sum_j a_j \times \ln(Pe_j) + \frac{1}{2} \times \sum_i \sum_j b_{i,j} \times \ln(Pe_i) \times \ln(Pe_j)$$

Old vintages:

$$Pq = \sum_i Pe_i \times Eq_i$$

Pq production cost/ton material Eq factor intensity (fixed)
Pe input prices λ cost-reducing technological
                                        progress coefficient

For derived values of price elasticities, see table 7.4.

4. Mineral price (see section 7.2.5)

$$\frac{Pg}{Py} = a \times \lambda^t \times \exp(\beta \times Qg_{\text{cum}})$$

Pg mineral price β exhaustion coefficient
Qg_{cum} cumulated mine λ technological progress
            production   coefficient

5. Scrap price (see section 7.2.5)

$$\frac{Ps}{Py} = \lambda^t \times \left[a - \beta \times \ln\left(1 - \frac{Q_s}{W}\right) \right]$$

Ps scrap price β exhaustion coefficient
Q_s secondary material λ technological progress
       production   coefficient

6. International trade of materials and scrap (see section 7.2.6)

$$\frac{X}{Dx} = Sx \times \left[\frac{Pq}{Pqx}\right]^\beta$$

$$\frac{M}{D-M} = Sm \times \left[\frac{Pq}{Pqx}\right]^\gamma$$

X export Pq domestic price
M import Pqx foreign price
D domestic demand β, γ substitution elasticity
Dx foreign demand Sx, Sm long-run market share
                                    (see equation (7))

For parameter values, see table 7.5.

7. Long-run market shares

$$Sx = Sx(-1) \times \left[\frac{Pq}{Pqx}\right]^{\sigma}$$

(*Sm* and *Sps* have analogous specifications)

For more information on the model equations see Mannaerts (2000).

Notes

1. The business-as-usual scenario features extrapolation of historical technological trends, average economic growth, and real price developments (with the exception of energy and materials prices). Real energy prices are assumed to remain constant, and real material prices are endogenous.

2. The second packing material agreement was a voluntary agreement between the Dutch government and industries to reduce the use and increase the recycling of packing materials between 1997 and 2000.

3. The gap between observed and simulated scrap prices in the 1960s arises because the (already outdated) Bessemer steel production with high scrap input has not properly been taken into account in the model.

4. Avoided emissions in a particular country resulting from import of energy-intensive products are referred to as the "environmental leakage" of that country.

5. Recycling of the proceeds through a company tax reduction will generate almost the same results, because for the energy-intensive basic industries, the costs of energy taxation will far exceed the revenues recycled by the company tax.

References

Armington, P. S. (1969). A Theory of Demand for Products Distinguished by Place of Production. International Monetary Fund Staff Papers, International Monetary Fund, Washington, DC.

Blok, K., J. G. Beer, M. T. Wees, and E. Worrell. (1994). ICARUS 3—The Potential of Energy Efficiency Improvement in the Netherlands from 1990 to 2000 and 2015. Paper no. 94013, Department of Science, Technology and Society, Utrecht University.

de Bruyn, S. M., and J. B. Opschoor. (1994). Is the economy ecologising? Discussion paper no. TI94-65, Tinbergen Institute.

EEA (European Environmental Agency). (2003a). *Annual European Community Greenhouse Gas Inventory, 1990–2001, and Inventory Report 2003*. Copenhagen: EEA.

EEA (European Environmental Agency). (2003b). *Environmental Signals 2002*. Copenhagen: EEA.

Gielen, A., and N. van Leeuwen. (1998). A note on Armington and the law of one price. In M. Brockmeier, J. Francois, T. Hertel, and P. M. Schmitz (eds.), *Economic Transition and the Greening of Policies*, 91–108. Kiel: Wissenschaftsverlag Vauk.

Gielen, D., T. Gerlach, and A. J. M. Bos. (1998). Matter 1.0: A Markal Energy and Materials System. Model Characterisation. ECN Report C-98-065, Energy Research Centre of the Netherlands, Petten.

Jänicke, M. (2001). Towards an end to the "era of materials"? Discussion of a Hypothesis. In M. Binder, M. Jänicke, and U. Petschow, *Green Industrial Restructuring*, 45–58. Berlin: Springer-Verlag.

Jänicke, M., H. Mönch, T. Ranneberg, and U. E. Simonis. (1989). Economic structure and environmental impacts: East-West comparisons. *Environmentalist* 9: 171–182.

Konijn, P., S. de Boer, and J. van Dalen. (1995). Material flows and input-output analysis. Statistics Netherlands (CBS), Report no. 689-95-EIN.PNR, Zoetermeer, the Netherlands.

Koopmans, C. C., D. W. te Velde, W. Groot, and J. H. A. Hendriks. (1999). *NEMO: Netherlands Energy Demand Model. A Top-Down Model Based on Bottom-Up Information.* CPB Research Memorandum no. 155, CPB Netherlands Bureau for Economic Policy Analysis, the Hague.

Mäenpää, I., and A. Juutinen. (1996). FMS3 Model System and Forest Sector. Working paper no. 1, Department of Economics, University of Oulu, Oulu, Finland.

Mannaerts, H. J. B. M. (2000). STREAM: A Partial Equilibrium Model for Material Flows in the Economy. CPB Research Memorandum no. 165, CPB Netherlands Bureau for Economic Policy Analysis, the Hague.

Shephard, R. (1970). *Theory of Costs and Production Functions.* Princeton: Princeton University Press.

Simonis, U. E. (1989). Ecological modernisation of industrial society: Three strategic elements. In F. Archibugi and P. Nijkamp (eds.), *Economy and Ecology: Towards Sustainable Development*, 119–138. Dordrecht, the Netherlands: Kluwer Academic.

Slade, M. E. (1981). Recent advances in econometric estimation of materials substitution. *Resources Policy* 7(2): 103–109.

Varian, H. R. (1984). *Microeconomic Analysis.* New York: Norton.

von Weiszäcker, E., A. B. Lovins, and L. H. Lovins. (1997). *Factor Four: Doubling Wealth, Halving Resource Use.* London: Earthscan.

Wilting, H. C., W. F. Blom, R. Thomas, and A. M. Idenburg. (2001). Dimitri 1.0: Beschrijving en toepassing van een dynamisch input-output model (Description and application of a dynamic input-output model). RIVM report: 778001005, Bilthoven.

8 DIMITRI: A Model for the Study of Policy Issues in Relation to the Economy, Technology, and the Environment

Annemarth M. Idenburg and Harry C. Wilting

8.1 Introduction

The realization of economic growth with far fewer undesirable consequences for the environment is one of the long-term challenges of the
Dutch government (Ministry of Economic Affairs 1997). Since the 1980s
this challenge has resulted in a sequence of national environmental policy
plans (NEPPs). These plans have succeeded only partially in reducing
environmental pressure in the Netherlands (National Institute for Public
Health and the Environment 2001a, 2001b). The emissions of several
pollutants, for example, nitrogen, phosphate and sulfur dioxide, as well
as many toxins have decreased substantially during the last twenty years.
Conversely, CO_2 emissions and noise due to traffic are still increasing.
The policy successes that have been achieved are due mainly to technological improvements in process efficiencies, abatement techniques, and
the prohibition of certain toxins. There is little support for implementation of policies that might affect lifestyles or economic structures (National Institute for Public Health and the Environment 2000).

Unfortunately, there are no simple solutions to resolve the persistent
environmental problems in the Netherlands. The latest NEPP (Ministry
of Housing, Spatial Planning and the Environment 2001) distinguishes
seven major environmental problems the country is facing. These problems concern the effects of Dutch production and consumption both
within the Netherlands and abroad. In particular, Dutch cattle breeding
and paper production have large impacts on biodiversity in other countries. The solution of these problems requires, according to the latest
NEPP, technological, economic, social-cultural, and institutional changes.
Therefore, transition management has been introduced as a policy strategy for dealing with environmental problems. The main road toward
sustainable development is still paved with technological improvements.

But these might also lead to radical changes in the economic system. And on the other hand, the successful implementation of these improvements requires some system adjustment of a social-cultural, institutional, and economic nature. Transition management implies stimulation of the required changes as well as the removal of existing bottlenecks. Knowledge and technology development and transfer are important aspects of transition management.

The Netherlands Environmental Assessment Agency of the National Institute for Public Health and the Environment (MNP-RIVM) is a governmental research institute that provides policymakers with information and analyses required for the development of new policies and evaluates the effectiveness of implemented and suggested policies. This cycle of scientific input, policy development, and scientific evaluation is reflected in a sequence of publications. Each NEPP is preceded by an environmental outlook study. Each year the MNP-RIVM publishes an environmental balance that evaluates the effectiveness of the country's environmental policies. Consequently, changes in policies have an impact on the analyses performed by the MNP-RIVM. And subsequently, these changes will have an impact on tools used by the MNP-RIVM.

One of these tools, DIMITRI (Dynamic Input-Output Model to Study the Impacts of Technology-Related Innovations), was developed to answer questions about interrelationships among economy, technology, and the environment. More precisely, it has to investigate

1. the impact of (changes in) final demand on production in the Netherlands and elsewhere;
2. environmental pressure and energy and material use in the Netherlands and elsewhere, related to this production;
3. the impact of new technologies on production, environmental pressure, and energy and material use in the Netherlands and elsewhere.

To do this DIMITRI, a mesoeconomic model, operates at the level of production sectors, focusing on production and related environmental pressure in the Netherlands and for the Dutch population. Production for the Dutch population takes place partly outside the national borders of the Netherlands. The use of a multiregional input-output model enables an analysis of changes, among sectors and among regions, resulting from structural and technological changes. The dynamic nature of the model is especially important for investigating the impact of new technologies. The model is used together with other analysis tools. Consumption patterns, for instance, are studied with a different model (Rood et al. 2001). Issues relating more to the economy, such as employment, income, and monetary production, are dealt with by the CPB (Netherlands Bureau for Economic Policy Analysis). Policy issues related to the production of materials are also dealt with in cooperation with the CPB, in particular with the STREAM model, which is described in chapter 7. Furthermore we would like to mention here that DIMITRI is a model in progress.

DIMITRI analyzes not only the consequences of changes in direct or operational inputs to production, but also the effects of shifts from operational inputs toward capital inputs, and vice versa. The dynamics of the model is limited to the description of physical production. The model does not include a more sophisticated description of economic behavior.

The characteristics of the model are illustrated here by three cases. First, the model is used to analyze changes in energy use by Dutch economic sectors between 1980 and 1997. By running the model with different technological scenarios, it is possible to discern the effect of efficiency changes within sectors and structural changes within the economy. Second, an analysis is performed for energy use related to Dutch production and consumption by economic sectors outside the Netherlands. Both cases are ex post studies and of a typical environmental balance nature. The third case is of an environmental outlook nature. It focuses on the effects of a transition to a new technology. The introduction of novel protein foods in the Netherlands between 1995 and 2030 will result in shifts among economic sectors and among regions.

8.2   Background

Economic input-output analysis, originally developed by Leontief, is a useful tool in studying the relations among economy, technology, and environment. From the 1930s, Leontief (1941) applied the input-output model to the U.S. economy to study the relations among economic sectors. This pioneering work has stimulated many researchers to enter the field of input-output analysis. One of the first applications of economic input-output analysis in relation to physical flows is input-output energy analysis (IOEA). IOEA originated in the 1970s with the pioneering work of Wright (1974) and Bullard and Herendeen (1975) (see also chapters 5 and 6 for discussions of this type of work), who almost simultaneously published articles on the calculation of sectoral energy intensities. Since then, IOEA has developed as a main stream in energy analysis in addition to traditional process analyses (Peet 1993; Wilting 1996). Nowadays, input-output analysis has also proved its worth for investigating other flows of natural resources from the environment through the economy (industrial metabolism). Duchin (1992, 1994) has already showed the role that input-output analysis might play in the field of industrial ecology. Some recent examples of the application of input-output analysis in the field of material use and emissions are studies on paper and wood (Hekkert 2000) and plastics (Joosten 2001). Another interesting application of input-output analysis concerns the analysis of ecosystems (Hannon 1995).

The starting point in input-output analysis is an input-output table, which describes the flows of goods and services through an economy in financial terms. From this input-output table, a "technological matrix" can be easily derived. The columns of the matrix can be seen as descriptions of the technologies in the individual sectors. The "technology of a sector" refers to the way in which intermediate inputs, capital, and value added are combined in the sector for production. So the columns of the matrix are the links between descriptions of technologies at the micro level and the macroeconomic system. Changes in column coefficients over time give insights into technological change (for an overview of the early literature on technological change in combination with input-output analysis, see, e.g., Rose 1984; and chapter 4 offers an application in the field of industrial ecology). In addition to investigating technological change in the past in series of input-output tables, input-output analysis also enables the study of the future impacts of new technologies on the economy and environment through the adjustment of input coefficients. The input-output framework has proved to be ideal for incorporating technical information into economic analyses (Veeneklaas 1990; Idenburg 1993). Duchin and Lange (1994) carried out an example of this type of analysis. They investigated the potential of technological measures to fulfill economic objectives (in terms of GDP per capita) and environmental objectives (in terms of CO_2, NO_x, and SO_2 emission reductions).

Generally, a static input-output model is sufficient to determine the impacts of new technologies on economy and environment. However, when technology changes imply a substitution between operational inputs and capital inputs, investments should be endogenous. Organic farming, for instance, will require less chemicals but more machinery compared to traditional farming. In addition a dynamic approach is also required to investigate the implementation of new technologies. In contrast to static input-output models, in which investments are part of final deliveries, dynamic input-output models include the stocks and flows of capital goods explicitly. Duchin and Szyld (1985) described such a dynamic input-output model for the United States. Leontief and Duchin (1986) extended this basic model to investigate the effects of new technologies on employment. For the same purpose, Kalmbach and Kurz (1992) developed a dynamic input-output model for Germany. At the MNP-RIVM, DIMITRI has been developed. This is a dynamic input-output model for the Dutch economy based on the model worked out in Duchin and Szyld (1985) and Duchin and Lange (1992). It will be obvious that we do not agree with Ruth, Davidsdottir, and Amato on the relevance of dynamic input-output analyses to policymakers (see chapter 5).

8.3   DIMITRI

This section presents the Dynamic Input-Output Model to Study the Impacts of Technology Related Innovations, or DIMITRI. Figure 8.1 is a schematic representation of the model, and appendix 8A gives a mathematical description of DIMITRI.

The core of the model concerns the determination of production per sector on the basis of final demand and the technological matrix. The

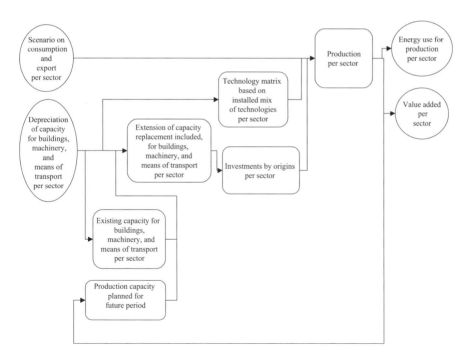

Figure 8.1
Schematic representation of DIMITRI. The scheme applies to each region in the model.

final demand is partly exogenous to the model and partly endogenous. The exogenous part concerns household consumption and exports and is based on scenarios that might be the outcome of different models used by the MNP-RIVM. The endogenous part of final demand concerns the investments in buildings, machines, and transport that are necessary for maintaining production capacity. These investments depend on the yearly depreciation of capacity, the degree of capacity utilization, and the expected production in future years. Furthermore, the amount of new investments in combination with technology scenarios determines the changes in the technological matrix, which depicts the mix of technologies installed in previous years. The technology scenario contains the production characteristics and required investments in goods for extended and replaced capacity. This means that the model investigates "What if, starting from year t, all extended and replaced production capacity will be using technology x?" The choice of technology x over technology y is beyond the scope of the model. Energy use and emissions per sector depend on production per sector and energy and emission intensities. These intensities belong to the technology scenarios. Similarly, value added per sector is determined by using sectoral primary input coefficients. Currently environmental pressure is represented only by energy use and the production of emissions. However, we are planning to incorporate land use and water use in a similar way. Also the physical data presented in chapter 4 could be incorporated into the model. As noted before, DIMITRI is a model in progress.

The advantage of an input-output model is its flexibility; the number of sectors, the type of emissions or the primary input types, and the number of regions can easily be adjusted. Its drawback is its strong reliance on the national accounts for data. Emissions and the use of natural resources such as water can be incorporated if related to the sectoral classification as used in the system of national accounts. Fortunately an increasing number of countries are extending their national accounts with environmental data.

The scope of the model is twofold. First, it calculates emissions and energy use related to production in the Netherlands for Dutch final demand, such as consumption and exports. Second, it calculates energy use and emissions related to production in the Netherlands and abroad for Dutch household consumption. The latter option concerns environmental pressure along the entire production chains of Dutch consumption items. Not all countries will have the same production technology installed as the Netherlands. To handle discrepancies among countries with respect to

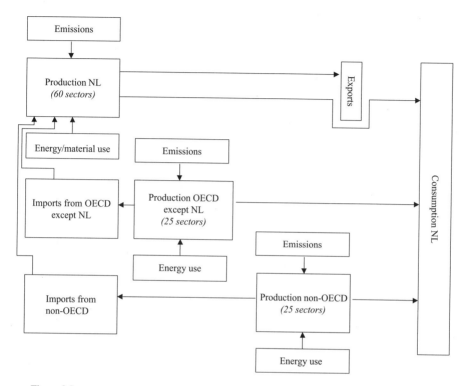

Figure 8.2
Overview of the regional structure in DIMITRI. Production takes place in the Netherlands and in two foreign regions, each with its specific energy and material use and emissions.

the technologies installed, the model contains two foreign regions covering the world. The first region covers the OECD countries excluding the Netherlands; the second region covers the non-OECD countries (in regard to their average technologies). The calculation of production, energy use, and so on, in the Netherlands as well as in the foreign countries is based on the diagram in figure 8.1. Figure 8.2 shows the structure of DIMITRI in regard to the flows of goods and services to the Netherlands from abroad.

The model is a simplification of a multiregional model, since it contains no feedback loops via the exports of the Netherlands to foreign regions. Furthermore, the trade flows between the foreign regions are not taken into account. For each region it is assumed that the required production takes place totally in that region. Production in the foreign regions concerns only production required for the delivery of goods and services to the Netherlands. The flows in a base year are the starting point in

DIMITRI. The model shows the direction in which these flows can develop into the scenarios that concern technology and demand.

8.4 DIMITRI versus Other Dynamic I/O Models

As described, input-output analysis is often used to estimate the (environmental) effects of new technologies. In most cases, these analyses are based on static input-output models. There exist only a few dynamic input-output models. Well-known empirical dynamic input-output models are the Leontief-Duchin model (1986), and the Duchin-Lange model (1992) (which are both based on the [theoretical] Duchin-Szyld model [1985]), and the Kalmbach-Kurz model (1990, 1992). A different type of dynamic input-output model is proposed by Los (2001), who incorporates important properties of endogenous growth theory within a dynamic input-output approach to investigate the spillover effects of research and development. One of the main differences between this model and other dynamic input-output models is the endogenous nature of technological progress. This explains why Los's approach is less suitable for our purpose. We want to be able to investigate the impact of the implementation of a specific technology on material flows. The description of such a technology should thus be based on technical information, not derived from model simulations. DIMITRI is therefore based on the Duchin-Lange model, but there are a few differences between DIMITRI and the other dynamic input-output models previously mentioned. The most important differences are related to the modeling of investments, the way capacity is planned, and the adjustment of input coefficients.

8.4.1 Investment Goods

Duchin and Szyld distinguished between investments for extension and those for replacement. They assumed that the replacement of capacity is carried out with old technologies, whereas the extension of capacity is carried out with new technologies. The Kalmbach-Kurz model and DIMITRI do not distinguish between investments for extension and replacement. The Kalmbach-Kurz model treats all investments as a mix of old and new technologies. For DIMITRI, we assume that all investments are carried out with new technologies.

Investments are, of course, an input of capital goods. In standard input-output notation, this input of capital goods is modeled as the product of a matrix $\mathbf{B}$ with capital coefficients (the amount of capital required for the extension of the capacity by one unit) and a variable representing the required capacity extension (replacements included or not). There is no straightforward way to derive this matrix **B** from statistics, as is done for the matrix **A** with technical coefficients related to the operational inputs of production. One of the reasons is the lag period between the production of capital goods and the installation. Another reason is the lack of statistics on capacity of production. Therefore, the models discussed deal differently with this matrix **B**. The Kalmbach-Kurz model has a fixed lag period of two time periods for all capital goods. This means that the amount of capital goods produced at period t is based on the required capacity in period $t + 2$. Leontief and Duchin, and Duchin and Lange, distinguish different lag periods for different types of capital goods but without going into details. DIMITRI, however, distinguishes two lag periods for three types of capital goods: construction (five years), machinery and means of transportation (both two years).

8.4.2 Planning of New Capacity and Investments
Dynamic input-output models of the Duchin-Szyld kind often show unstable behavior, presumably originating from the way capacity is planned. Los (2001) also faces stability problems with his model. Fleissner (1990), for example, studied the stability and sensitivity properties of the Duchin-Szyld model. However, there is not a generally accepted explanation on what causes these instabilities or, for that matter, how to resolve the problem. In case of the early versions of DIMITRI, the instability caused cyclic fluctuations in the production of investment-goods-producing sectors. Because they increased in each cycle, these fluctuations were especially problematic for long-term simulations.

The modeling of capacity planning is weak in production models in general partly because of the lack of (good) data on capacity and capacity utilization. In most cases, investments are modeled regardless of the amount of required capacity. Investment decisions are, for instance, based on savings (Norén 2001) or on factor prices and past changes of outputs (Meade 2001). If we had included the dynamic input-output price model, this would have been possible for DIMITRI as well. However, DIMITRI does not include this model, so we chose to follow the approach used by Duchin-Szyld and Kalmbach-Kurz to derive investments from the planning of capacity, which in these models is based on growth rates of production. We based the planning of capacity on the implementation in the MESEMET (Macro Economic Semi Equilibrium Model with Endogenous Technology) model (van Bergeijk et al. 1995), in which sectoral decisions for expansions are based on the weighted growth rate of production in the previous four periods. The growth rates for the most recent periods have the highest weights. The other models previously discussed do not use weighted growth rates. To limit exorbitant fluctuations in planned capacity, all models use a fixed sector-specific maximum admissible annual rate of extension of capacity.

The yearly investments concern replacement of existing capacity that is depreciated and newly required capacity based on the planning of new capacity. DIMITRI includes a restriction on these investments. They take place depending on the degree of capacity utilization per sector.

8.4.3 Adjustment of Input Coefficients

DIMITRI uses a dynamical calculation of the technological matrices for each period. Carter (1967) has already proposed such a dynamic adjustment of the input coefficients. This approach is different from the procedure employed by Leontief and Duchin (1986), who, as is done in most input-output studies, constructed a technological matrix for a certain future period. For the intervening periods between the base year and the future year, technological matrices were determined by interpolating. So, in these cases, the technology mix for the final year of a model simulation is determined or predicted exogenously. Within DIMITRI, the final technology mix is not fixed beforehand but is the outcome of a model simulation. The technology used for new capacity is then defined exogenously. The model computes the pace of introduction. It turns out, however, that it is easier to determine the technical coefficients of a mix of technologies than to define the technical coefficients of a new "pure" technology. This is, of course, especially true for ex post studies, since input-output tables give the mix of technologies, yet it is almost impossible to recapture the introduction of new technologies in the past. But for ex ante studies as well, it turns out that there is more information on the expected technology mixture for future years than on the introduction and description of new pure technologies. As a result of this, the empirical application of DIMITRI is, at this point, less different from that of the other dynamic input-output studies than indicated by its theoretical potential.

An alternative manner of dealing with technological change in an input-output framework is to add new sectors (Veeneklaas 1990; van den Broek 2000). The shift from old to new technology is then modeled by increasing the final demand of the new sector at the expense of the final demand of the old one. This method is applied with DIMITRI for the third case study described in the next section.

8.5   Case Studies

We now discuss three case studies to show how DIMITRI has been applied. The first two cases are ex post studies and are of a typical environmental balance nature. The third case concerns the effects of a new technology and is of an environmental outlook nature.

The first case concerns the identification of changes in production and energy use in the Netherlands. By trying to specify the factors that caused these changes, it is possible to indicate the effectiveness of applied environmental policies.

The second case focuses on the environmental impact of Dutch consumption and production on other countries. Dutch policy shows an increasing interest in issues concerning the effects of economic activities in the Netherlands on global environment and biodiversity (Ministry of Housing, Spatial Planning and the Environment 2001). The case illustrates, on the one hand, the impact of Dutch economic activities on the global environment. On the other hand, it illustrates the relative environmental performance of the Netherlands. By using the multiregional character of the model, it can be shown that the Dutch have a "footprint" abroad, but conversely, others too have a part of their footprint in the Netherlands.

The third case shows how DIMITRI may play a role in the simulation of innovations or transitions. This type of analysis is performed to support and evaluate transition management as a new environmental policy strategy.

8.5.1   Case 1: Analysis of Production and Energy Use in the Netherlands in 1980–1997

The first case illustrates how a dynamic input-output model like DIMITRI can be used for investigating the effects of changing production structures (technologies) on the economy and environment. The case concerns an analysis of the Dutch economy for the period 1980–1997. The calculations were carried out with a first empirical version of DIMITRI for the Dutch economy (consisting of thirty economic sectors), which was built on the basis of the theoretical model. For running the empirical model, we required large amounts of data, which we partly obtained from the Netherlands Bureau of Statistics (CBS). The main part of the data, for example, that concerning capacities, capacity utilization, and capital coefficients, had to be constructed. We calibrated the model on the basis of investments and production in the construction sector. A simulation of the past, for the period considered, enabled us to validate the model. Wilting et al. (2001) present a comparison of the model outcomes for production and energy use with the realized data obtained from the statistics for a set of sectors.

In this case, we calculated the effects of technological change on sectoral production and energy use for two simulations. The starting point for the calculations was the production structure of the Dutch economy in 1980, depicting the mix of technologies implemented in the years before 1980. In the first simulation (changed technology), all new investments are based on the technological matrix of 1997 (which depicts a mixture of technologies implemented in the periods before 1997). The model calculates dynamically the production structures for the period 1980–1997 in this simulation. The second simulation (fixed technology) concerns the imaginary situation that no technological change occurs: The 1980 technology mix is fixed for the whole period. In fact, the calculation described is a form of structural decomposition analysis (SDA) of production in the period considered. See chapter 4 for a genuine structured decomposition analysis.

Figure 8.3 shows values for production in agriculture and business services for both simulations mentioned (expressed in 1980 prices). The fixed-technology simulation depicts only the development in the production of agriculture and business services due to changes in final demand of all sectors. The differences between these values and the values based on the changed-technology simulation indicate effects of changing production structures (depicted in the technological matrix). The figure shows that the importance of business services in the production structure of the Dutch economy increased strongly in the period 1980–1997. This means

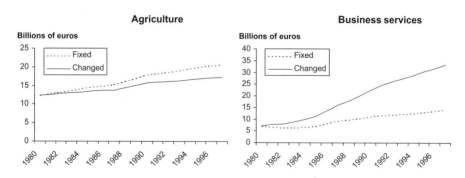

Figure 8.3
Production (in billions of euros) for the agriculture and business services sectors, calculated with DIMITRI, for two simulations for the period 1980–1997.

that in all sectors, on average, the inputs coming from the service industry increase. On the other hand, there is a slight decline in the importance of agriculture in the production structure in the period considered. This means that in general, in the technologies of the economic sectors, the share of the inputs coming from agriculture decreases.

The demand for energy in individual sectors is used as an example for illustrating the effects of technological change on the environment. Energy use in a sector is the result of the physical production in that sector and the energy efficiency of production. Figure 8.4 shows the developments in energy use calculated with DIMITRI in the agriculture and business services sectors for three simulations: changed technology and two variants of fixed technology. In the first variant of the fixed-technology simulation (fixed $t + e$), production structure and energy efficiency are fixed at the 1980 values for the whole period 1980–1997. The developments in energy use per sector are the same as the developments in production in figure 8.3 (fixed simulation). In the second variant (fixed t), only the production structure is fixed. So this simulation, compared with the previous one, shows the improvements in energy efficiency that occurred in the period considered. The changed-technology simulation shows that energy use in the agricultural sector remains at the same level during the period 1980–1997 because of changing production structures and energy efficiencies. In the business services sector, energy efficiency improvements moderate the growth in energy use (as a consequence of an increasing importance of this sector in total production).

This case shows that the demand for agricultural products decreased and the demand for services increased as a result of changes in the Dutch

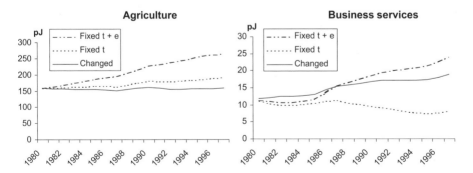

Figure 8.4
Energy use (in petajoules [pJs]) in the agriculture and business services sectors, calculated with DIMITRI, for three simulations for the period 1980–1997.

production structure during the period 1980–1997. Both sectors show significant improvements in energy efficiency, but in the case of the business services sector, the energy savings are almost canceled out as a result of the fast expansion of production. The outcomes of analyses like those carried out in this case provide insights into the factors that underlie energy use in economic sectors. So these outcomes can be employed in the evaluation of applied policy, and they also indicate directions for policymaking.

8.5.2   Case 2: Analysis of the International Aspects of Dutch Production and Consumption

The previous case concerned the Dutch economy and energy use related to production activities in the Netherlands. Since the Netherlands is a relatively small country with an open economy and a long history of trading, these activities are not intended only for the benefit of the Dutch population. They also comprise production of goods and services for production and consumption in other countries. Conversely, goods and services consumed by Dutch households are partly produced in foreign countries, and so they cause energy use and environmental pressure abroad. This case compares both cases: energy use in the Dutch production system, limited by the geographical borders of the country, and energy use related to production for consumption by Dutch people, which appears on a global scale. The latter is the embodied energy of Dutch consumption. (See chapter 6 for a discussion of embodied energy.) So this case shows the international dimension of the activities in the Netherlands, and of the Dutch, depicted in imports and exports.

The energy use for both approaches is calculated by using a version of DIMITRI comprising the Netherlands and two foreign regions: the OECD (excluding the Netherlands) and the non-OECD countries. Figure 8.2 has already shown the relations between these regions. Appendix 8B gives an overview of the countries on which the regions are based and a global description of the main data sources used.

We allocate energy use for production in the Netherlands to two final demand categories, namely, exports and domestic final demand. The latter category comprises not only consumption by Dutch households, but also other domestic final demand, like government consumption. We assume that all domestic final demand is ultimately intended for Dutch population. The energy use in the foreign regions concerns just the energy that is used for production related to consumption by the Dutch population.

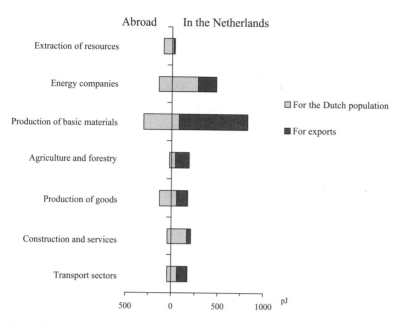

Figure 8.5
Energy use in the Netherlands (for the Dutch population and exports) and abroad (for the Dutch population) for seven sectors (1995).

Figure 8.5 shows the outcomes of the calculations for the year 1995. To illustrate the relevance of the economic structure in the Netherlands for the energy use in the country, energy use is given for seven groups of production sectors. More than 60 percent of energy use for production in the Netherlands is destined for exports. So only a minor part of energy use in Dutch production sectors is for the Dutch population. In contrast, more than half of energy use for production destined for the Dutch population takes place in other countries. More than 70 percent of this energy use abroad takes place in other OECD countries.

Figure 8.5 shows that particularly the energy-intensive sectors at the beginning of the chain, namely, the heavy chemical and primary metal industries, are responsible for a fair amount of export of the embedded energy. An important explanation for the size of these sectors is the location of the country and the availability of natural gas in the Netherlands (see chapter 7). The agricultural, transport, and service sectors in the Netherlands use more energy for exports than is used for the Dutch population in the same sectors abroad. More than 50 percent of the energy use abroad intended for the Dutch population takes place in the energy and basic materials sectors.

Targets for reducing energy use are set at a national scale. For countries like the Netherlands with an open economy, a shift of production across borders could help to meet those targets, but without changes in consumption patterns, this does not necessarily have to decrease global energy use. If production technologies abroad are less efficient than in the Netherlands, global energy use may even rise. The efficiencies of processes and production systems differ from country to country. Not only does the state of and knowledge about technology in the various countries explain the differences, but also characteristics of climate, soil properties, population density, and spatial relations can be country specific and may have an impact on production efficiencies. To gain insight into differences in production technologies over regions, we analyze these differences from the perspective of energy use related to production for the Dutch population that takes place abroad. The energy use is calculated for four variants:

1. Each foreign region has its own production technology and its own energy efficiency (this corresponds to the total energy use depicted in figure 8.5, to the left of the vertical axis).
2. Each foreign region has the Dutch production technology and energy efficiency.
3. Each foreign region has the Dutch production technology and its own energy efficiency.
4. Each foreign region has its own production technology and the Dutch energy efficiency.

Figure 8.6 shows the energy use calculated for the four variants. We realize that the outcomes should be interpreted with care, since the calculations assume that the sectors are comparable among regions and that they produce the same output. Since the sectors are aggregated, this may not always be the case. Nevertheless, we think it is justified to make some general conclusions concerning differences in production efficiencies among regions.

A comparison of the variant 1 outcomes with those of variant 2 shows that energy use in the OECD is lower if production is carried out with the Dutch production technology and energy efficiencies. The calculated energy use for the OECD is at the same level in variants 2 and 3, which differ only in the energy efficiencies used in the calculation. This implies that overall energy efficiencies in the OECD and the Netherlands are at

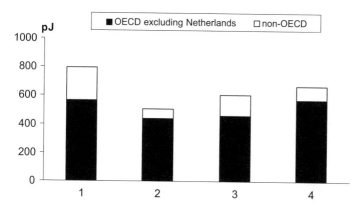

Figure 8.6
Energy use (in petajoules [pJs]) for production for the Dutch population in two foreign regions in 1995, calculated for four variants (see text). Energy use in variant 1 matches the total energy use for the Dutch population abroad summed up over all sectors in figure 8.5.

the same level. Comparing variant 1 with variant 3 (or variants 2 and 4) for the energy use in the OECD shows that production technology in the Netherlands results in a lower energy use than that in the OECD. Comparison of the variant 1 energy use in the non-OECD countries with that in the other variants shows that energy use in the non-OECD countries is the lowest if production is carried out with Dutch technologies and efficiencies. This means that both production technology and energy use in the non-OECD countries are less efficient than in the Netherlands.

The case shows the international character of energy use related to Dutch production and consumption activities. Since there are differences in production technologies among countries, a shift of production across borders could help to reduce global energy use by optimizing over chains. Policy may use the outcomes of such calculations in decisions concerning stimulating or discouraging some economic activities within the borders of the country. However, besides economic and ecological considerations, social considerations too should be taken into account.

8.5.3   Case 3: Analysis of a New Technology—Novel Protein Foods
Meat features largely in the current Western diet, but the production of meat is one of the most polluting processes in the food supply (Kramer 2000). One of the options for a more sustainable food supply may be a large-scale replacement of meat by what are known as "novel protein foods" (NPFs), foodstuffs based on vegetable or microbial proteins with

a much lower pollution level per functional unit of product. By using DIMITRI, the effects on the economy and environment of the replacement of meat with NPFs were calculated.

The calculations were carried out with a simplified version of DIMITRI consisting of forty-two production sectors. For the purpose of this case, the model was extended with two new sectors. The first sector added concerns the production of raw materials for making NPFs and is related to the agricultural sector. The second sector added is a more industrial sector in which the processing of raw materials to final products takes place. For these new sectors, input-output coefficients and energy and emissions intensities were derived using detailed information on NPF production and technologies. These coefficients were derived for a number of years to take into account the development of technologies for new NPF options. To calculate production and environmental pressure abroad, the model consists of a foreign region with the same forty-two production sectors as for the Netherlands. It is assumed that production technology abroad is the same as in the Netherlands. This version of DIMITRI therefore differs from the one used in case 2.

Calculations are carried out for several variants of an economic scenario for the period 1995–2030 developed by the Bureau for Economic Policy Analysis (1996). The results refer to effects on the economy as well as to effects on the environment both in time and location. The following aspects, which defined the different variants, influence the outcomes to a considerable extent:

1. *The percentage of meat replacement* For the period 1995–2030, curves are formulated depicting the percentage of meat replacement for each year. The results discussed in this chapter concern a comparison of a variant without meat replacement and a variant in which replacement occurs on a large scale (40 percent meat replacement in 2030). Meat replacement by NPFs is about 0.5 percent in the base year 1995.

2. *The location of raw material production and of final product processing (in the Netherlands or abroad)* The results are based on the variant that 50 percent of NPF production (both raw materials and final products) takes place in the Netherlands.

3. *The consumer price of NPFs related to meat* The consumer price of NPFs is varied between a price equal to that of meat and a price of a quarter of that of meat. In the latter case, consumers save money that may be spent on other consumption items. The saved money is assigned

Table 8.1
Changes in value added per sector in the NPF variant compared to the variant without meat replacement in 2030 (in millions of euros)

Sector	Change[a]
Livestock industry	−583−−208
Slaughterhouses and meat-packing industry	−767−−276
Other sectors	−838−+66
NPF: Raw materials	+13−+121
NPF: Processing	+71−+636

[a] Variance due to assumptions concerning the ratio of prices of NPF and meat products and concerning exports.

to other consumption items through the use of a model concerning consumer expenditures (Rood et al. 2001).

4. *Meat replacement for exports* In some variants, meat replacement concerns only domestic consumption, and in others it concerns the exports as well. In the latter case, we assume no compensation through exports of other products (if the price of NPFs is lower than that of meat).

Table 8.1 shows the changes in value added per sector for the NPF variant compared with the variant without NPFs. The table shows the combined effects of the variance in aspects 3 (consumer prices) and 4 (exports) on the value added.

High meat replacement increases the value added of the NPF sectors (raw material production and processing). Conversely, value added particularly decreases for sectors involved in meat production, like slaughterhouses and the meat-packing and livestock industries. Total GDP in the year 2030 is about 0.05 to 0.3 percent lower in the NPF variant compared with the meat variant. In particular, if the value of the exports decreases, GDP decreases, since we assume no compensation for the exports. This decrease may be a temporary effect, since in the long term, the saved labor and capital will be used for activities elsewhere in the economy. Production capacity in the NPF sectors has to increase rapidly if there is a fast introduction of the NPFs. Since the required investments are only a small fraction of total investments in the Netherlands, the expansion can be carried out without significant economic costs.

Figure 8.7 shows some of the positive effects on the environment in the Netherlands of a high percentage of meat replacement. In 2030, emissions of methane and ammonia in the livestock industry will be about 20 percent lower in the NPF variant (meat replacement in the Netherlands

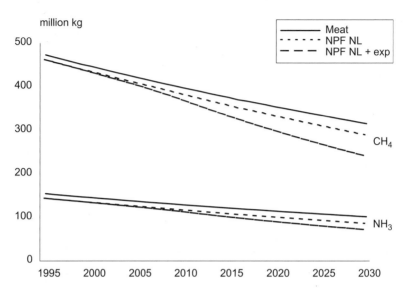

Figure 8.7
Methane (CH_4) and ammonia (NH_3) emissions in the livestock industry in the Netherlands
in the period 1995–2030, according to three variants (see text).

and for the exports) compared with the meat variant as a result of lower
production in the livestock industry. If meat replacement concerns only
Dutch consumption, the reductions in emissions are far smaller: on the
order of 5 percent. Meat replacement has only a fringe effect (less than
0.1 percent) on energy use and CO_2 emissions in the Netherlands.

Summarizing, then, a high percentage of meat replacement has a positive effect on the environment and only a small negative effect on the
economy. The economic and environmental effects have to be weighed to
conclude that replacement of meat with NPFs is a sustainable option.

8.6   Discussion

The case studies presented in this chapter have demonstrated how the
MNP-RIVM uses input-output analysis in investigating policy issues
concerning the interrelationships among consumption, production, technology, and the environment. One should bear in mind that DIMITRI is
one of many tools used by the MNP-RIVM. The model is not meant to
provide an answer on all issues related to the environmental impact of
consumption, production, and technology. It has been developed to investigate the environmental impact of (changes in) final demand and of

new technologies in the Netherlands and elsewhere. DIMITRI is a dynamic input-output model, but we point out that not all policy questions require dynamic modeling. For many situations, a static model suffices, by analyzing one base year or a base year and a future year. Dynamic models are essential when studying investments required for new technologies or exploring paths without well-defined technological developments in future years. Furthermore, dynamic models provide insights into the pace at which production structures may change.

Case 1 showed how DIMITRI has been applied for investigating the effects of changing production structures. In all fairness, we have to admit that this type of analysis does not require a dynamic input-output model. A year-to-year structural decomposition analysis would have yielded similar results (de Haan 2001). However, DIMITRI can also be used to perform this type of analysis for future years, as is shown in case 3. The year-to-year changes presented in both cases could be derived only with a dynamic model. But neither case has explicitly addressed the surplus value of a dynamic analysis compared with a static analysis. A comparative study would provide more empirical insight into the importance of a dynamic model for environmental impact analyses.

Case 2 focused on the multiregional character of the model. There is a significant difference between energy use for Dutch production and energy use for Dutch consumption. Moreover, the case also shows the importance of including regional-specific information on production characteristics.

We would also like to mention the limitation of DIMITRI for investigating policy issues. DIMITRI describes production flows within the economy. Changing prices and thereby changing supply and demand are beyond the scope of the model. This means that DIMITRI does not cover important aspects of the dynamics of an economic system. Consequently, outcomes of the model should be interpreted with care.

A second drawback is the amount of data required for the model. Reliable data on sectoral capacities and capacity utilization are of great importance for a dynamic input-output model. Furthermore, DIMITRI has a large demand for sectoral emission data in the Netherlands and abroad and data on sectoral technologies. Meeting this demand is not simple. But case 3 shows that an extension to include emissions can proceed smoothly if the required data are available. Furthermore, data used for one case study can be used easily for another, even when the model structure itself has to be adjusted. This is a major advantage of the general nature of the input-output framework.

Appendix 8A: Model Equations

Production
The model describes the relations between production, final demand, and investments in time. The production is determined as usual in dynamic input-output models:

$$\mathbf{x}(t) = [\mathbf{I} - \mathbf{A}(t)]^{-1}[\mathbf{y}(t) + \mathbf{z}(t)], \qquad (8A.1)$$

where

$\mathbf{x}(t)$ is a vector of total outputs in period t;
$\mathbf{I}$ is the identity matrix;
$\mathbf{A}(t)$ is a matrix with technological coefficients for period t (see equation (8A.10));
$\mathbf{y}(t)$ is a vector of final deliveries (excluding investments) for period t (the final deliveries are exogenous);
$\mathbf{z}(t)$ are investments (by origin) in period t.

The main purpose of the model is the calculation of the impacts of new technologies on environmental characteristics like the use of energy and materials and emissions into the environment. Moreover, the consequences for economic characteristics such as GDP and employment are of interest. Both environmental and economic characteristics are closely related to sectoral production, and they can be accounted for by distinguishing the items as separate value-added categories. Value added is calculated as follows:

$$\mathbf{v}(t) = \mathbf{V}(t)\mathbf{x}(t), \qquad (8A.2)$$

where

$\mathbf{v}(t)$ is a vector of total value-added per category in period t;
$\mathbf{V}(t)$ is a matrix with value-added coefficients per category per sector for period t.

Energy use, material use, and emissions are modeled in a similar way. In these cases, matrix $\mathbf{V}(t)$ has to be extended with coefficients concerning energy use (in petajoules), material use (in kilograms), or emissions (in relevant units) per unit of sectoral production for period t.

Investments

Sectors plan their investments on the basis of existing capacity, depreciation of capital goods, and the expected or planned capacities for the subsequent periods. We assume that investments for capital replacement and expansion are carried out with the same technology available at the moment of investment. The investments by origin are determined using:

$$\mathbf{z}(t) = \sum_k \mathbf{B}^k(t)\mathbf{c}^{\Delta,k}(t), \tag{8A.3}$$

where

$\mathbf{B}^k(t)$ is a matrix with capital requirements for capital good k per unit of production installed in period t;

$\mathbf{c}^{\Delta,k}(t)$ is the extension of the capacity (inclusive of the replacement of depreciated capacity) of capital good k installed in period t.

DIMITRI distinguishes among three types of capital goods, namely, buildings, machines, and means of transport. The matrix of capital coefficients, $\mathbf{B}$, describes the required capital goods per unit of capacity per sector.

Extension and Replacement of Capacity

The capacity to be installed newly in every period consists of the replacement of existing capacity that will be depreciated in the subsequent periods and the extension of capacity. The latter is based on expectations of future capacities per sector (see equation (8A.6)). The planning of new capacity is a number of periods ahead, depending on the type of capital good involved in the sector. This distinction among sectors is made because the planning and realization of new buildings generally takes longer than the ordering and introduction of new machines or means of transport. We assume that the installation of new buildings requires five periods and that the acquisition of machines and new means of transport takes two periods.

New capacity is installed only when this is necessary. By using the degree of capacity utilization as a measure, the model determines the new capacity that has to be installed. If this degree is below 0.5, there is an enormous overcapacity, and no new capacity will be installed. A capacity utilization degree between 0.5 and 0.9 enables only some replacement of depreciated capacity, and a capacity utilization degree above 0.9 allows all required new capacity to be installed. The new capacity per type of capital good i is

$$
c_i^{\Delta,k}(t) = \begin{cases} c_i^{a,k}(t) + c_i^{u,k}(t), & \text{if } b_i^k(t-1) > 0.9 \\ (2.5 \cdot b_i^k(t-1) - 1.25) \cdot c_i^{a,k}(t), & \text{if } 0.5 < b_i^k(t-1) \le 0.9, \\ 0, & \text{if } b_i^k(t-1) \le 0.5 \end{cases}
$$

(8A.4)

where

$c_i^{\Delta,k}(t)$ is the newly installed capacity of capital good k in sector i in period t;

$b_i^k(t)$ is the degree of capacity utilization of capital good k in sector i in period t (i.e., the quotient of production and capacity for each type of capital good);

$c_i^{a,k}(t)$ is the depreciation of capital good k in sector i in period t (see equation (8A.8));

$c_i^{u,k}(t)$ is the extension of capacity of capital good k in sector i in period t (see equation (8A.5)).

DIMITRI aims at maintaining a capacity utilization degree of one. The model permits the degree to exceed this level, although in actual practice this is not possible.

Extension of Capacity
According to equation (8A.4), the newly installed capacity is based on depreciated capacity and the extension of capacity. The extension of capacity is modeled on the basis of the difference between the planned capacity and the existing capacity. Assuming that this amount will be installed in equal shares over the planning period, the next year's extension will be one-fifth for buildings and one-half for machinery and means of transport. (Note that the remaining four-fifths or one-half will not be installed.) The extension of capacity is planned regardless of the previous year's planning. The extension of capacity in sector i for capital good k in period t is

$$
c_i^{u,k}(t) = \max\left(0, \frac{c_i^{*,\theta_k}(t) - c_i^k(t)}{\theta_k}\right),
$$

(8A.5)

where

$c_i^{*,\theta_k}(t)$ is the capacity of sector i planned in period t for period $t + \theta_k$ (see equation (8A.6));

$c_i^k(t)$ is the capacity of capital good k in sector i in period t (see equation (8A.9));

θ_k is the number of periods required to realize a capital good of type k; θ_k is five for buildings and two for machinery and means of transport.

The capacity to be installed new has to be positive, which means that a lowering of the capacity, for example, by selling part of the capital stock to other sectors, is not allowed. So capital stock can be decreased only by depreciation.

Planning of Future Capacity
As already discussed, sectors have to plan their capacity for subsequent periods to determine how much new capacity has to be installed each period. As usual in dynamic input-output modeling, the planned capacity per sector depends on the developments in the production of that sector in previous periods. Each period, the capacity for subsequent periods is planned on the basis of the developments in the production in the preceding periods. In DIMITRI, the planned capacity in period t for period $t + \tau$ $(\tau > 0)$ is

$$c_i^{*,\tau}(t) = \min\left(1 + \alpha_i, \left\{ \begin{array}{l} 1 + 0.4 \cdot gx_i(t-1) + 0.3 \cdot gx_i(t-2) \\ + 0.2 \cdot gx_i(t-3) + 0.1 \cdot gx_i(t-4) \end{array} \right\} \right)^{\tau+1}$$
$$\times \; x_i(t-1), \qquad\qquad\qquad\qquad\qquad\qquad (8A.6)$$

where

α_i is a maximum growth rate for sector i (exogeneous);

$gx_i(t)$ is a growth rate of total output in sector i in the period between t and $t - 1$, i.e. $(x_i(t) - x_i(t-1))/x_i(t-1)$;

$x_i(t)$ is a total output of sector i in period t.

Each period, the planned capacity is compared with the actual capacity (see equation (8A.5)). In this way, previous prognostications for future capacities may be adjusted.

Depreciation

Capacity is depreciated on the basis of depreciation periods per type of capital good.

In general, the share $S_i^{\tau,k}(t)$ describes the share of the capacity of capital good k in sector i implemented in period $t - \tau$ (and for the first time in use in period $t - \tau + 1$):

$$S_i^{\tau,k}(t) = 0, \qquad\qquad\qquad\qquad \text{if } \tau > \text{dep}_{i,k},$$

$$S_i^{\tau,k}(t) = \frac{S_i^{\tau-1,k}(t-1)c_i^k(t-1)}{c_i^k(t)}, \quad \text{if } 1 < \tau \le \text{dep}_{i,k}, \qquad (8A.7)$$

$$S_i^{\tau,k}(t) = \frac{c_i^{\Delta,k}(t-1)}{c_i^k(t)}, \qquad\qquad\quad \text{if } \tau = 1,$$

where

$S_i^{\tau,k}(t)$ is the share in period t of the capacity of capital good type k in sector i installed in period $t - \tau$ (which will be depreciated in period $t + \text{dep}_{i,k} - \tau$);

$\text{dep}_{i,k}$ is the sector i–specific depreciation period of capital good of type k.

So, $S_i^{\text{dep}_{i,k},k}(t)$ is the share of the capacity of capital good type k installed in period $t - \text{dep}_{i,k}$, in use for the first time in period $t - \text{dep}_{i,k} + 1$, and to be depreciated in period $t + 1$. The depreciated capacity is

$$c_i^{a,k}(t) = S_i^{\text{dep}_{i,k},k}(t)c_i^k(t). \qquad (8A.8)$$

All capital goods placed in a certain period are depreciated entirely at the end of the economic life of the capital goods. Intermediate falling out of capital goods is thus not taken into account.

Present Capacity

The capacity in period $t + 1$ is the sum of the present capacity minus the depreciated capacity and the newly installed capacity in period t for each type of capital good i:

$$c_i^k(t+1) = (1 - S_i^{\text{dep}_{i,k},k}(t))c_i^k(t) + c_i^{\Delta,k}(t) \qquad (8A.9)$$

Technological Matrix

The installed technology is a mix of technologies implemented in previous periods. As a result of depreciation and new investments, the installed technology in all sectors will change every period. After installing new technologies, the technological matrix should depict the new installed mix of technologies. This is achieved using the following formula:

$$\mathbf{A}_{ji}(t) = \sum_{\tau=1}^{\mathrm{dep}_{i,\kappa_i}} \mathbf{A}_{ji}^{N}(t-\tau) S_i^{\tau,\kappa_i}(t),$$ (8A.10)

where

κ_i is a type of capital good that specifies the technology per sector (means of transport for the transport sector, machines for other sectors);

$\mathbf{A}_{ji}^{N}(t)$ is the matrix of technical coefficients belonging to the technology of sector i installed in period t.

Since the value added per unit of production also depends on the installed technology, the derivation of the coefficients of $\mathbf{V}(t)$ is similar to the derivation of coefficients of the technological matrix in equation 8A.10.

Appendix 8B: Regions and Data

The DIMITRI model, as used in case 2, distinguishes between two regions covering the world. Table 8B.1 shows the groups of countries on which the regions are based.

Dynamic input-output models require a large amount of data. Table 8B.2 shows the most important sources for DIMITRI. In most cases, the original data needed some adjustments to fit the DIMITRI format.

Table 8B.1
Countries per region: GTAP countries

Region	Countries
OECD	Australia, Canada, Denmark, Germany, Finland, Japan, Rest of the European Union, Sweden, United Kingdom, United States
non-OECD	Argentina, Brazil, China, India, Indonesia, South Africa, Taiwan, Thailand, Turkey

Table 8B.2
Sources of data

Subject	Source
Economic input-output data, the Netherlands	National accounts: Statistics Netherlands (1998)
Economic input-output data, foreign regions	Global Trade Assistance and Protection (GTAP): McDougall, Elbehri, and Truong (1998)
Energy use, the Netherlands	Energy statistics: Statistics Netherlands (1996)
Energy use, foreign regions	Emission Database for Global Atmosphere Research (EDGAR): Olivier et al. (1996)
Emissions, the Netherlands	Environmental balances and environmental outlook: National Institute for Public Health and the Environment, institute databases
Emissions, foreign regions	EDGAR: Olivier et al. (1996)
Sectoral imports in the Netherlands from foreign regions	Import statistics: Statistics Netherlands (unpublished)

Note: For more information on data, see Wilting et al. (2001).

A more extensive discussion of the data can be found in Wilting et al. (2001).

References

Bullard, C. W., and R. A. Herendeen. (1975). The energy cost of goods and services. *Energy Policy* 3(4): 268–278.

Bureau for Economic Policy Analysis. (1996). *Omgevingsscenario's Lange Termijn Verkenning, 1995–2020*. Working paper no. 89, Centraal Planbureau, Den Haag.

Carter, A. P. (1967). Incremental flow coefficients for a dynamic input-output model with changing technology. In T. Barna (eds.), *Structural Interdependence and Economic Development: Proceedings of an International Conference on Input-Output Techniques, Geneva, September 1961*, 277–302. London: MacMillan.

de Haan, M. (2001). A structural decomposition analysis of pollution in the Netherlands. *Economic Systems Research* 13: 181–196.

Duchin, F. (1992). Industrial input-output analysis: Implications for industrial ecology. *Proceedings of the National Academy of Sciences* 89: 851–855.

Duchin, F. (1994). Input-output analysis and industrial ecology. In B. B. Allenby and D. J. Richards (eds.), *The Greening of Industrial Ecosystems*, 61–68. Washington, DC: National Academy Press.

Duchin, F., and G.-M. Lange. (1992). Technological choices and prices, and their implications for the US economy, 1963–2000. *Economic Systems Research* 4: 53–76.

Duchin, F., and G.-M. Lange, with K. Thonstad, and A. Idenburg. (1994). *The Future of the Environment: Ecological Economics and Technological Change*. New York: Oxford University Press.

Duchin, F., and D. B. Szyld. (1985). A dynamic input-output model with assured positive output. *Metroeconomica* 27: 269–282.

Fleissner, P. (1990). Dynamic Leontief models on the test bed. *Structural Change and Economic Dynamics* 1: 321–357.

Hannon, B. (1995). Input-output economics and ecology. *Structural Change and Economic Dynamics* 6: 331–333.

Hekkert, M. P. (2000). *Improving Material Management to Reduce Greenhouse Gas Emissions*. Utrecht, the Netherlands: Faculteit Scheikunde, Universiteit Utrecht.

Idenburg, A. M. (1993). *Gearing Production Models to Ecological Economic Analysis*. Enschede, the Netherlands: Faculteit Bestuurskunde, Universiteit Twente.

Joosten, L. A. J. (2001). *The Industrial Metabolism of Plastics—Analysis of Material Flow, Energy Consumption and CO_2 Emissions in the Lifecycle of Plastics*. Utrecht, the Netherlands: Faculteit Scheikunde, Universiteit Utrecht.

Kalmbach, P., and H. D. Kurz. (1990). Micro-electronics and employment: A dynamic input-output study of the West German economy. *Structural Change and Economic Dynamics* 1: 371–386.

Kalmbach, P., and H. D. Kurz. (1992). *Chips & Jobs, Zu den Beschäftigungswirkungen programmgesteuerter Arbeitsmittel*. Marburg, Germany: Metropolis Verlag.

Kramer, K. J. (2000). *Food Matters: On Reducing Energy Use and Greenhouse Gas Emissions from Household Food Consumption*. Groningen, the Netherlands: Rijksuniversiteit Groningen.

Leontief, W. W. (1941). *The Structure of the American Economy, 1919–1929*. 2nd ed. New York: Oxford University Press.

Leontief, W. W., and F. Duchin. (1986). *The Future Impact of Automation on Workers*. New York: Oxford University Press.

Los, B. (2001). Endogenous growth and structural change in a dynamic input-output model. *Economic Systems Research* 13(1): 3–34.

McDougall, R. A., A. Elbehri, and T. P. Truong. (1998). *Global Trade Assistance and Protection: The GTAP 4 Data Base*. Lafayette, IN: Center for Global Trade Analysis, Purdue University.

Meade, D. S. (2001). *The LIFT Model*. Working paper, INFORUM, available at http://inforumweb.umd.edu/.

Ministry of Economic Affairs. (1997). *Nota Milieu en Economie; op weg naar een duurzame economie* (Mimeo on environment and economics: Toward a sustainable economy). Den Haag: SDU.

Ministry of Housing, Spatial Planning and the Environment. (2001). *Een wereld en een wil, werken aan duurzaamheid. Nationaal Milieubeleidsplan 4* (Where there's a will, there's a world: Working on sustainability). Den Haag: Ministry of Housing, Spatial Planning and the Environment.

National Institute for Public Health and the Environment. (2000). *Milieubalans 2000; het Nederlandse milieu verklaard* (Environmental Balance 2001: Explaining the Dutch environment). Alphen aan den Rijn: Samsom.

National Institute for Public Health and the Environment. (2001a). *Milieubalans 2001; het Nederlandse milieu verklaard* (Environmental Balance 2001: Explaining the Dutch environment). Alphen aan den Rijn: Kluwer.

National Institute for Public Health and the Environment. (2001b). *Dutch Environmental Data Compendium 2001*. Available at http://www.rivm.nl/environmentaldata/.

Norén, R. (2001). Endogenous Disinvestment Activities and the Transformation to a New Equilibrium: A Computable General Equilibrium Approach. Scandinavian Working Papers in Economics no. 569, Department of Economics, Umeå University, available at http://swopec.hhs.se/umnees/abs/umnees0569.htm.

Olivier, J. G. J., A. F. Bouwman, C. W. M. van der Maas, J. J. M. Berdowski, C. Veldt, J. P. J. Bloos, A. J. H. Visschedijk, P. Y. J. Zandveld, and J. L. Haverlag. (1996). *Description of EDGAR Version 2.0: A Set of Global Emission Inventories of Greenhouse Gases and Ozone-Depleting Substances for All Anthropogenic and Most Natural Sources on a Per Country Basis and on 1 Degree × 1 Degree Grid*. RIVM rapport 771060002, RIVM, Bilthoven.

Peet, J. (1993). Input-output methods of energy analysis. *International Journal of Global Energy Issues* 5(1): 10–18.

Rood, G. A., J. P. M. Ros, E. Drissen, K. Vringer, T. G. Aalbers, G. Speek. (2001). *Model-structuur voor de milieudruk door consumptie* (Model structure for environmental pressure due to consumption). National Institute for Public Health and the Environment (RIVM) rapport 550000002, RIVM, Bilthoven.

Rose, A. (1984). Technological change and input-output analysis: An appraisal. *Socio-Economic Planning Sciences: An International Journal* 18: 305–318.

Statistics Netherlands (CBS). (1996). *De Nederlandse Energiehuishouding, jaarcijfers 1995* (Dutch energy statistics of the year 1995). Voorburg: St CBS.

Statistics Netherlands (CBS). (1998). *Input-Output Tables for the Netherlands, 1995: Domestic and Imports*. Voorburg: CBS.

van Bergeijk, P. A. G., M. A. van Dijk, R. C. G. Haffner, G. H. A. van Hagen, R. A. de Mooij, and P. M. Waasdorp. (1995). *Economic Policy, Technology and Growth*. BTE (Policy Studies Technology and Economics) no. 30. Den Haag: Ministry of Economic Affairs.

van den Broek, R. (2000). *Sustainability of Biomass Electricity Systems*. Delft, the Netherlands: Eburon.

Veeneklaas, F. R. (1990). *Dovetailing Technical and Economic Analysis*. Rotterdam, the Netherlands: Erasmus Drukkerij.

Wilting, H. C. (1996). *An Energy Perspective on Economic Activities*. Groningen, the Netherlands: Rijksuniversiteit Groningen.

Wilting, H. C., W. F. Blom, R. Thomas, and A. M. Idenburg. (2001). *DIMITRI 1.0: Beschrijving en toepassing van een dynamisch input-output model*. National Institute for Public Health and the Environment (RIVM) rapport 778001005, RIVM, Bilthoven.

Wright, D. J. (1974). Goods and services: An input-output analysis. *Energy Policy* 2(4): 307–315.

IV Waste Management and Recycling

9 Modeling Market Distortions in an Applied General Equilibrium Framework: The Case of Flat-Fee Pricing in the Waste Market

Heleen Bartelings, Rob B. Dellink, and Ekko C. van Ierland

9.1 Introduction

Current waste management policies are insufficient to obtain a significant reduction in waste generation from both industries and households. Although governments have made a great effort to reduce waste generation, the actual amount of waste generated is still rising, mostly as a result of economic growth. Governments have failed to achieve a decoupling of waste generation from economic growth, partly because of market distortions present in the waste market. In the Netherlands, these market distortions include (1) flat-fee pricing,[1] (2) virgin-material-biased regulations, and (3) contracts between municipalities and waste treatment facilities that require municipalities to supply minimum quantities of waste. All of these market distortions can cause a type of market failure: namely, that waste generation is higher than socially optimal and that not enough waste is recycled, with the result that more environmental costs than optimal are incurred (Miedema 1983).

Several studies have been conducted to analyze the effects of market distortions in the waste market and to suggest possible solutions. Most of these studies use a partial approach. Wertz (1976) was the first to analyze the effects of a user charge on solid waste disposal. Miedema (1983) analyzed the effects of other distortionary characteristics of the waste market, like virgin-material-biased tax policies, virgin-material-biased regulations, and indirect subsidization of virgin materials. Other, more recent studies are Jenkins (1993), Hong, Adams, and Love (1993), Miranda et al. (1994), Morris and Holthauzen (1994), and Sterner and Bartelings (1999). The overall conclusion of these empirical studies is that the demand for waste services is sensitive to user fees, and that the introduction of user fees can cause a substantial reduction in waste generation, especially if the fees are combined with programs that increase the public awareness of the waste problem. However, thoughtless construction of waste-handling tariffs may not have the desired effect and can encourage illicit dumping, burning, or other improper disposal (e.g., Fullerton and Kinnaman 1995).

Although most studies agree that a flat-fee pricing system is not optimal, they differ in their opinion on what the optimal policy to minimize cost of disposal should be. Studies like Miedema (1983), Jenkins (1993), Strathman, Rufolo, and Mildner (1995), and Linderhof et al. (2001) propose to introduce a "downstream" tax, for example, a unit-based pricing system.[2]

Other studies, like Fullerton and Kinnaman (1995, 1996), Palmer and Walls (1997), Fullerton and Wu (1998), and Choe and Frasier (1999), favor an "upstream" tax, like a deposit refund system or a "waste tax" on particular consumption products to internalize the waste treatment costs in the price of those products. They fear that a downstream tax will be nonoptimal because of huge implementation and enforcement costs. They especially fear an increase in illegal disposal. Households may start to dump their waste in their neighbors' bins, dispose of it at work, illegally dump waste, or burn it themselves. Such behavior leads to large social costs and has been identified as one of the most serious obstacles to the introduction of unit-based pricing for waste collection. Both Dobbs (1991) and Fullerton and Kinnaman (1995) demonstrate that if illegal disposal is a possibility, it may be optimal to have a negative tax on waste disposal; that is, legal waste disposal should be subsidized. In such a case, policymakers would be better off implementing other policy instruments to reduce waste generation. Fullerton and Kinnaman (1996) estimate that about 28 percent of the decrease in waste generation resulting from unit-based pricing may be caused by increased illegal disposal. Empirical studies, like Jenkins (1993) and Miranda and Aldy (1998), also report instances of increased illegal dumping. These results, however, are contradicted by those of other empirical studies. For example, Miranda et al. (1994), Strathman, Rufolo, and Mildner (1995), Nestor and Podolsky (1998), Podolsky and Spiegel (1998), Sterner and Bartelings (1999), and Linderhof et al. (2001) found no significant evidence of increased illegal disposal as a result of unit-based pricing.

In this chapter the question of whether an upstream or a downstream tax is the optimal policy tool for reducing waste generation, considering the effects waste management policies will have on the rest of the economy, is addressed. We have developed a general equilibrium model to analyze the efficiency of a downstream tax, namely, a unit-based pricing scheme, and an upstream tax, namely, an advance disposal fee (or waste tax). In this model we will consider the entire product life cycle from production, packaging, sale, and use to disposal. Thus it is possible to analyze the full impact of waste management tools and determine the cost-effectiveness of these tools. Because of the potential high social cost and transaction costs imposed by the introduction of unit-based pricing, we will include these costs in the analysis. In contrast to previous studies, in our model, municipalities are given an explicit role as the collectors of municipal solid waste. The method of solid waste collection, the pricing of waste collection, and the subsequent choice of waste treatment options lies solely with the municipality. Municipalities, therefore, can have a significant effect on the social costs of waste treatment.

The chapter is structured as follows: Section 9.2 describes the model and gives some insights into how different policy options can be included in an applied general equilibrium model. Section 9.3 presents a stylized example based on numerical data from the Netherlands in 1996 and shows how a unit-based pricing system can inadvertently promote waste leakage. Section 9.4 concludes and gives policy recommendations.

9.2   Description of the Model

9.2.1   Introduction
In this section, an applied general equilibrium model of the waste market is presented, together with two submodules of the model. In section 9.2.3, the model with endogenous prices for waste collection is described; in section 9.2.4, the model including the flat-fee pricing scheme is introduced; and in section 9.2.5, the model with an upstream tax is presented.

Like, for example, Fullerton and Kinnanman (1995), Fullerton and Wu (1998), and Calcott and Walls (2002), we have chosen to build a general equilibrium model, as this provides the ideal tool for analyzing the interactions among the waste treatment sector, the consumption sector, the production sector, the recycling sector, and the extraction sector. By including the whole economy, it is possible to analyze the effects of waste management tools to the fullest extent. The choice between re-cycling and generating waste, for example, will have significant conse-quences for the production sector and the extraction sector, and thus the consumption sector.

There are several ways to present or formats for presenting a general equilibrium model: for example, the "Arrow-Debreu format," the "com-putable generalized equilibrium format," the "open economy format," the "full format," and the "Negishi format." Some of these formats are written in terms of excess demand, others in terms of welfare programs. Extensive information about the strengths and weaknesses of each of these formats can be found in Ginsburgh and Keyzer (1997). It should be stressed, however, that a format is just a way of presenting a model. A different format still describes the same model and will result in the same equilibrium solution.

In this chapter, the Negishi format is chosen as the preferred tool for building a general equilibrium model. The advantage of this format is that it makes it relatively easy to incorporate externalities and non-convexities (see also Ginsburgh and Keyzer 1997). Hence, this format is especially suitable for incorporating into the model externalities of waste treatment and market distortions like flat-fee pricing. In contrast to, for example, the CGE format, the Negishi format is able to calculate the equilibrium solution in the case of price rigidities without requiring additional proof that a general equilibrium solution has been found. Moreover, the Negishi format is particularly suitable for incorporating multiple consumers, given that it can maximize several utility functions at the same time.

The Negishi format can, however, be written only in the primal form, which is a weakness of the Negishi format. This means that in the model, only production sets exist. Prices are calculated exogenously from the model. In the primal format, the equilibrium solution is found by one or several mathematical programs using some iterative procedure on parameters to find a fixed-point solution. In the dual form, which is used, for example, in the computable generalized equilibrium format, net supply and input demand are explicit functions of prices. The model is solved through a system of nonlinear equations. The advantage of the dual form over the primal form lies in the way in which the model is solved. As the dual form is based on a system of nonlinear equations, the computation and parameter estimation are normally far less difficult than the computation in the primal form. Thus the dual form will find an equilibrium solution much faster than the primal form. Nevertheless, we feel that this disadvantage does not offset the strong points of the Negishi format.

9.2.2   The Subsidy-cum-Tax Scheme

Most municipalities in the Netherlands have chosen to charge a fixed amount of money for waste collection, known as "flat fee." In a flat-fee pricing scheme, the amount of money paid for waste collection is independent of the amount of waste actually generated. The perceived price for waste collection (in economic terms, the marginal perceived costs of generating waste) equals zero in such a case. If the price of a good equals zero, the equilibrium demand for that good can no longer be determined through the normal demand and supply functions. Especially in the general equilibrium framework, in which it is assumed that some equilibrium price will make sure that demand equals supply, the zero price poses a problem. Hence, in order to implement a zero price in a general equilibrium model, we need some indirect approach. It is possible to implement a zero perceived price by using subsidies that compensate households for their cost of waste generation.

In a flat-fee scheme, households pay a fixed lump-sum transfer, based on the flat fee, for waste collection to the government. This lump-sum transfer takes away part of the income of the households. Total expenditure of the households declines. The expenditure pattern, i.e. the percentage of income the households spend on a certain product will, however, not be affected.

In our model, private households demand waste collection services and pay an equilibrium price for these services. However, in order to introduce the zero perceived price, the government reimburses these costs to the consumers in the form of a subsidy, which exactly equals the equilibrium price for every unit of waste collection services. Thus, the perceived price of waste collection for the households equals zero. If the revenue from the lump-sum transfer is lower than the amount spent on the subsidy, the government expenditure decreases (in this case, there is a net subsidy on waste generation). If the revenue of the flat fee is higher than the total costs, government expenditure increases. The idea of the subsidy-cum-lump-sum-transfer scheme is illustrated in figure 9.1. The next subsection will show how such a subsidy-cum-tax scheme can be implemented in a general equilibrium model.

9.2.3   Description of the Model with a Unit-Based Price for Waste Collection

In a simplified economy two types of actors are distinguished: households and firms. Households consume goods and supply endowments; firms produce goods and use endowments and intermediate goods. Two types of consumer households are distinguished: private consumers and a government consumer. Five different types of production sectors are distinguished, producing among them eight unique goods. These sectors are (1) an extraction sector, producing virgin material; (2) a production sector, producing agricultural goods, industrial goods, and services; (3) a recycling sector, producing recycling services; (4) a collection sector,

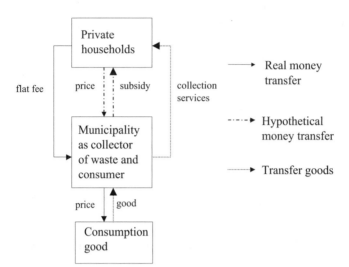

Figure 9.1
The subsidy-cum-tax scheme.

producing collection services; and (5) a waste treatment sector, producing incineration services and landfilling services. The hypothetical economy is shown in figure 9.2.

In this hypothetical economy, private households consume the consumer goods: agricultural goods, industrial goods, and services. The government consumes only services. Consumption of agricultural and industrial goods will cause waste generation. Waste will have to be either recycled or collected by the municipality. We assume that collected rest waste is not separated and recycled after collection but is sent immediately to an incineration plant or landfill unit. Although this puts some constraints on the model, we feel that this assumption is justified. We are mostly interested in the choice the consumer makes: The consumer can, for example, choose to separate organic waste, paper, or glass from rest waste. The consumer will have to incur costs to recycle these materials. It will, for example, cost the consumer time and storage space. This is modeled as if the consumer buys "recycling services." By buying recycling services, consumers generate recyclable waste; this waste is then sent to a recycling unit, where it is turned into recycled material.

Consumers will have the opportunity to prevent waste by recycling more waste or, to a lesser extent, by substituting waste-extensive goods (i.e., services) for waste-intensive goods (i.e., agricultural and industrial goods). In reality, consumers have the possibility of two kinds of substi-

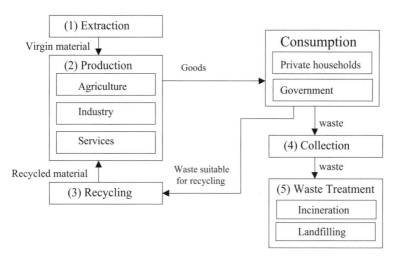

Figure 9.2
Representation of the hypothetical economy.

tution: namely, substitution within a sector and substitution between sectors. Substitution within a sector would make it possible to substitute one product for another that is basically the same except for waste intensity. Substituting between sectors means changing consumption patterns. For example, in Oostzaan, a municipality in which a unit-based pricing scheme was recently introduced, households reported that they not only started to buy products with less packaging, an example of substitution within a sector, but also started to use diaper services instead of disposable diapers, an example of substitution between sectors (Linderhof et al. 2001). However, waste prevention through substitution within a sector would add a certain degree of complexity to the model, as different products within the same sector and their associated (varying) "waste intensity" would have to be explicitly modeled. In our opinion, this trade-off between accuracy and transparency of the model is not easy to make, but in this case, we have chosen to include only the more straightforward channel of waste prevention via substitution between sectors. We realize that as a consequence, we may underestimate the possibility of waste prevention and therefore underestimate the effects of introducing either a unit-based price or an upstream waste tax. However, this assumption will not affect the comparison between the effectiveness of a unit-based price or an upstream waste tax, since the substitution possibilities will be identical in these two scenarios.

We assume that only the consumer sector generates waste. Neither the government nor the firms generate waste. This assumption is made because the focus is on policies affecting the generation of household waste, not on policies affecting the generation of industrial waste. Implicitly this means that we treat the municipal solid waste market and the industrial waste market as two separate markets and assume that changes in the industrial waste market do not affect the municipal solid waste market. It is quite justifiable to make this assumption because of the structure of the waste market in the Netherlands. In the Netherlands treatment of municipal solid waste is strongly regulated, and there is hardly any competition between treatment of municipal solid waste and treatment of industrial waste. Incinerators, for example, are obligated to treat as much municipal solid waste as their capacity allows. They are allowed to treat industrial waste only if their incineration capacity is larger than the quantity of waste offered by the municipalities. Therefore, the price of incineration paid by the municipalities is not affected by the quantity of industrial waste generated. Offering more industrial waste to an incinerator will affect only the price of incineration paid by the industry and not the price paid by the municipalities.

All the firms in the model produce goods or services with the use of capital and labor as inputs. The extraction sector produces virgin material, which is sold to the production sector of consumer goods. The recycling sector sells recycling services to the consumer and recycled material to the production sector of consumption goods. Besides capital and labor, the production sector of consumption goods uses virgin materials and recycled materials as inputs. The collection sector sells collection services to the private households. It uses capital and labor and waste treatment services as input. Finally, the waste treatment sector sells waste treatment services to the collection sector. It consists of two producers: a producer of incineration services and a producer of landfilling services.

9.2.3.1 Consumer Utility Function   In the Negishi format, total welfare is maximized subject to relevant balance and production possibility constraints (Ginsburgh and Keyzer 1997). The total welfare function is shown in equation (9.1). Total welfare (TW) equals the sum of weighted utilities (u_i) over consumer i ($i = 1, \ldots, n$); that is,

$$TWF(\alpha_i) = \max \sum_i \alpha_i u_i(x_i^g). \qquad (9.1)$$

Consumers derive utility from the consumption of consumer goods (x_i^g), where $g = 1, \ldots, G$. The utility of each consumer is weighted by a factor α_i, values of which are known as Negishi weight.[3]

Consumers generate waste by consuming products. Waste generation is dynamic; not all products will turn into waste immediately when they are consumed. Durable goods, for example, can function properly for several years. Over an infinite time scheme, every good will turn into waste. However, at any particular point in time, only part of the good will have turned into waste. To take this dynamic aspect into account in a comparative static model, waste is determined as a fraction β of the consumption product.[4] Total waste generation per consumer (w_i) is equal to a fixed percentage of total consumption; that is,

$$W_i = \sum_g \beta^g x_i^g. \tag{9.2}$$

All waste that is generated has to be dealt with. Private households can choose either to recycle the waste by demanding waste-recycling services (x_i^r) or to let the waste be collected by demanding waste collection services (x_i^w).

9.2.3.2 Production Functions The eight types of producers use two primary production factors—namely, capital (k) and labor (l)—and four intermediate inputs—namely, virgin material (m^v), recycled material (m^r), incineration services (w^i), and landfilling services (w^l)—to produce goods. All producers generate commodities y_j within their given production set Y_j; that is,

$$y_j \in Y_j. \tag{9.3}$$

The production set for the three consumption goods (agricultural goods, industrial goods, and services) is given by a nested Leontief-CES (constant elasticity of substitution) production function, which depends on the input of capital, labor, virgin material, recycled material, and waste treatment services:[5]

$$Y_j = A_j[\min\{CES(k_j, l_j; \sigma^{kl}), CES(w_j^{is}, w_j^{ls}, CES(m_j^v, m_j^r; \sigma^{vr}); \sigma^{wm})\}],$$

$$\text{for } j = 1, 2, 3, \tag{9.4}$$

where A stands for the technology level. The production set for the producer of recycled material is given by a nested CES function, which depends on the input of capital labor and recyclable waste:

$$Y_j = A_j[CES\{CES(k_j, l_j; \sigma^{kl}), X^r; \sigma^{pr}\}], \quad \text{for } j = 5, \tag{9.5}$$

where X^r is the total amount of recyclable waste generated by the private households. The production set for the producer of collection services given by a nested Leontief-CES function, which depends on the input of capital, labor, incineration, and landfilling, is

$$Y_j = A_j[\min\{CES(k_j, l_j; \sigma^{kl}), CES(w_j^{is}, w_j^{ls}; \sigma^{il})\}], \quad \text{for } j = 6. \tag{9.6}$$

The production sets of all other production sectors are given by CES functions that depend only on the input of capital and labor.

9.2.3.3 Balance Equations   As in all general equilibrium models, one of the conditions is that the demand for commodities (consumed goods and primary factors) be equal to the supply of these commodities (produced goods and endowments). This is ensured by the following balance equations.

First of all, total demand for consumption good g by consumer i and producer j must be equal to or less than total supply (y^g) of consumer good g, where g is an index of the three consumer goods (agricultural goods, industrial goods, and services). The shadow prices of the commodities can be determined from the balance equations by calculating the shadow price of the balance constraint. In the following equations, this is symbolized by $\perp$ and a price variable p.

$$\sum_i x_i^g + \sum_j x_j^g \leq y^g, \quad \perp p^g. \tag{9.7}$$

Total demand of all firms j for the intermediate goods "virgin material" (m_j^v), and "recycled material" (m_j^v) must be equal to or less than total supply of these materials (y). Since virgin material and recycled material are intermediate goods only, that is, not demanded by the consumers, the only demand comes from firm j:

$$\sum_j m_j^v \leq y^v, \quad \perp p^v, \tag{9.8}$$

$$\sum_j m_j^r \le y^r, \quad \perp p^r. \tag{9.9}$$

Total demand for "recycling services" (x^{rs}) and "waste collection services" (x^w) by consumer i must be equal to or less than the total supply of these services:

$$\sum_i x_i^{rs} \le y^{rs}, \quad \perp p^{rs}, \tag{9.10}$$

$$\sum_i x_i^w \le y^w, \quad \perp p^w. \tag{9.11}$$

Total demand for the intermediate good "waste treatment services" (w_j^n) (incineration or landfilling) must be equal to or less than total supply of waste treatment services:

$$\sum_j w_j^n \le y^n, \quad \perp p^n. \tag{9.12}$$

Total demand for primary factors must be equal to or less than the total supply of primary factors $(\bar{k}, \bar{l})$. Total supply of capital and labor is equal to the sum of the initial endowments of each consumer:

$$\sum_j k_j \le \sum_i \bar{K}_i, \quad \perp p^k, \tag{9.13}$$

$$\sum_j l_j \le \sum_i \bar{L}_i, \quad \perp p^l. \tag{9.14}$$

Prices for all commodities are calculated as the marginal value of the associated balance equations. In the model the budget constraint for each consumer must hold. The consumer obtains income by selling production factors capital and labor and spends its income on the three consumer goods, recycling services, and waste collection services:

$$\sum_g p^g x_i^g + p^{rs} x_i^{rs} + p^w x_i^w = p^k \bar{K}_i + p^l \bar{L}_i. \tag{9.15}$$

9.2.4 Description of the Model with a Flat-Fee Price Scheme for Waste Collection

To implement the subsidy-cum-tax scheme, as discussed in subsection 9.2.2, we extend the objective function, equation (9.1), with a subsidy term.[6] This subsidy term works like a benefit on the allocation of production output. Maximum social welfare now depends on the weighted utility of consumer I, on the one hand, and on the total benefits of the subsidy (ξW), on the other, where W stand for the total amount of waste generated and ξ for the cost of the subsidy per unit of waste:

$$TWF(\alpha) = \max \sum_i \alpha_i u_i(x_i^g) + \xi X^w, \tag{9.16}$$

$$x_i^g \geq 0, w_i \geq 0, r_i \geq 0 \quad \forall i, y_j \quad \forall j,$$

The subsidy costs are added to the social welfare function solely to change the perceived price of waste collection. Doing so does not imply that introducing subsidies would positively influence the social welfare of a region. The social welfare that is calculated with this model is not comparable to the social welfare calculated with the model presented in equation (9.1). The subsidy is present in the welfare function for technical reasons and specific to the Negishi format of the model. If the model were written in another format, the subsidy would not have to be made explicit in the welfare function.

The subsidy wedge (ξ) is defined as the difference between the equilibrium price for waste collection (p^w) and the perceived price (p_c^w). In our case, the perceived price of waste collection equals zero, so the subsidy wedge equals the real price of waste collection.

The balance equation for waste collection services (equation (9.8)) is rewritten as follows:

$$X^w \leq y^w, \quad \perp p^w, \tag{9.17}$$

$$\sum_i x_i^w \leq X^w, \quad \perp p_c^w. \tag{9.18}$$

One new variable is introduced: total waste demand (X^w). In equation (9.17), the shadow price of waste collection is calculated. This price equals marginal production costs. In equation (9.18), the shadow price of waste collection, as consumers perceive it, is calculated. This price equals the equilibrium price minus the subsidy.[7]

The Negishi weights (α_i) are determined in such a way that every consumer is exactly at its budget constraint. The budget constraint for private households is as follows:

$$\sum_g p^g x_i^g + p^{rs} x_i^{rs} + p_i^w x_i^w + F_i = p^k \overline{K}_i + p^l \overline{L}_i, \quad \text{for } i = 1. \tag{9.19}$$

Private households spend their income on the consumption of consumer goods, recycling services, and collection services (keep in mind that p_c^w is zero, so the costs of consumption of waste collection services are equal to zero) and pay a flat fee (F) to the government for the collection of waste.

The budget constraint of the government is as follows:

$$\sum_g p^g x_i^g + S = p^k \overline{K}_i + p^l \overline{L}_i + F_i, \quad \text{for } i = 2. \tag{9.20}$$

The government spends its income on consumer goods and subsidy costs (S). Since the government does not generate waste, it need not spend any income on the collection of waste. We assume that the government owns primary factors and earns income both from selling these primary factors and from benefits of the lump-sum transfer.

The size of the subsidy costs depends on the total amount spent on the subsidy for waste collection, which is calculated as follows:

$$S = \xi \sum_i x_i^w. \tag{9.21}$$

The total transfer equals the subsidy wedge (ξ) multiplied by the total demand for waste collection services. The subsidy wedge is calculated as follows:

$$\xi = p^w - p_c^w. \tag{9.22}$$

The subsidy wedge is equal to the real price of waste collection minus the perceived price of waste collection.

9.2.5 Description of Model with an Upstream Tax for Waste Collection Services

In the model with an upstream tax, the price of waste collection is internalized in the price of the good. A small tax on each of the three consumption goods is introduced. Introducing a tax in the Negishi format is quite similar to introducing a subsidy. First of all, the social welfare function must be adjusted. The new social welfare function is as follows:

$$TWF(\alpha) = \max \sum_i \alpha_i u_i(x_i^g) + \xi^w X^w + \sum_g \xi^g X^g, \tag{9.23}$$

$$x_i^g \geq 0, w_i \geq 0, r_i \geq 0 \quad \forall i, y_j \quad \forall j,$$

where ξ^g is the tax wedge and X^g is the total demand for good g.

The balance constraints for consumption goods also have to be changed, just as in section 9.2.3.3. It is important to realize that only the private households pay an upstream tax for waste collection. Neither the producers, who demand goods as intermediate deliveries, nor the government, which is assumed not to generate waste, have to pay this tax. The new balance constraints are as follows:

$$X^g + \sum_j x_j^g + x_{gov}^g \leq y^g, \quad \perp p^g, \tag{9.24}$$

$$x_c^g \leq X^g, \quad \perp p_c^g. \tag{9.25}$$

In the first balance constraint, the real equilibrium price for the consumption goods can be calculated. Both the producers and the government pay this price while consuming these goods. In the second balance constraint, the price, including the upstream tax, is calculated. Only the private households pay this price.

Again the Negishi weights are calculated in such a way that all consumers meet their budget constraints. The budget constraint for the private household equals

$$\sum_g p_i^g x_i^g + p^{rs} x_i^{rs} + p_i^w x_i^w + F_i = p^k \overline{K}_i + p^l \overline{L}_i, \quad \text{for } i = 1, \tag{9.26}$$

and the budget constraint for the government equals

$$\sum_g p^g x_i^g + S = p^k \overline{K}_i + p^l \overline{L}_i + F_i + T, \quad \text{for } i = 2, \tag{9.27}$$

where T equals the total gains from the upstream tax.

9.3 A Numerical Example

The model discussed in the previous section is now applied in a stylized example with numerical data from the Netherlands. The economic data used in the numerical example are based on national accounts for the Netherlands in 1996 (Statistics Netherlands 1998). These data are aggregated to four sectors (agricultural goods, industrial goods, services, and extraction) and two production factors (capital and labor) and supplemented with detailed data of the waste sectors (recycled material, recycling services, collection, incineration, landfilling, fee, and subsidy) based on AOO (1998) and RIVM (1998). To keep the model as simple as possible, we have chosen to give the government an income dependent on capital endowments instead of an income from taxes on labor and consumer goods.[8] Income dependent on capital endorsements has been added to the social-accounting matrix.

9.3.1 Parameter Values Used in the Numerical Example

The social-accounting matrix displayed in table 9.1 describes the initial equilibrium. Supply or producers' output and consumer endowments are given positive values; demand or producer inputs and consumption are given negative values.[9] All prices are normalized to unity except the price of waste collection. Waste collection basically has two prices: the perceived price and the social price of waste collection. The perceived price equals the total fee divided by the total demand for waste collection. The social price equals the total fee plus the total amount paid by the government for waste collection divided by the total demand. We have chosen to normalize the perceived price for waste collection, which means that the social price for waste collection (which is shown in table 9.1) is higher than unity.

Government expenditure is kept constant at benchmark level to avoid problems concerning public expenditure incidence.[10] This means that the government has a constant income to spend on consumption of services. If the income of the government changes as a result of policy measures, private households will compensate for the change through a lump-sum transfer.

All production sectors are characterized by a CES production function. Substitution elasticities for the different sectors are given in table 9.2. As mentioned before, each producer uses the primary production factors capital and labor. The three production sectors of consumer goods (agriculture, industry, and services) also use intermediate inputs for production. The use of primary factors and intermediate inputs is strictly complementary. Only the producer of industrial goods uses recycled material. It can fully substitute between recycled materials and virgin materials.

The substitution parameters for households are shown in table 9.3. Utility of private households depends on consumption of agricultural goods, industrial goods, and services. Among goods, a substitution elasticity of unity is assumed (Cobb-Douglas utility function). The government consumes only services. The initial Negishi weights are determined on the basis of the initial income (sales of endowments). Since the income of the private households is far larger than the government income,

Table 9.1
Benchmark social-accounting matrix (expenditures in hundreds of millions of Dutch guilders, 1996)

	Agri-culture	Industry	Services	Extrac-tion	Rec. material	Rec. services	Collec-tion	Incin-eration	Landfill	Cons.	Gov.	Price
Agriculture	400	-180	-10	0	0	0	0	0	0	-210	0	1.00
Industry	-120	2,190	-480	0	0	0	0	0	0	-1,590	0	1.00
Services	-40	-570	5,270	0	0	0	0	0	0	-4,460	-200	1.00
Extraction	-10	-318	-20	348	0	0	0	0	0	0	0	1.00
Rec. material	0	-4	0	0	4	0	0	0	0	0	0	1.00
Rec. services	0	0	0	0	0	2.5	0	0	0	-2.5	0	1.00
Rec. waste	0	0	0	0	-2.5	2.5	0	0	0	0	0	1.00
Collection	0	0.0	0	0	0	0	9.5	0	0	-9.5	0	1.00
Incineration	0	0.0	0	0	0	0	-6	6	0	0	0	1.05
Landfill	0	0.0	0	0	0	0	-2	0	2	0	0	1.00
Capital	-180	-468	-1,970	-273	-0.5	-4	-1	-5	-1.5	2,700	200.5	1.00
Labor	-50	-650	-2,790	-75	-1	-1	-1	-1	-0.5	3,572	0	1.00
Fee	0	0	0	0	0	0	0	0	0	-9.5	9.5	1.00
Subsidy	0	0	0	0	0	0	0	0	0	10	-10	1.00

Note: "Rec. material" stands for the production sector of recycled material; "Rec. services" stands for the production sector of recycling services; "Rec. waste" stands for waste suitable for recycling; "Fee" stands for the flat fee consumers pay to the government for waste collection; "Subsidy" stands for the total amount of money that the government gives for waste collection as a subsidy to the consumers; "Cons." stands for private households; and "Gov." stands for the government.

Table 9.2
Substitution elasticities for the production sectors

	Agri-culture	Industry	Services	Extrac-tion	Rec. material	Rec. services	Collec-tion	Incin-eration	Landfill
Substitution elasticity, capital and labor (σ^{kl})	0.8	0.8	0.8	0.8	0.8	0.8	0.8	0.8	0.8
Substitution elasticity, primary and intermediate inputs (σ^{pi})	0.0	0.0	0.0	—	—	—	—	—	—
Substitution elasticity, materials and other intermediate inputs (σ^{wm})	1.0	1.0	1.0	—	—	—	—	—	—
Substitution elasticity, recycled material and virgin material (σ^{vr})	—	∞	—	—	—	—	—	—	—
Substitution elasticity, primary factors and recycled waste (σ^{pw})	—	—	—	—	0.125	—	—	—	—
Substitution elasticity, landfilling and incineration (σ^{il})	—	—	—	—	—	—	0.2	—	—

Note: "Rec. material" stands for the production sector of recycled material; "Rec. services" stands for the production sector of recycling services.

Table 9.3
Additional parameters for households in the benchmark

		Cons.	Gov.
Elas1	Substitution elasticity between consumer goods	1.0	
α	Negishi weights	96.9	3.1

Note: "Cons." stands for private households; "Gov." stands for the government.

the Negishi weight for the private households is far larger that for the government.

In the base case scenario, collection and treatment of municipal solid waste costs about 1.2 billion guilders. Consuming either agricultural goods or industrial goods generates waste. One unit of agricultural goods contains a smaller percentage of waste than one unit of industrial goods. The percentage of waste present in a unit of agricultural goods is equal to 0.46, and the percentage of waste present in a unit of industrial goods is equal to 0.69. Of the waste generated, about 20 percent is recycled and about 80 percent is collected for waste treatment (either landfilling or incineration). Most of the waste collected is incinerated (75 percent); the rest is landfilled. Private households pay a total of 9.5 million guilders in the form of flat fees for collection of waste. This is slightly lower than the real cost of waste collection, which equals 10 million guilders.

9.3.2 Policy Scenarios

The model specified in the previous section is used to analyze the effects of different policy options on the generation of waste and the cost of waste treatment. A base case scenario plus four others are distinguished: (1) unit-based price, (2) upstream tax, (3) unit-based price plus transaction costs, and (4) upstream tax plus transaction costs.

The base case (first) scenario is described by the data presented in section 9.3.1 without added policies. The four other scenarios will all be compared with the base case scenario.

In the first scenario, labeled the "unit-based price scenario," the flat-fee pricing scheme of the base case is replaced by a unit-based pricing scheme. Private households now bear the full costs of waste collection. This means that their costs of waste collection have risen, since private households pay only 95 percent of the total costs in the benchmark scenario.

In the second scenario, labeled the "upstream tax scenario," a small upstream tax on agricultural and industrial goods is introduced. This tax

Table 9.4
Policy scenarios

Scenario	Fee collection	Waste tax	Transaction costs
Benchmark	Flat fee	No tax	No
Unit-based price	Unit-based fee	No tax	No
Upstream tax	No fee	Tax on consumption good	No
Unit-based price plus transaction costs	Unit-based fee	No tax	Yes
Upstream tax plus transaction costs	No fee	Tax on consumption goods	Yes

internalizes the cost of waste collection and treatment in the price of the product. Private households do not have to pay anything for the collection of waste.

In the third scenario, labeled the "unit-based price plus transaction costs scenario," the unit-based pricing system is again introduced. Two major disadvantages of unit-based pricing recognized in the literature are (1) the high transaction costs of introducing such a system and (2) the potential costs arising from illegal disposal of waste. Fullerton and Kinnaman (1996) show, for example, that the transaction costs of unit-based pricing are extensive and may offset the benefits of introducing such a system. Illegal disposal also poses a serious problem. Several studies, like Fullerton and Kinnaman (1995), Miranda and Aldy (1998), and Calcott and Walls (2002), show that consumers will illegally dispose of waste after unit-based pricing is introduced. As illegal disposal generates high social costs, it is important that these costs be considered in the analysis. In this scenario, therefore, some transaction costs of introducing a unit-based price are taken into account. These transaction costs represent both transaction costs of introducing unit-based pricing, like costs of installing weighing scales in garbage trucks, and increased social costs due to illegal disposal. By changing the available technology (A) in the production function, we can introduce transaction costs. We assume that a more expensive technology has to be used, which means that less output can be generated with the same amount of input.

In the fourth and final scenario, labeled the "upstream tax plus transaction costs scenario," the upstream tax is combined with the transaction costs of introducing an upstream tax. This means that the tax will be slightly higher than in the upstream tax scenario. The different elements of each scenario are summarized in table 9.4.

Table 9.5
Changes in the main variables under the unit-based price scenario relative to the flat-fee scenario (expenditures in hundreds of millions of Dutch guilders, 1996)

Variable	Flat fee		Unit-based price	
	Cons.	Gov.	Cons.	Gov.
Agricultural goods	210.00	—	209.43 (−0.268%)	— —
Industrial goods	1,590.00	—	1,581.79 (−0.517%)	— —
Services	4,460.00	200.00	4,468.92 (0.200%)	199.92 (−0.040%)
Recycling services	2.50	—	2.66 (6.645%)	— —
Waste collection	9.50	—	9.27 (−2.376%)	— —
Utility	3,097.72	200.00	3,097.77 (0.002%)	199.92 (−0.040%)

Note: "Cons." stands for private households; "Gov." stands for the government.

9.3.3   Results

9.3.3.1   Unit-Based Pricing Scheme Scenario   In the first scenario, a unit-based fee is introduced for waste collection. Households now pay the equilibrium price for waste collection. This means that generating more waste will result in higher costs for waste collection.

Table 9.5 shows the changes in the main variables. The government no longer bears the costs of collection. This means that its relative income that can be used for the consumption of services increases. Therefore, to keep government expenditure constant, as discussed in section 9.2, private households get a positive lump-sum transfer from the government. Private households now bear the full cost of waste collection, but because of the positive lump-sum transfer, there is little change in their available income. Households have an incentive to prevent waste and to recycle more waste. They substitute the waste-intensive agricultural and industrial goods for the waste-extensive services. Since the perceived price for waste collection has risen, there is some substitution between recycling and waste collection. Demand for recycling services increases, and demand for collection services decreases. Since the consumer's income is hardly affected by the policy change, the utility of the private households remains almost unchanged.

9.3.3.2   Upstream Tax Scenario   In the second scenario, an upstream tax is introduced. The tax is quite small and is meant to cover only the real cost of waste collection. Because the consumption of agricultural goods and industrial goods causes waste generation, these two goods are taxed.

Table 9.6
Changes in the main variables under the upstream tax scenario relative to the flat-fee scenario (expenditures in hundreds of millions of Dutch guilders, 1996)

Variable	Flat fee		Upstream tax	
	Cons.	Gov.	Cons.	Gov.
Agricultural goods	210.00	—	209.55 (−0.214%)	— —
Industrial goods	1,590.00	—	1,583.44 (−0.412%)	— —
Services	4,460.00	200.00	4,467.11 (0.159%)	199.95 (−0.027%)
Recycling services	2.50	—	2.50 (−0.039%)	— —
Waste collection	9.50	—	9.45 (−0.490%)	— —
Utility	3,097.72	200.00	3,097.76 (0.001%)	199.95 (−0.027%)

Note: "Cons." stands for private households; "Gov." stands for the government.

The results of the upstream tax scenario are shown in table 9.6. Because households pay a higher price for agricultural goods and industrial goods, the demand for these goods has declined. The demand for services has increased, because the price of services has not been affected. Since less agricultural and industrial goods are consumed, the amount of waste generated decreases slightly. The utility of the consumers is hardly affected by the measure. We see that compared with the unit-based price scenario, the upstream tax is less effective in minimizing the waste problem. Moreover, there is no substitution between recycling and collection.

9.3.3.3 Unit-Based Price Plus Transaction Costs Scenario An often-heard complaint about the unit-based pricing scheme is the huge transaction costs involved in introducing such a scheme. In most models, these costs are left out of the analysis. In the third scenario, we take into account both the unit-based pricing scheme and some of the transaction costs involved in introducing such a scheme. To cover these extra costs, private households will have to pay a higher fee compared with the unit-based pricing scenario.

Table 9.7 shows the results of this scenario for the main variables. Consumption is slightly decreased, because the cost of waste disposal has increased. Private households have to spend more income on waste disposal. Because the costs of waste collection have increased, consumers start to substitute demand for recycling services for demand for waste collection services. Compared with the second scenario, we see that more waste is recycled, whereas the utility of both the private households and the government is lower. Implementing a unit-based pricing scheme seems inefficient, given the results presented in table 9.7. No government should

Table 9.7
Changes in the main variables under the unit-based price plus transaction costs scenario relative to the flat-fee scenario (expenditures in hundreds of millions of Dutch guilders, 1996)

Variable	Flat fee		Unit-based price plus transaction costs	
	Cons.	Gov.	Cons.	Gov.
Agricultural goods	210.00	—	209.34 (−0.312%)	— —
Industrial goods	1,590.00	—	1,580.79 (−0.579%)	— —
Services	4,460.00	200.00	4,468.98 (0.201%)	199.90 (−0.049%)
Recycling services	2.50	—	3.14 (25.454%)	— —
Waste collection	9.50	—	8.80 (−7.403%)	— —
Utility	3,097.72	200.00	3,097.26 (−0.015%)	199.90 (−0.049%)

Note: "Cons." stands for private households; "Gov." stands for the government.

implement a policy that lowers the total welfare of the country. However, one should keep in mind that no environmental damage is included in the model. Collecting and treating waste will cause environmental damage. Recycling, on the other hand, causes far less environmental damage. If the state of the environment were included in the social welfare function, it might well be that this policy scenario would perform relatively well in terms of an increase in social welfare.

9.3.3.4   Upstream Tax Plus Transaction Costs Scenario   In the fourth scenario, an upstream tax on consumption goods is introduced. We assume that all transaction costs of introducing such a tax are borne by the private households. This means that we can introduce the transaction costs by increasing the total tax. This tax is slightly higher than in the third scenario, in order to cover the transaction costs.

In table 9.8, the results of this scenario are shown. The higher tax does not change the results very much. Somewhat less waste is generated. The demand for agricultural goods and industrial goods decreases slightly, and the demand for services, the only good without a tax, increases. The results show a minor decrease in the demand for both recycling services and collection services. Because of the costs of implementing the tax, the utility of both consumers decreases.

9.3.4   Sensitivity Analysis

9.3.4.1   Substitution Elasticity among Consumer Goods   The effectiveness of the upstream tax and, to a lesser extent, the unit-based pricing scheme depends on the substitution elasticity among the different con-

Table 9.8
Changes in the main variables under the upstream tax plus transaction costs scenario relative to the flat-fee scenario (expenditures in hundreds of millions of Dutch guilders, 1996)

Variable	Flat fee		Upstream tax plus transaction costs	
	Cons.	Gov.	Cons.	Gov.
Agricultural goods	210.00	—	209.55 (−0.261%)	— —
Industrial goods	1,590.00	—	1,583.44 (−0.477%)	— —
Services	4,460.00	200.00	4,467.11 (0.160%)	199.95 (−0.018%)
Recycling services	2.50	—	2.50 (−0.046%)	— —
Waste collection	9.50	—	9.45 (−0.569%)	— —
Utility	3,097.72	200.00	3,097.20 (−0.017%)	199.95 (−0.018%)

Note: "Cons." stands for private households; "Gov." stands for the government.

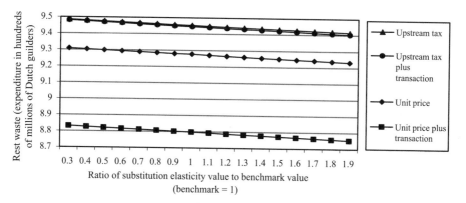

Figure 9.3
Sensitivity analysis of the substitution elasticity among different consumption goods: Impacts on the generation of rest waste.

sumer goods. If the demand for consumer goods is more elastic, it can be expected that consumers will substitute more services for industrial and agricultural goods, since consumption of services does not generate waste.

A sensitivity analysis is now performed for the substitution elasticity among the different consumption goods. The substitution elasticity is changed from low to high in a number of (equidistant) steps, resulting in a range of outcomes from a very inelastic demand to an elastic demand. The effects of parameter changes on the variables *rest waste* and *recyclable waste* are calculated for all four scenarios.

Figure 9.3 shows the impact of the substitution elasticity on the generation of rest waste, which is collected by the municipality.[11] For each scenario, the demand for waste collection is slightly sensitive to the

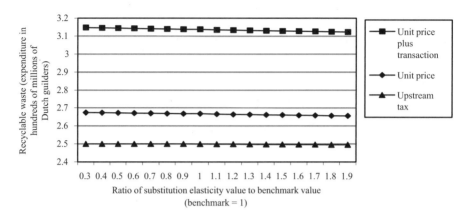

Figure 9.4
Sensitivity analysis of the substitution elasticity among different consumption goods: Impacts on the generation of recyclable waste.

substitution elasticity. If the demand for consumption goods is elastic, private households will substitute services for agricultural and industrial goods and generate less waste. From figure 9.3, it is clear that the unit-based pricing scheme is more efficient in reducing waste generation than the upstream tax.

Figure 9.4 shows that the demand for recycling services is hardly affected by changes in the substitution elasticity. A unit-based price gives households an incentive to increase recycling. However, if less waste is produced, recycling will also decline slightly. An upstream tax stimulates private households to generate less waste but does not stimulate recycling. Therefore, the demand for recycling services is not affected and stays at benchmark level.

9.3.4.2 Transaction Costs   The total costs of implementing a policy will greatly determine the effectiveness of that policy. In the scenarios presented earlier, we assumed that transaction costs involved in implementing the various policies would make the collection of waste 11 percent more expensive. However, the transaction costs involved could also be far larger. To analyze how sensitive the results are to a change in transaction costs, we have performed a sensitivity analysis. Higher transaction costs have been implemented by varying the technology parameter between 0.4 and 1.0. A lower technology parameter means that more inputs are necessary to produce the same amount of output. The benchmark-level technology parameter is equal to unity. Changing the value of the tech-

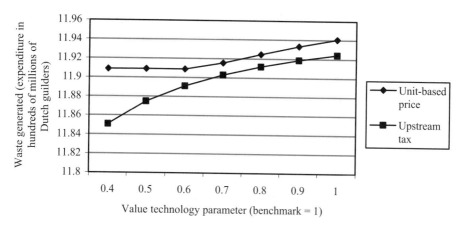

Figure 9.5
Sensitivity analysis of the transaction costs of implementing policy: Impacts on the generation of total waste.

nology parameter to a value higher than unity is not necessary, since the firms would then be able to produce output with fewer inputs, that is, the equivalent of negative transaction costs. The results for the total amount of waste generated (both recyclable waste and rest waste) are shown in figure 9.5.

As can be seen in figure 9.5, the upstream tax is more efficient at preventing waste than the unit-based price. If the transaction costs are large, more waste will be prevented with the upstream tax than with the unit-based price. However, the unit-based price stimulates not only the prevention of waste, but also the recycling of waste.

The results for the individual categories *waste recycling* and *waste collection* are shown in figure 9.6. The larger the transaction costs become, the more waste will be recycled in the unit-based price scenario. If the costs become very large, all waste generated will be recycled.[12]

Finally, the effects on the utility of the private households are shown in figure 9.7. The effects on utility are nearly equal for both the unit-based price policy and the upstream tax policy, if the transaction costs are small. However, when the transaction costs become larger, the upstream tax will have a far larger negative effect on utility than the unit-based pricing scheme. This effect is caused by the fact that if the transaction costs become too large in the unit-based pricing scheme, private households will start recycling all their waste, thereby eliminating all costs connected to the collection of waste.

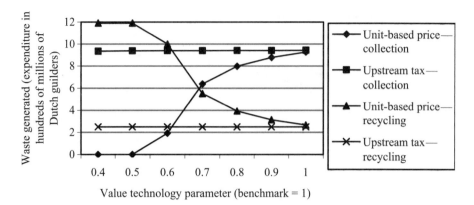

Figure 9.6
Sensitivity analysis of the transaction costs of implementing policy: Impacts on the generation of recyclable waste and rest waste.

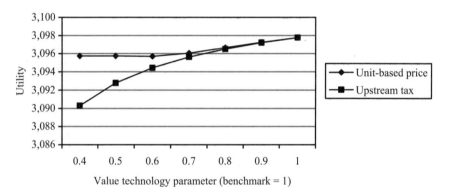

Figure 9.7
Sensitivity analysis of the transaction costs of implementing policy: Impacts on the utility of private households.

9.4   Conclusion

In this chapter, we have demonstrated how a simple model simulating the waste market and incorporating market distortions can be built using an applied general equilibrium framework. Since one of the characteristics of the waste market is a flat-fee pricing scheme for waste collection, the marginal costs or the actual price for waste collection that the consumers perceive is equal to zero. Special attention has therefore been given to modeling goods with a zero price. Introducing such a market distortion has strong effects on the results of the model, as was shown in the application of the model in a numerical example.

Flat-fee pricing for waste collection causes an inefficiently high generation of waste. The effects of two possible pricing methods for waste collection have been analyzed: an upstream tax and a unit-based pricing
scheme. The upstream tax can stimulate private households to generate
less waste if the demand for consumption goods is elastic. If the demand
is inelastic, the policy change will have hardly any effect. The unit-based
pricing scheme stimulates private households to both prevent and recycle
waste.

In contrast to other studies like Fullerton and Kinnaman (1995)
and Fullerton and Wu (1998), the results of the model presented in this
chapter clearly show that even considering the costs of illegal disposal
and high transaction costs, the unit-based pricing scheme is more efficient
at tackling the waste problem than an upstream tax. Private households
get a strong incentive to prevent and recycle more waste. The biggest
disadvantage of an upstream tax is that it hardly provides any incentive
to increase recycling; it provides only an incentive to reduce generation
of waste. This price incentive, however, is far too small to significantly change consumption patterns. Like Calcott and Walls (2002) and
Shinkuma (2003), we can therefore conclude that the introduction of unit-
based pricing should always be part of the optimal set of policy instruments to reduce waste generation.

The results also show, however, that both a unit-based pricing scheme
and an upstream tax also negatively affect utility and social welfare. Particularly if transaction costs are taken into account, the utility of private
consumers and social welfare will be negatively affected by either type of
measure. In this model, however, the environmental gains due to reduced
waste generation and treatment are not taken into account. Further research should show whether these environmental gains would be large
enough to offset the loss of utility. If the environmental gains of waste
prevention and more recycling are not large enough to offset this decrease
in utility, neither policy option is efficient.

In this chapter, different polices have been compared in a relatively
easy example. For a more extensive assessment of waste management
policies in the Netherlands, more data should be gathered. It is also necessary to resolve certain modeling issues, such as including environmental
aspects and including substitution possibilities among products within a
sector.

In the analysis presented in this chapter, we have assumed that private
households would pay all the transaction costs of introducing a policy.
But in reality, this may not be possible. In particular, the social costs of
illegal disposal and the costs of preventing illegal disposal may imply that

the unit-based pricing scheme is less desirable than an upstream tax. However, policymakers should bear in mind that the upstream tax is far less efficient, especially since it does not stimulate consumers to start recycling waste.

Appendix 9A: Solving a Negishi Format

The Negishi model calculates an equilibrium through an iterative process. First, the equilibrium is determined by solving the maximization model:

$$TW = \max \sum_i \alpha_i u_i(x_i, r_i, w_i), \tag{9A.1}$$

subject to the balance constraint

$$\sum_i \sum_g x_i^g + \sum_i r_i + \sum_i w_i \leq \sum_i \omega_i + \sum_j y_j, \quad \perp p. \tag{9A.2}$$

The Negishi weights are initialized as follows:

$$\alpha_i = \frac{h_i}{\sum_i h_i}. \tag{9A.3}$$

This means that the Negishi weight of consumer i is determined by the initial share this consumer has in total income. If the share of consumer i in the total income is large, the Negishi weight of that consumer will be large, and vice versa. It is assumed that the utility functions of both consumers are homothetic and commodity endowments are strictly proportional. Homotheticity ensures that the composition of a utility-maximizing commodity is unaffected by the level of income. Because of this assumption, the social demand, that is, the sum of individual demands, is proportional to the level of the total income, independent of its distribution. The competitive equilibrium prices and therefore the resultant allocation of resources are independent of income distribution. Thus, the problem of income distribution is assumed away (Negishi 1972).

After the model is solved, the shadow price of each commodity is calculated. Then these shadow prices are used to calculate the income deficit of each consumer, that is, the difference between total income and total expenditure of each consumer, labeled "loss":

$$loss_i = p\omega_i - \sum_i px_i^g + pr_i + pw_i. \tag{9A.4}$$

If the loss for each consumer is equal to zero, then the equilibrium solution has been found. If the loss is not equal to zero, then the Negishi weights are adjusted as follows (Ginsburgh and Keyzer 1997):

$$\alpha_i = \alpha_i + \beta \frac{loss_i}{\sum_i h_i}. \tag{9A.5}$$

If a consumer has a surplus income, that is, its income is larger than its expenditure, then the Negishi weight will be increased. In the next iteration, the consumption by this consumer will be greater, because of the larger Negishi weight. This iterative procedure will result in a set of unique equilibrium Negishi weights and prices.

Appendix 9B: Definition of Model Indices, Parameters, and Variables

Indices

Label	Entries	Description
G	1, 2, 3	goods (agriculture, industry, and services)
h	1, 2	material (recycled and virgin)
i	1, 2	consumers
j	1, ..., 8	goods and services
n	1, 2	waste treatment services (incineration and landfilling)
z	1, ..., 10	commodities (goods, services, capital, and labor)

Parameters

Symbol	Description
α	Negishi weight
β	waste percentage
σ^{kl}	substitution elasticity between labor and capital
σ^{pi}	substitution elasticity between primary and intermediate inputs
σ^{wm}	substitution elasticity between materials and other intermediate inputs
σ^{vr}	substitution elasticity between virgin and recycled material
σ^{pw}	substitution elasticity between primary factors and waste treatment services

σ^{il}	substitution elasticity between incineration and landfilling services
ξ	subsidy wedge
A	technology parameter
F	flat fee for waste collection
$\bar{K}$	endowment of capital
$\bar{L}$	endowment of labor
LST	lump-sum transfer to keep income of government constant
P	price
p_t	price including subsidy
S	transfer cost subsidy
T	gains upstream tax
Y^0	initial income

Variables

Symbol	Description
k	capital use
l	labor use
m	use material
TWF	total welfare
U	utility
w	use waste treatment services
W	total generation of waste
x	consumption
X	total consumption
Y	production

Notes

1. In a flat-fee pricing system, households pay a fixed amount of money to the municipality to collect their waste. The amount of money that they pay is not related to the actual amount of waste they generate.

2. In a unit-based pricing system, private households pay a variable fee for waste collection. The fee can, for example, depend on the actual amount of waste generated, the volume of waste generated, or the number of persons in a household.

3. These Negishi weights are determined in such a way that each consumer's budget constraint holds. This means that consumers cannot spend more money on goods and services then they receive from sales of primary inputs (capital and labor). The value of the Negishi weights is exogenous to the model. How these Negishi weights are determined and how the equilibrium solution is found is described in appendix 9A. (For more information, see Ginsburgh and Keyzer 1997.)

4. Implicitly this means that part of the used material will accumulate in the stock of durable goods. This stock is not constant: New materials will enter the stock, and other materials will leave the stock as waste. Therefore, at any given point in time, the material inflow in the model does not have to be equal to the material outflow.

5. The notation $z = CES(x, y; \sigma)$ reflects the following function:
$z = (x^{(\sigma-1)/\sigma} + y^{(\sigma-1)/\sigma})^{\sigma/(\sigma-1)}$.

If a good is produced with production factors that are completely complementary ($\sigma \to \infty$), a Leontief production function can be used as a special case of the CES production function. The standard notation for a Leontief production function is: $z = \min(x, y)$.

6. See Ginsburgh and Keyzer (1997) for details of this procedure.

7. Note that mathematically speaking, the introduction of the total waste demand variable is irrelevant. The equation $W = \sum_i w_i$ can be substituted in the balance equation in the equilibrium solution. However, the distinction of W enables the separation of the social (equilibrium) price for waste collection and the perceived price.

8. Although it is less realistic, we feel that this assumption is justified, since our focus is on waste generation and recycling by private households. The government's source of income will not influence the results in any way. For example, an equal tax rate on all produced goods will not affect the relative demand for any of these goods.

9. The column of each producer sums to zero to ensure that the zero profit condition holds: Value of inputs equals value of outputs. The column of each consumer sums to zero to ensure that the budget constraint holds: Each consumer spends exactly its income on the consumption of goods and services. Each row must sum to zero to ensure that each market clears: Total demand for each commodity must equal total supply.

10. When government expenditure changes, this will likely have an effect on the relative prices in the economy. The relative prices of factors of production and the relative prices of final consumption and production goods will change. Changes in relative prices will affect the income of private households directly through changed prices for production factors and indirectly through changing prices of consumption goods. Thus, changing public expenditure has some effects on the income distribution of private households and the utility of private households (Brown and Jackson 1990).

11. The value of the substitution elasticity is calculated as the value of the benchmark substitution elasticity multiplied by a certain factor, the value of which is shown on the x-axis in the figure.

12. For reasons of simplicity, we have assumed that a 100 percent recycling rate is possible, which is probably not a very realistic assumption. However, in this example, we want only to show the general working of the waste market. Limiting the recycling rate will have no effect on the overall trend of increased recycling.

References

AOO (Afval Overleg, Orgaan). (1998). Waste treatment in the Netherlands, 1998 [in Dutch]. Report AOO 98-05, Afval Overleg, Orgaan, Utrecht, the Netherlands.

Brown, C. V., and P. M. Jackson. (1990). *Public Sector Economics*. Oxford: Basil Blackwell.

Calcott, P., and M. Walls. (2002). Waste, Recycling, and Design for the Environment: Roles for Markets and Policy Instruments. Discussion paper no. 00-30REV, Resources for the Future, Washington, DC.

Choe, C., and I. Fraser. (1999). An economics analysis of household waste management. *Journal of Environmental Economics and Management* 38: 234–246.

Dobbs, I. M. (1991). Litter and waste management: Disposal taxes versus user charges. *Canadian Journal of Economics* 24: 221–217.

Fullerton, D., and T. C. Kinnaman. (1995). Garbage, recycling and illicit burning or dumping. *Journal of Environmental Economics and Management* 29: 78–91.

Fullerton, D., and T. C. Kinnaman. (1996). Household demand for garbage and recycling collection with the start of a price per bag. *American Economic Review* 86: 971–984.

Fullerton, D., and W. Wu. (1998). Policies for green design. *Journal of Environmental Economics and Management* 36: 131–148.

Ginsburgh, V., and M. Keyzer. (1997). *The Structure of Applied General Equilibrium Models.* London: MIT Press.

Hong, S. R., M. Adams, and H. A. Love. (1993). An economic analysis of household recycling of solid waste: The case of Portland, Oregon. *Journal of Environmental Economics and Management* 25: 136–146.

Jenkins, R. R. (1993). *The Economics of Solid Waste Reduction: The Impact of Users Fees.* Aldershot, England: Elgar.

Linderhof, V., P. Kooreman, M. Allers, and D. Wiersma. (2001). Weight-based pricing in the collection of household waste: The Oostzaan case. *Resource and Energy Economics* 23: 359–371.

Miedema, A. K. (1983). Fundamental economic comparisons of solid waste policy options. *Resources and Energy* 5: 21–43.

Miranda, M. L., and J. E. Aldy. (1998). Unit pricing of residential municipal solid waste: Lessons from nine case study communities. *Journal of Environmental Management* 52(1): 79–93.

Miranda, M. L., J. W. Everett, D. Blume, and A. R. Barbeau Jr. (1994). Market-based incentives and residential municipal solid waste. *Journal of Policy Analysis and Management* 13(4): 681–698.

Morris, G. E., and D. M. Holthausen. (1994). The economics of household solid waste generation and disposal. *Journal of Environmental Economics and Management* 26: 215–234.

Negishi, T. (1972). *General equilibrium theory and international trade.* Amsterdam: North-Holland.

Nestor, D. V., and M. J. Podolsky. (1998). Assessing incentive-based environmental policies for reducing household waste disposal. *Contemporary Economic Policy* 16(4): 27–39.

Palmer, K., and M. Walls. (1997). Optimal policies for solid waste disposal taxes, subsidies, and standards. *Journal of Public Economics* 65: 193–205.

Podolsky, M. J., and M. Spiegel. (1998). Municipal waste disposal: Unit-pricing and recycling opportunities. *Public Works Management and Policy* 3: 27–39.

RIVM (National Institute for Public Health and the Environment). (1998). *Environmental Balance Sheet, 1998.* Bilthoven, the Netherlands: RIVM.

Shinkuma, T. (2003). On the second-best policy of household's waste recycling. *Environmental and Resource Economics* 24: 77–95.

Statistics Netherlands. (1998). *National Accounts of the Netherlands.* Voorburg: Statistics Netherlands.

Sterner, T., and H. Bartelings. (1999). Household waste management in a Swedish municipality: Determinants of waste disposal, recycling and composting. *Environmental and Resource Economics* 13: 473–491.

Strathman, J. G., A. M. Rufolo, and G. C. S. Mildner. (1995). The demand for solid waste disposal. *Land Economics* 71(1): 57–64.

Wertz, K. L. (1976). Economic factors influencing household production of refuse. *Journal of Environmental Economics and Management* 2: 263–272.

10 International Trade, Recycling, and the Environment

Pieter J. H. van Beukering

10.1 Introduction

Recycling is generally considered as an important strategy for alleviating the pressure of society on the environment. Natural resources are saved, emissions decrease, and the burden of solid waste is reduced. At the same time, recycling creates employment and attracts investments. In recent years, many countries have experienced large increases in recycling. The rationale behind this development varies between the developed and the developing world. In the North, the increase in recycling is assumed to have resulted mainly from higher disposal costs, increased public concern about health and the environment, and public policy to stimulate recycling. Recycling in the South is thought to be driven mainly by economic opportunities and employment, largely as a result of the lower labor costs.

In addition to domestic causes, international trade has played an important role in the expansion of the recycling sector. In recent decades, international trade of recyclable materials has increased significantly. As is the case with any commodity, international trade of recyclable materials allows countries to exercise their comparative advantages to bring about a more efficient allocation of resources. This trend of increased trade in recyclable materials raises the question of how important international trade has been in stimulating recycling across the globe.

Few economic studies have been devoted to examining the internationalization of markets for recyclable materials and products. Most available studies have focused on an empirical analysis of the relationship between trade and recycling (Grace, Turner, and Walter 1978; Johnstone 1998; Hoffmann 1999). This neglect of recycling market internationalization in the literature is unjustified, because the growth rate of international recycling nowadays tends to exceed the development of agricultural and consumer product markets. In contrast with economic studies, environmentalists have noticed the changes in international recycling (McKee 1996; van der Klundert 1997; Greenpeace 1997; Kellow 1999). Rather than analyzing international recycling as an economic process, however, they tend to present international trade of recyclables as a form of ecological dumping and assume that it is.

The aim of this chapter is to determine the economic and environmental significance of the simultaneous increase in trade and recycling of materials. A number of questions emerge: (1) What are the specific patterns in trade and recycling? (2) What are the underlying causes of these patterns? (3) What are their economic and environmental effects? (4) How can these effects be measured and modeled? and (5) How should public policy respond to potentially negative externalities resulting from international recycling? Special attention will be devoted to differences in recycling in developed and developing countries.

10.2   Global Patterns in Trade and Recycling

It is important to differentiate between the materials and the products that go through different stages in the production sequence of extraction, production, consumption, and waste management. This sequence has been referred to as the material-product chain (Opschoor 1994; Kandelaars 1999). To analyze the trends in trade and recycling, a good understanding of the flows in the M-P chain is required. The following types of materials are especially relevant:

• Primary commodities or virgin materials: raw materials that have been extracted from natural resources. Examples are iron ore and wood pulp.
• Secondary commodities or recyclable materials: raw materials that have been recovered after production or consumption. Examples are iron scrap and wastepaper.
• Final commodities: materials suitable for conversion directly into final products. Examples are crude steel and printing paper. It is assumed that final commodities can be produced from both primary and secondary commodities.

The materials considered in the present analysis include aluminum, copper, lead, nickel, paper, iron, tin, zinc, and plastics. These materials have been selected because of the availability of relatively long time series (1970–1997) for related primary, secondary, and final commodities. Production and consumption data have been derived from various sources.[1] Several global trends can be identified.[2]

10.2.1 Recycling Is Increasing

Over the last three decades, the recycling of most commodities has increased. It is difficult to evaluate and compare different materials in absolute terms. For example, over the period 1970–1997, the recycling of lead has increased from approximately one to three million metric tons (tonnes).[3] The relative recycling rate of iron and steel is, however, much higher; steel recovery increased from 253 to 375 million tonnes. Relative measures have been used to evaluate changes in recycling. One indicator is the global "recycling rate," which is defined as the total amount of secondary materials consumed (or produced) as a share of the total volume produced (or consumed) of a particular final commodity. When the stocks of secondary commodities are constant, the total amount of recovered materials equals the total amount of utilized materials at a global level. Therefore, one can use either consumption or production of secondary commodities as the numerator of the recycling rate.

Figure 10.1 depicts the development of the five-year average recycling rates for six materials between 1970 and 1997. For all these materials, recycling is increasing, although at different rates. Commodities that show high growth rates in recycling are aluminum (increasing from 18 to 26 percent), lead (from 26 to 50 percent) and paper (from 25 to 39 percent). Zinc recycling has also increased (from 8 to 11 percent) but has remained rather low in an absolute sense. The recycling rates of iron (increasing from 44 to 51 percent) and copper (steady at around 39 percent) are relatively high but do not show high rates of change.

Various factors determine the variations in the growth of recycling rates. *Economic growth* tends to increase both the supply of and the demand for secondary materials in the total supply of commodities (Radetzki and van Duyne 1985). The supply of secondary materials increases during periods of economic growth, inter alia, because more old buildings are demolished and consumers replace cars and home appliances more often (Moison 1997). Levels of demand increase because of higher levels of per capita income, resulting in larger requirements for certain raw materials. This is the case for aluminum and paper. Iron and copper are characterized by low growth in final demand and therefore show relatively little progress in recycling.

Technical and institutional limits may constrain recycling. Figure 10.1 shows that iron and copper have the highest recycling rate and the lowest

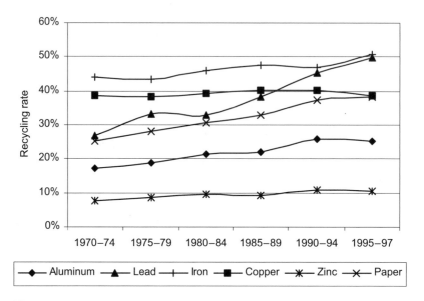

Figure 10.1
Global recycling rates for the period 1970–1997. The global recycling rate is the global quantity of recycled materials utilized as a share of total global quantity of final production, shown here in five-year averages.

growth rate in 1970. These materials seem to come within reach of the maximum level of recycling that is determined by technological and institutional factors. These same constraints can also have an impact at the level of final commodities. For example, the supply of copper scrap is constrained because the potential supply of secondary copper is "frozen" in existing equipment and buildings (Linnemann and Kox 1995). In the case of lead, a positive effect has emerged from technological changes relating to its final uses. Although the overall demand for lead has decreased, the growing use of lead in car batteries has reduced the average costs of lead scrap recovery, leading to a positive impact on the recycling of lead (Tilton 1999). Aluminum recycling has been boosted as a result of new technologies to recover the by-products of the recycling process (Thomas and Wirtz 1994).

Substitution between commodities plays an important role. Copper and iron have gradually been replaced by substitutes, such as glass fiber, plastics, and aluminum (Linnemann and Kox 1995; Pei and Tilton 1999). This replacement has functioned as a negative incentive to further increase recycling. Substitution between primary and secondary materials has also changed as a result of technological developments. For example, compared with a decade ago, a much larger share of newsprint can be produced from wastepaper (Michael 1998). Similarly, a technological change in the metal-processing industry has enabled the exploitation of low-grade and oxide ores, which the old technology could not treat. This technological advance has stimulated the substitution of iron ore for iron scrap and copper ore for copper scrap (Tilton 1999).

Environmental policies have a major impact on recycling. An argument often used by policy makers to promote recycling is the prevention of mineral resource depletion. Research has demonstrated, however, that the availability of mineral resources has not declined over time (Tietenberg 1996). Depletion has taken place at much slower rates than foreseen, and new resource reserves are being discovered at an almost continuous rate (Frechette 1999). Another motivation to promote recycling, especially in small countries, has been the need to reduce landfilling and waste disposal. For example, wastepaper is widely recovered to reduce the use of landfill space (Weaver et al. 1995). Lead is increasingly being recovered because of its hazardous characteristics (Berger 1998). Environmental policies can also have an indirect effect on recycling: for example, by taxing energy more heavily. Given the relatively high energy intensity of primary production, the competitiveness of recycling grows as a result of higher energy prices (Bartone 1990).

10.2.2 Growth in Secondary Commodity Trade

International trade of most secondary materials has increased at an even faster rate than their production. The combined trade volume of secondary aluminum, lead, zinc, copper, and paper increased from 2.5 million tonnes in 1970 to 21.5 million tonnes in 1997. Iron and steel dominate the international recycling market: The iron and steel scrap trade increased from 20 to 37 million tonnes over that same time period. To compare and evaluate the trade of secondary commodities on a global scale, the indicator "trade rate" is used. The trade rate is defined as the total amount exported (imported) internationally as a share of the total amount globally produced (consumed).

Figure 10.2 shows the five-year average trade rates for a number of secondary materials. The rates for four of the six materials show exceptionally high growth. For example, the trade rate of secondary copper increased from 15 percent in 1970 to 48 percent in 1997. The trade rate of wastepaper and aluminum also increased significantly: namely, from 7 to 17 percent in the case of the former and from 11 to 31 percent in the case of the latter. These trends indicate that for most secondary materials, the

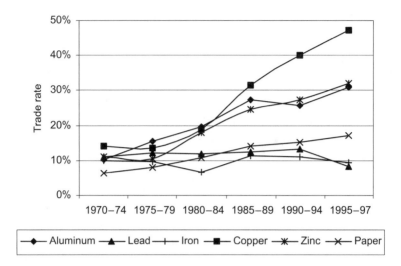

Figure 10.2
Global trade rate of secondary materials for the period 1970–1997. The global trade rate is
defined as the total quantity of traded secondary materials as a share of total quantity of
recovered secondary materials, shown here in five-year averages.

local market has changed into a global market. The development of trade
in iron scrap shows dynamics similar to that in iron recycling. The trade
rate remains constant at approximately 10 percent. Trade in lead scrap
also lags behind the increasing trade rates of other secondary materials.

As a result of trade liberalization and changing production and consumption patterns, the international trade of primary and final commodities has also expanded. The growth of trade in secondary commodities is,
however, much more pronounced than trade in primary commodities. As
table 10.1 shows, the share of secondary materials in the total market for
raw commodities has grown for most materials since 1970. Copper scrap,
lead scrap, and wastepaper have the largest share in the international
market for raw commodities. Nickel, iron, zinc, and plastics are relatively
insignificant in the overall international market. The share of aluminum
scrap grew most rapidly. Iron scrap is the only commodity that has become less important as a traded commodity over the last three decades.

Several factors may have contributed to the change in trade of secondary materials. The most important incentive to trade is the difference in
recycling costs and benefits among different countries. In the last three
decades, markets for secondary materials in developed countries have
been characterized by oversupply. At the same time, a shortage of high-
quality secondary materials for product manufacture has prevailed in

Table 10.1
Market shares of traded secondary commodities in the international market for raw commodities[a]

	Average 1970–1974	Average 1995–1997
Aluminum	4%	13%
Copper	24%	37%
Lead	12%	22%
Nickel	8%	8%
Paper	13%	27%
Plastics[b]	2.9%	3.5%
Iron	14%	7%
Tin	13%	20%
Zinc	2%	4%

[a] The raw commodity market is defined as the sum of the markets for primary and secondary commodities.
[b] The share that is given in the first column is the average for 1980–1984. Plastics include polyethylene, polypropylene, polystyrene, and polyvinyl chloride.

developing countries that have attempted to widen their industry base (Savage and Diaz 1996). This has led to relatively low prices of recovered materials in the North and relatively high prices for these commodities in the South. A typical example of this oversupply of secondary materials is the wastepaper market in Europe in the early 1990s (Byström and Lönnstedt 1995). For certain types of wastepaper, even negative prices have been recorded. Another example that illustrates the impact of cost differences on the trade flows of secondary materials is the trade of computer scrap. Kellow (1999) reports how computer scrap is collected in Australia and exported to the Philippines for manual disassembly by cheap labor. Certain components of old computers, such as diodes and switches, are exported to China for reuse. Printed circuit boards are sent back to Australia, where gold, silver and copper are recovered through high-tech metallurgical processes. The remaining materials are used in the Philippines.

The gradual decline of transaction and transport costs has promoted the trade of secondary materials. To protect the home market, and to prevent increased waste disposal, trade policies, particularly in developing countries, originally tended to discriminate against imports of secondary commodities (Navaretti, Soloaga, and Takacs 1998). Negotiations in the context of the World Trade Organization (WTO) have gradually reduced these trade barriers. Improvements in information technologies and the establishment of national and international networks have also made it much easier for buyers and sellers to meet (Buggeln 1998). For example, in the late 1990s the Chicago Board of Trade (CBOT) opened an international exchange for recyclable materials (CBOT 1999). Whether these information technologies will be successful in encouraging trade in recyclables has yet to be seen.

Finally, an important question raised by figure 10.2 is why the trade in lead and iron scrap remains at low levels of approximately 10 percent. These low levels persist because both scrap metals are generally of too low value per unit of volume to be transported over long distances. Moreover, lead scrap is often considered hazardous. In anticipation of the ratification of the Basel Convention, many governments have already introduced constraints on the import of lead scrap (Elmer 1996).

10.2.3 Specialization in the Global Recycling Market

The international markets for secondary materials differ from primary and final commodity markets in that for most secondary materials, imports occur mainly in developing countries. Figure 10.3 portrays changes over time for each of the four directions in which materials can move between developed and developing countries.[4] The curves represent trade of primary, secondary, or final commodities between regions over time. Each point on the lines represents an unweighted average share in the world trade across the various commodities (aluminum, copper, lead, nickel, paper, plastics, iron, tin, and zinc). By definition, the four directions add up to 100 percent for each period and commodity.

Figure 10.3 indicates several developments. Of the four trade directions, trade among developed countries is dominant (figure 10.3a). Over time, however, the importance of North-North trade has gradually decreased, especially for secondary commodities. Developing countries have become more important customers for exports from developed countries (figure 10.3b). For the commodities considered, developing countries have been able to expand their export share on the world market only by trading with developing countries. The share of trade from developing countries to developed countries has fallen. The share of primary commodities exported from developing to developed countries in figure 10.3c still exceeds the flow in the opposite direction, but the gap has been reduced substantially. Developing countries are characterized as major net importers of secondary commodities. Figure 10.3d shows that the decreasing levels of trade in secondary commodities among developed countries shown in figure 10.3a has resulted in a shift toward trade among developing countries. On the whole, the import share of traded secondary commodities in developing countries has increased from 10 to 38 percent.

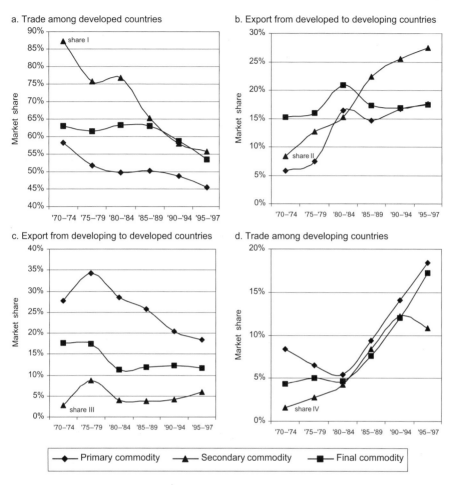

Figure 10.3
Trade flows between and within developed and developing regions. Vertical axes indicate the trade of a commodity between two regions as a share of the global trade in that commodity. Therefore, the sum of the four quadrants for each year represents the total trade of that commodity in a particular year. An example of this rule is given for secondary materials in 1970–1974 by the respective shares I, II, III, and IV, which add up to 100 percent.

For primary and final commodities, the import share has increased, respectively, from 15 to 36 percent and from 20 to 35 percent.

In summary, three main trends can be observed. First, recycling has increased at a global level. In both developed and developing countries, secondary materials have gained importance as a basic raw material. Second, trade in secondary commodities has grown faster than trade in other commodities. Third, over the last three decades, the North-South trade in secondary commodities has increased the most among trade flows. Developed countries have specialized in the recovery of secondary commodities, whereas developing countries have specialized in the utilization of secondary materials.

10.3   Causes of Trade and Recycling

What causes these rapid changes in trade and recycling? First, one of the few empirical studies that has focused on the underlying causes of trade and recycling is van Beukering and Bouman (2001), in which regression analysis was conducted for paper and lead commodities in developed and developing countries. The results show that population density is positively related to materials recycling. Land scarcity limits the possibility of landfilling, thereby creating an incentive for increased recovery. For hazardous materials, such as lead scrap, however, environmental regulations discourage recycling in densely populated areas. Second, another finding is that an increase in waste in higher-income countries has promoted recycling. This can be explained by the economies of scale of waste recovery and increased environmental awareness at the household level. It is easier to recover a large share of waste if one produces large quantities, because one puts some sort of infrastructure for doing so in place, especially if one is environmentally motivated. Third, economic growth and recycling have a negative relationship. A possible explanation is that rapidly growing economies utilize easily accessible primary materials rather than the more "laborious" recyclable materials. Moreover, there is a time lag between the rapid increase in consumption of materials and the discarding of materials. Finally, openness of an economy is found to promote the utilization of secondary materials in that economy. An explanation is that recyclers in open economies are more flexible in arranging their input mix from domestic and foreign sources and therefore perform in a more cost-effective manner. Moreover, open economies are generally well equipped with transport, communication, and industrial infrastructure, which have spin-off effects on the recycling industry. In a few cases, however, increased foreign competition creates strenuous market conditions for the recoverer of waste, possibly leading to bankruptcies in the waste recovery sector.

Van Beukering and Bouman's study identifies the factors that influence recycling activities but does not necessarily explain the cause of international trade of recyclables. Several trade theories can explain the observed patterns and trends in trade and recycling (van Beukering et al. 2000). Rather than providing an all-encompassing explanation of international recycling markets, each trade theory enlarges a specific feature of international recycling. According to the Heckscher-Ohlin theorem, developed countries export the abundant factor "secondary materials," whereas developing countries use cheap labor to convert secondary materials into products. Trade theories incorporating environmental elements show that if developing countries have a comparative advantage in environment-intensive products and materials, they import relatively more low-grade commodities and waste materials (Rauscher 1994; Ulph 1999). According to demand-oriented theories, secondary commodities flow to developing countries because consumers in these countries demand lower-quality materials and products (Linder 1961). According to neo-technology theories, new (recycling) technologies originate and mature in developed countries, after which they gradually diffuse to developing countries (Posner 1961; Vernon 1966; Porter 1990). Recycling activities based on using advanced technology therefore lag behind in the South. Location theories suggest that because the recycling industry is strongly dependent on forward and backward linkages to other activities in the vicinity of the recycling industry, geographical concentration of industrial clusters is important for a prosperous recycling sector (Venables 1999; Krugman 1999). Only when the recycling industry is strategically located in relation to the supplying and demanding agents can economies of scale in this sector be realized.

10.4 Effects of Trade and Recycling: Case Studies

Globalization of the recycling market leads to an increased interdependence of economic activities. This interdependence affects every stage in the life cycle: extraction, production, consumption, and solid waste management. To analyze interdependencies, the concept "international material-product chain" provides a useful tool for analyzing developments in the international trade of materials and products in a comprehensive manner (see figure 10.4). Taking into account the variations in

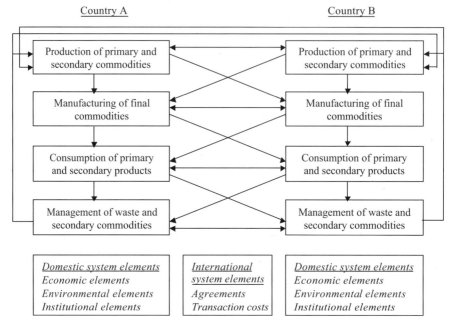

Figure 10.4
A conceptual model for the analysis of international material-product chain.

the factor requirements of the different segments in the I-M-P chain and the local availability of these factors, the I-M-P chain illustrates vertical and horizontal integration among the various segments of the chain at national and international levels. As a result, observed developments in international trade can be explained more accurately. Moreover, private (financial) and external (environmental) costs and benefits of increased trade and recycling can be linked to the I-M-P chain so that the allocation of welfare effects among countries, sectors, and industries can be determined.[5]

Using the concept of the international M-P chain, three case studies have been conducted in which economic and environmental effects of the trade in secondary materials were studied. These case studies, as presented hereafter, cover three different recyclable commodities (wastepaper, waste plastic, and used truck tires) in three different regions (India, Thailand, and the United States), which are analyzed with three different model types (static optimization, static multilocation optimization, and two-region simulation). The commodities have been selected on the basis of their varying degree of environmental impacts. The regions have been

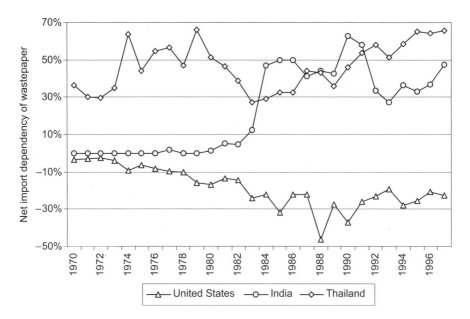

Figure 10.5
Import dependency of wastepaper (net import share of total domestic consumption of wastepaper).

chosen on the basis of policy measures taken that constrain the trade of the recyclable commodities. The variety of models is the result of the specific requirements and conditions of the case studies. The sequence of the case studies displays an increasing degree of comprehensiveness, developing from nonspatial to spatial, from static to dynamic, and from one- to two-region models. Field surveys have been conducted for all three case studies. On the basis of the qualitative information from the field surveys and the quantitative findings from the model studies, several conclusions on the effects of trade and recycling can be drawn.

10.4.1   Wastepaper Trade in India
The Indian paper recycling industry is one of the most import-dependent paper industries in the world. Figure 10.5 shows the development of net traded wastepaper as a share of the total wastepaper consumed, which rapidly increased in the mid-1980s. Only Thailand is generally more dependent than India on wastepaper imports. The United States is a net exporter of wastepaper on the world market. The figure also shows the sudden reduction of net import dependency of wastepaper in the early

1990s, after the Indian government implemented a 100 percent import duty on wastepaper. Once this duty was reduced significantly in 1993, imported wastepaper recovered.

Using a static optimization model, a case study was carried out to investigate the role of the wastepaper trade in the paper cycle in India (van Beukering and Duraiappah 1998). The study indicates that wastepaper imports perform as a cheap and environmentally extensive commodity for the Indian paper sector. If imports of wastepaper are constrained, alternative domestic inputs to produce paper are required. Imported wastepaper does not necessarily substitute for domestic waste paper but replaces fibers from agro-residues such as bagasse and straw, and to a lesser extent, wood pulp. Environmental pressure increases with trade constraints on wastepaper because agro-residues and wood pulp are more-polluting inputs. Yet the competition among domestic wastepaper recoverers is hardly reduced by the exclusion of wastepaper imports. Because of a quality difference—imported wastepaper has a longer fiber than domestic wastepaper—there is limited competition between imported and domestic wastepaper. The local complaint that foreign dumping of wastepaper has a crowding-out effect seems therefore unfounded. In fact, by combining domestic and imported wastepaper, Indian papermakers can produce products of a higher quality than if they depended solely on local wastepaper. Thereby, Indian papermakers are more capable of better meeting the demand requirements for paper. Rather than imposing trade barriers, the Indian government should therefore focus on reducing environmental externalities in the domestic pulp and paper industry.

10.4.2   Waste Plastic Trade in China

China is a major importer of waste plastics (see table 10.2). This foreign dependency was clearly emphasized when, in response to various incidences of illegal importing of contaminated and nonrecyclable waste plastics from Europe and the United States, the Chinese government banned imports of waste plastics in the late 1990s. As a result of the sudden lack of inputs, Chinese plastics recyclers struggled to survive.

Based on a static multilocation optimization model, a case study was conducted to examine the environmental and economic role of the trade and recycling of waste plastics in China (van Beukering 1999). The results show that installing barriers to international trade is not beneficial for the economy and the environment in China. The main bottleneck in the Chi-

Table 10.2
Main traders of waste plastics in 1994 (based on trade value)

Imports of waste plastics			Exports of waste plastics		
Rank	Country	Market share (%)	Rank	Country	Market share (%)
1	Hong Kong	54.9	1	United States	44.8
2	United States	17.4	2	Japan	23.2
3	China	13.2	3	Germany	7.2
4	European Union	4.8	4	Saudi Arabia	5.1
5	Canada	4.2	5	Canada	4.5
6	India	1.4	6	Mexico	3.8
7	Korea Republic	0.6	7	China	3.8
8	Malaysia	0.6	8	Netherlands	2.4
9	Singapore	0.6	9	China Taiwan	1.7
10	Mexico	0.5	10	Korea Republic	1.0

Source: UNCTAD (1996).

nese plastics cycle is the lack of production capacity in the primary industry and lack of input in the secondary industry. This problem is only
magnified by reduction of imports. Solutions have to be found domesti-
cally. The model demonstrates that at present, the manufacturing sector
is using a higher proportion of primary input (i.e., resin) in its final goods
than what is economically and technically desirable. The costs of meeting
the Chinese demand for plastics are unnecessarily high. This situation is
caused by the limited supply of secondary resin, which in turn is mainly
due to insufficient availability of recyclable waste plastics. Therefore, the
supply of domestic and imported waste plastics should be encouraged.
Using domestic waste plastics in the recycling industry avoids long-
distance transport activities and reduces the significant waste burden
created by postconsumer plastics in China. Increasing plastics recycling
requires that additional attention be given to the domestic recovery sector
of waste plastics. However, because of the limited quality of local waste
plastics, it is unlikely that the recovery sector will be able to meet the total
demand for waste plastics. Therefore, imports of waste plastics should be
allowed to fill the gap between the demand and supply of waste plastics in
China. To anticipate the risk of an increase of trade in nonrecyclable
plastics, the monitoring of quality standards for imported waste plastics
should be improved. This policy approach requires international collaboration. In general, exporting countries are better equipped to avoid undesirable exports of waste plastics than importing countries are to control
the problem through import monitoring.

Table 10.3
World trade in tires (in tonnes)

Commodity	1991	1997
New pneumatic tires	1,894,089	4,635,631
Old pneumatic tires	297,966	907,233
Share of old tires in world trade	13.6%	16.4%

Source: UNCTAD (2000).
Note: Old tires are defined as the sum of retreaded and used tires.

Table 10.4
Direction of world trade in old tires (in tonnes)

Direction	1991	1997
From North to North	72.0%	58.7%
From North to South	13.2%	20.2%
From South to North	11.8%	13.9%
From South to South	3.0%	7.1%

Source: UNCTAD (2000).

10.4.3   Trade of Used Truck Tires in Europe

International markets are important throughout the life cycle of tires. Natural rubber can be produced only in tropical areas, whereas high-quality tires are manufactured only in a few industrialized countries. The waste management stage is increasingly subject to international trade. Yet information on the commodity and regional patterns of international trade in used tires and tire-derived rubber waste is scant (Hoffmann 1999). Table 10.3 demonstrates how, in both absolute and relative terms, trade in old tires has become more important in the 1990s.

Table 10.4 reports the movement of international trade between developed and developing countries. Although the South expanded imports of old tires from 16 percent in 1991 to 27 percent in 1997, the North-North trade flow still dominates global trade, accounting for 72 percent of the imports. The share of Eastern European countries in European imports of old tyres expanded from 19 percent in 1992 to 47 percent in 1997.

A large number of used truck tires are exported from Western to Eastern Europe. Major reasons for this trade flow are the high disposal fees in Western Europe and excess demand for used tires for retreading in Eastern Europe. This trade is claimed to cause undesirable environmental damage in the longer term. European policymakers aim to solve this problem through the harmonization of environmental legislation and through trade measures. To examine these two approaches, a case study on used truck tires has been conducted (van Beukering and Janssen 2001). It examines the economic and environmental effects of the proposed measures. A two-region simulation model was developed that integrates the complete life cycle, incorporates environmental impacts, and accounts for learning-by-doing effects. The two regions represent Western and Eastern Europe.

The results show that the improvement of domestic environmental legislation, especially in Eastern Europe, is more effective than introducing a trade ban on used tires. The environmental effects caused by the trade of used truck tires from Western to Eastern Europe are small compared with the overall environmental damage caused by truck tires. Through implementation of domestic measures, such as the enforcement of strict laws on minimum tread depth and harmonization of disposal fees, the causes of undesirable trade practices are addressed rather than the symptoms. As a result, the trade in used tires for retreading, desirable because retreaders in Eastern Europe lack tire castings of good quality, whereas the demand for retreaded tires in Eastern Europe is significant, is left unharmed.

10.5   General Conclusions

The high labor intensity and the low technological capital requirements of the recycling industry cause it to operate particularly well in developing countries. An additional advantage of recycling activities in low-income countries is the existence of a high local demand for cheap recycled products, which reduces transport costs and externalities. Moreover, large quantities of recyclable waste allow for substantial economies of scale in the recovery sector. As a result, developed countries, whose citizens have relatively high incomes and are generally more environmentally conscientious and therefore participate actively in waste recovery programs, are efficient recyclers of secondary materials. In other words, many developing countries have a comparative advantage in utilizing particular nonhazardous secondary materials, whereas developed countries are specifically talented in recovering these secondary materials. As is the case with any commodity, the international trade in recyclable materials allows countries to exercise their comparative advantages so as to realize a more efficient international allocation of recyclable waste materials. This means that international trade in recyclable materials generates welfare gains for all trading parties, in the absence of market failures.

The extent to which the international trade in recyclables can play a role in achieving welfare gains is determined mainly by whether imported and domestic secondary materials are substitutable or complementary inputs. It depends, among other things, on whether the imports of secondary materials crowd out the local recovery sector and support the domestic recycling industry. If there are economies of scale in the recycling industry of a particular material, both domestic and imported materials may be required to maintain continuity in secondary production. Imported secondary materials can complement local materials in cases in which the quality of the domestic materials is a problem for the recycling industry.

The expansion of the market for secondary materials from a local to a global scale increases the flexibility of recyclers and thereby leads to greater efficiency in the secondary industry. Similarly, globalization has encouraged the recovery of more materials than would be feasible for the local market, because the recovery sector of secondary materials has a substantially larger market of potential customers. The recovery sector in developed countries has been found to exert only a minimal crowding-out effect on the recovery sector in developing countries. For producers that utilize secondary materials, the expansion of the market for recyclable inputs implies a more stable supply and a wider choice in terms of the quality of inputs.

If the international trade in secondary materials leads to higher levels of recycling, the environmental benefits are likely to outweigh the environmental costs, because recycling generally alleviates the pressures of society on the environment. Natural resources can be saved, emissions can be reduced, and the burden of solid waste can be reduced. An indirect environmental effect found is that importing recyclers operate on a larger scale than recyclers who depend solely on local materials. Small recyclers are unable to take measures to avoid environmental damage. Larger recyclers, on the other hand, have sufficient scale to implement abatement technologies. Moreover, by being part of international networks, trading recyclers also have better access to newer and cleaner production technologies.

10.6   Policy Recommendations

National governments and international organizations are in a position to develop and implement policies to overcome the market failures observed in international and national recycling markets. Here, some policy recommendations aimed at international and national recycling markets are presented.

Several general principles derived from the case studies presented in this chapter may be helpful in improving international recycling markets. To avoid the importing of nonrecyclable and hazardous materials, monitoring of the international trade in secondary materials needs to be improved. In this context, the Basel Convention on the Control of Transboundary Movements of Hazardous Waste plays a crucial role, as it introduced the concept of "prior informed consent." This is a control system that requires an exporting country to ask an importing country prior for written authorization to ship secondary materials. Despite the existence of the Basel Convention, incidents of illegal exports of hazardous and nonrecyclable wastes, under the disguise of recyclable materials, still occur, although on a much smaller scale than before. Besides causing significant environmental damage, these incidents severely damage the image of the international trade in recyclable materials. To prevent such illegal practices, unambiguous definitions of "hazardous waste" and "secondary materials" are required. The costs of monitoring and enforcement should predominantly be covered by the importers and exporters of secondary materials.

The creation and expansion of trade networks and markets for materials supports the international exchange of secondary materials. In the last decade, it has become much easier for buyers and sellers to meet, mainly as a result of improved information technologies and the establishment of national and international networks. Formal trading systems alleviate uncertainties in the international supply and prices of secondary materials. For example, the CBOT opened an international exchange for recyclables in the late 1990s. International cooperation is required in the development of these exchanges to agree on common standards, such as those concerning different qualities of secondary materials. As recognized in the Basel Convention, developed countries are responsible for financing the promulgation of such standards.

Restraining measures, such as uniform trade bans, are generally less cost-effective than supportive measures, such as technological assistance. The empirical cases presented in the chapter have shown that open economies have higher recycling levels than closed economies. One possible explanation is that because of the access to foreign markets for materials, recyclers in open economies also have better access to novel recycling technologies. Another disadvantage of constraining trade measures, in addition to limiting access to new recycling technologies, is that a ban on exports may discourage developed countries from assisting developing countries in the technological transfer of solid waste management techniques. From this perspective, it is desirable to take "positive" measures, for example, by mobilizing funds for technical cooperation between developed and developing countries in the recycling sector. This implies that the main advantage of the Basel Convention is not so much its potential to reduce international flows of materials as the effect it has on the capacity of the South to manage waste effectively.

As emphasized by theories on environmental policies, approaches to environmental externalities that are custom-made for specific countries or regions, based on domestic economic and environmental conditions, resource endowments, and social preferences, can be more effective than uniform multilateral approaches. In the context of recycling, this means that countries should avoid setting uniform recycling targets at an international level. For example, in the European Union, uniform recycling targets have been defined for all member countries. Such uniformity in targets ignores the economic, geographic, and cultural differences among European countries and therefore cannot be optimal. Similar conclusions hold for policies that do not distinguish among materials, because each material has different environmental and economic effects. Therefore, recycling and trade policies ideally should be country- and material-specific, unless the administrative costs exceed the economic benefits of such specificity.

The rationale behind international recycling is possibly also applicable to other types of international waste management, such as landfilling and the incineration of solid waste. Therefore, international organizations should investigate the option of liberalizing the international movement of waste materials between countries that are equipped to manage these materials for purposes other than recycling. If waste is allowed to be managed in countries other than where it originates, waste-processing companies will rationalize their operations. This approach requires transparent international guidelines, minimum standards, adequate monitoring, and "correct prices."

Several policy recommendations have been derived from the case studies discussed in this chapter that are aimed specifically at national recycling markets. If recycling has significant environmental and economic benefits, then the expansion of the capacity of the recycling industry has to be further promoted. Most countries acknowledge the truth of this assertion. The question remains of how such promotion should be accomplished.

Economic measures, such as tax exemptions, could be introduced. This might give a fledgling recycling industry in a particular country a competitive advantage over the established primary commodity sector. It is doubtful, however, that such measures would be effective in developing countries. Most recyclers currently do not pay taxes, so a tax reduction or exemption would have only a minimal effect on stimulating recycling. Subsidies might be helpful in the short term to support a country's recycling industry during its infancy but could cause inefficiencies in operations in the long term. Moreover, most governments in developing countries cannot afford such instruments. Therefore, the prime role of the government is to remove economic and technological obstacles for the recycling industry, rather than to directly intervene in the sector. Governments should avoid participating directly in recycling activities.

At present, the recycling sector suffers from a number of technological impediments. Governments could support research and development to overcome these problem areas in the recycling industry. First, the currently limited market of secondary material applications could be expanded. For example, products could be designed to contain a greater amount of recycled material. In addition to R&D support, governmental procurement policies that give priority to the purchase of products with a high recycled content could promote the adoption of these products. Second, improved product design might increase the recyclability of consumer products. This could be achieved, for example, by reducing the number of different materials in one product. Technology transfer from developed to developing countries could also be an effective strategy to overcome the limited market for secondary materials.

Despite the fact that recycling is often cleaner than primary production and waste disposal, production activities in the recycling sector itself may also generate substantial environmental problems. Governments could play an important role in the promotion of cleaner and more efficient recycling technologies. As a result of the high capital costs and the relatively less strict enforcement of environmental regulations in developing countries, the best available technologies are not yet widely implemented in those countries. Therefore, collaborative projects are required to promote technology transfer from developed to developing countries.

Notes

1. Food and Agriculture Organization of the United Nations (1999); Metallgesellschaft Aktiengesellschaft (annual). The trade data presented here are derived from the United Nations Statistical Office COMTRADE (commodity trade) databases at the International Computing Center in Geneva.

2. Data on trade and recycling in both developed and developing countries are not always reliable. Often, trade in recyclables is recorded under the label of new materials. Moreover, data sometimes hide the fact that a certain share of secondary "raw materials" are not recycled, but dumped.

3. Hereafter, instead of "metric ton(s)," the shorter term "tonne(s)" will be used.

4. As developed countries are both the most reliable reporters of trade statistics and the dominant exporters, it is plausible to choose export rather than import values as the basis of analysis. Import values tend to be slightly lower than export values and show more unreliability for secondary materials.

5. Besides taking into account the various processes within the industry and the waste management sector, transport between and within countries has also been accounted for in the I-M-P chain.

References

Bartone, C. (1990). Economic and policy issues in resource recovery from municipal wastes. *Resources Conservation and Recycling* 4: 7–23.

Berger, N. (1998). North-South Trade in Recyclable Waste: Economic Consequences of Basel. Centre for International Economic Studies Seminar Paper no. 98-03, University of Adelaide.

Buggeln, R. (1998). *Industrial Waste Exchanges: An Overview of Their Role.* Knoxville: Tennessee Materials.

Byström, S., and L. Lönnstedt. (1995). Waste paper usage and fibre flow in Western Europe. *Resources, Conservation and Recycling* 15: 111–121.

CBOT (Chicago Board of Trade). (1999). *Recyclables Exchange.* Available at http://www.cbot-recycle.com.

Elmer, J. W. (1996). The Basel Convention: Effect on the Asian secondary lead industry. *Journal of Power Sources* 59: 1–7.

FAO (Food and Agriculture Organization of the United Nations). (1999). *FAO Database Gateway.* Rome: FAO. Available at http://apps.fao.org/default.htm.

Frechette, D. L. (1999). Scarcity rents and the returns to mining. *Resources Policy* 25: 39–49.

Grace, R., R. K. Turner, and I. Walter. (1978). Secondary materials and international trade. *Journal of Environmental Economics and Management* 5: 172–186.

Greenpeace. (1997). *Plastic waste trade from Atlanta to Asia.* Available at http://www.archive.greenpeace.org/majordomo/index-press-releases/1996/msg00224.html.

Hoffmann, U. (1999). *Requirements for Environmentally Sound and Economically Viable Management of Lead in the Wake of Trade Restrictions on Secondary Lead by Decision III/1 of the Basel Convention: The Case of Used Lead Batteries in the Philippines.* Geneva: United Conference on Trade and Development.

Johnstone, N. (1998). The implications of the Basel Convention for developing countries: The case of trade in non-ferrous metal bearing waste. *Resource Conservation and Recycling* 23: 1–28.

Kandelaars, P. P. A. A. H. (1999). *Economic Models of Material-Product Chains for Environmental Policy Analysis.* Dordrecht, the Netherlands: Kluwer Academic.

Kellow, A. (1999). Baptists and bootleggers? The Basel Convention and metals recycling trade. *Agenda* 6(1): 29–38.

Krugman, P. (1999). What's new about the new economic geography? *Oxford Review of Economic Policy* 14(2): 7–17.

Linder, S. B. (1961). *An Essay on Trade and Transformation*. New York: Wiley.

Linnemann, H., and H. L. M. Kox. (1995). International Commodity-Related Environmental Agreements as an Instrument for Sustainable Development. Summary report, Economics Department, Vrije Universiteit, Amsterdam.

McKee, D. L. (1996). Some reflections on the international waste trade and emerging nations. *International Journal of Social Economics* 23(4–6): 235–244.

Metallgesellschaft Aktiengesellschaft. (annual). *Metallstatistik*. Frankfurt am Main: Metallgesellschaft Aktiengesellschaft.

Michael, J. A. (1998). Recycling, international trade, and the distribution of pollution: The effect of increased U.S. paper recycling on U.S. import demand for Canadian paper. *Journal of Agricultural and Applied Economics* 30(1): 217–223.

Moison, D. L. (1997). The myth and reality of the relationship between scrap metal supplies and prices. In *Book of Proceedings: The Recycling of Metals, ASM International Europe, Brussels (Third ASM International Conference, 11–13 June, Barcelona, Spain)*, 415–428.

Navaretti, G. B., I. Soloaga, and W. Takacs. (1998). When Vintage Technology Makes Sense: Matching Imports to Skills. Policy research paper no. 1923, World Bank, Washington, DC.

Opschoor, J. B. (1994). Chain management in environmental policy: Analytical and evaluative concepts. In J. B. Opschoor and R. K. Turner (eds.), *Economic Incentives and Environmental Policies*, 197–229. Dordrecht, the Netherlands: Kluwer Academic.

Pei, F., and J. E. Tilton. (1999). Consumer preferences, technological change, and the short-run income elasticity of metal demand. *Resource Policy* 25: 87–109.

Porter, M. E. (1990). *The competitive advantage of nations*. New York: Free Press.

Posner, M. V. (1961). International trade and technical change. *Oxford Economic Papers* 13(3): 323–341.

Radetzki, M., and C. van Duyne. (1985). The demand for scrap and primary metal ores after a decline in secular growth. *Canadian Journal of Economics* 18(2): 435–449.

Rauscher, M. (1994). On ecological dumping. *Oxford Economic Papers* 46: 822–840.

Savage, G. M., and L. F. Diaz. (1996). Future trends in solid waste management. In *The ISWA Yearbook 1995/96*, 22–28. London: James & James Science.

Thomas, M. P., and A. H. Wirtz. (1994). The ecological demand and practice for recycling aluminium. *Resources, Conservation and Recycling* 10: 193–204.

Tietenberg, T. (1996). *Environmental and Natural Resource Economics*. New York: Harper Collins.

Tilton, J. E. (1999). The future of recycling. *Resource Policy* 25: 197–204.

Ulph, A. M. (1999). Strategic environmental policy and foreign trade. In J. C. J. M. van den Bergh (ed.), *Handbook of Environmental and Resource Economics*, 433–448. Northampton, England: Elgar.

UNCTAD (United Nations Conference on Trade and Development). (1996). Trains database. UNCTAD Commodities Division, Geneva.

UNCTAD (United Nations Conference on Trade and Development). (2000). UN Commodity Trade Statistics database (UN COMTRADE). UNCTAD Statistics Division, Geneva.

van Beukering, P. J. H. (ed.). (1999). *Plastics Recycling in China: An International Life Cycle Approach*. CREED (Collaborative Research in the Economics of Environment and Development) report, prepared by the Institute for Environmental Studies (IVM) and the Chinese Academy of International Trade and Economic Cooperation (CAITEC). London: International Institute for Environment and Development.

van Beukering, P. J. H., and M. A. Bouman. (2001). Empirical evidence on recycling and trade of paper and lead in developed and developing countries. *World Development* 29(10): 1717–1739.

van Beukering, P. J. H., and A. Duraiappah. (1998). The economic and environmental impact of wastepaper trade and recycling in India: A material balance approach. *Journal of Industrial Ecology* 2(2): 23–42.

van Beukering, P. J. H., and M. A. Janssen. (2001). Trade and recycling of used truck tyres in Western and Eastern Europe. *Resources, Conservation and Recycling* 33: 235–265.

van Beukering, P. J. H., J. C. J. M. van den Bergh, M. A. Janssen, and H. Verbruggen. (2000). International Material Product Chain: An Alternative Perspective on International Trade and Trade Theories. Discussion paper no. TI2000-034/3, Tinbergen Institute, Amsterdam.

van der Klundert, A. (1997). *Policy aspects of urban waste management*. Issue paper presented at Programme Policy Meeting, Urban Waste Expertise Programme, Gouda, the Netherlands, April 1–8.

Venables, A. J. (1999). The international division of industries: Clustering and comparative advantage in a multi-industry model. *Scandinavian Journal of Economics* 101(4): 495–513.

Vernon, R. (1966). International investment and trade in the Product Cycle. *Quarterly Journal of Economics* 80: 190–207.

Weaver, P. M., H. L. Gabel, J. M. Bloemhof-Ruwaard, and L. N. van Wassenhove. (1995). Optimising environmental product life cycles: A case study of the European pulp and paper sector. Working paper no. 95/29/EPS/TM, INSEAD, Fontainebleau.

V Dynamics of Eco-Industrial Parks

11 Understanding the Evolution of Industrial Symbiotic Networks: The Case of Kalundborg

Noel Brings Jacobsen and Stefan Anderberg

(Denmark)

R32

Q53 Q58

11.1 Introduction

During recent decades, industrial environmental strategies have shifted from traditional "end-of-pipe" solutions to pollution prevention strategies, with the focus on cleaner production technology. It has been shown, however, that there are difficulties in reducing the environmental impact of industrial activities if one considers an industrial company in isolation from other companies (Nemerow 1997; Ayres 1999). No matter the level of improvement of technology, every production process generates, in addition to the useful product, some waste. If total material efficiency is to increase and overall resource use is to decrease, these wastes must find productive use. This is why networks based on exchange of by-products and resource sharing has attracted increasing attention during the last decade in many countries all over the world. This type of network activity —"industrial symbiosis"—seems often to have both economic and environmental benefits, and it may include mechanisms for making industry more environmentally sustainable. Therefore, it has been stressed as a model for local sustainable development (Erkman 1997; Dunn 1999). Here, we sometimes use the term "industrial symbiotic network" to stress that industrial symbiosis is one type of industrial network and that it resembles other types of collaboration among industrial companies.

Industrial symbiosis typically emerges spontaneously through the establishment of bilateral exchange relations among industrial manufacturers in an industrial district. This process may go very slowly, but as more actors get involved, an integrated network is established. On the one hand, this network generates concrete economic and environmental advantages; on the other, it develops a local collective competence—"a cooperation culture" (Schwartz 1995)—that may strengthen regional industrial development and ability to compete. The eco-industrial park in

Kalundborg, Denmark, is one of the most internationally well-known examples of a local network for exchanging waste products among industrial producers. Its development has been studied particularly by Nicholas Gertler and John Ehrenfeld (e.g., Ehrenfeld and Gertler 1997; Erkman 1997; Jacobsen 2002; and Chertow 2000). The Kalundborg symbiosis is an example of a spontaneous network that has evolved over a period of several decades. In many respects, it is unique, but still, insights derived in this case should be able to contribute to the development of eco-industrial parks elsewhere.

This chapter presents some preliminary observations and questions from ongoing studies of industrial symbiosis in Denmark. The aims of this research are not only to improve the understanding of the complex local development dynamics behind industrial symbiotic networks, but also to evaluate the potential of such networks, understood as a strategic model for local sustainable development. This second aim is partly motivated by the increased interest in implementing the idea of industrial symbiosis that can be observed in Denmark. Understanding how successful industrial collaboration evolves is of central importance, both for the realization of industrial symbiotic projects and for the strengthening of the regional economy and identity, which also are important aspects in connection with sustainable development. In this chapter, we discuss what can be learned from the well-known example of the symbiosis of Kalundborg that is of general value for the realization of local industrial symbiotic networks. For comparison and contrast, the ongoing efforts to develop an eco-industrial park in Avedøre Holme, a large industrial district around the major power plant in the Copenhagen area, is introduced.

We have studied the Kalundborg symbiosis from various angles. In addition to conducting research, based on interviews and written materials, Jacobsen has been employed at the Symbiosis Institute[1] in Kalundborg. For the analysis of Kalundborg, we introduce an analytical framework inspired by research on networks in connection with regional economic and technological development that includes physical preconditions and possibilities, local economic and environmental effects of exchange activities, and central mechanisms, including technological, institutional, organizational, economic, and mental elements, behind the development.

The focus of this chapter is limited to a discussion of some of the preconditions for the successful realization of local exchange opportunities for improving existing production systems. Questions about the desirability of local symbiotic networks for long-term sustainable development—for example, whether, and to what extent, local recycling and exchange activities may inhibit technological improvements and pollution prevention, and the problems with the (perhaps exaggerated) focus on the Kalundborg example—are also of interest to us but will be dealt with elsewhere.

11.2   An Analytical Framework for Industrial Symbiotic Networks

The study of regional and industrial environmental development has so far had very limited communication with research on regional and industrial development. One important reason for this is, perhaps, that regional science and economic geography have shown little interest in the natural environment, which has not been considered a central factor shaping industrial production. However, with the radically increased emphasis on the environmental performance of industrial companies during the 1990s, there are some signs that this is about to change.

As some authors (e.g., O'Rourke, Connelly, and Koshland 1996; Boons, Füssel, and Georg 2000) have observed, industrial ecologist visionaries often have a rather limited view of and interest in the social context of the industry. A factory, a company, and even more, a large corporation, are complex social phenomena. They are influenced by their history and particular culture, their location, their employees, their customers and competitors and the markets involving them, and their owners. To be able to understand the developments in an industrial company, and even to influence strategies and actions, it is necessary to be well-acquainted with its context. For example, even if some technological or organizational opportunities that present themselves to a company seem economically profitable and have short payoff times, this is not sufficient for the company's realization of these opportunities. There are often conflicts with priorities of more direct concern for the company or other barriers involved that might be of decisive importance. To improve the analysis of industrial environmental developments, and their future opportunities and constraints, it would be important to increase the understanding of and interest in industrial firms and regional economic development in industrial ecological research. Therefore, we would definitely welcome both an increased interest in environment and environmental activities among regional and industrial researchers and more inspiration from, for example, economic geography in industrial ecological research. In this chapter, we derive inspiration from the analysis of industrial networks in connection with regional development, in our search for understanding of the dynamics of the development of local industrial symbiotic networks.

From an industrial development perspective, industrial symbiosis is one type of industrial network. Industrial networking activities of various types have developed in many regions and are often stressed in economic geography as a central factor in connection with economically successful regions (Storper 1992a; Asheim 1996; Maskell et al. 1998), such as networks in metals, the textile and clothing industry in the "Third Italy" (Pyke and Becattini 1992), or the high-tech networks of Silicon Valley. Such networking activities have created an innovative and competitive production milieu in which communication and cooperation are central elements. During the past fifteen to twenty years, these industrial networks have been analyzed from varying perspectives, but central to the analysis has been an interest in how the networks have been formed, developed, and run, with the aim of understanding their dynamics and competitiveness (Storper 1992b). Many researchers have shown that learning and innovation that can be achieved through networking may form the basis for strong regional development (e.g., Maskell et al. 1998). There has also been interest in looking into problems of policy-driven interfirm collaborative projects (Huggins 2000). Edward Cohen-Rosenthal and Thomas McGaillard (1998) also stress the importance of inspiration from studies on business networks and regional development for developing policies to stimulate eco-industrial collaboration. Industrial symbiotic networks, on the one hand, show many of the same traits as the production-related networks that are the focus of regional development research, but on the other hand, they also add another dimension: namely, environmental performance. Even though environmental performance is a rather recent aspect in connection with industrial management, it is a frequent misunderstanding that striving for resource efficiency, the closing of loops, recycling and reuse of waste, development of new products from waste materials, and exchange of resources among firms are something genuinely new and revolutionizing. Such efforts have been quite a normal part of the development of many resource-intensive industries in many parts of the world for some time (e.g., Desrochers 2002).

It should be possible to use theoretical frameworks connected to the analysis of traditional production-related networks for analyzing the dynamics of industrial symbiotic networks. However, since resource efficiency is central to this type of networking, it is essential to complement these frameworks with aspects such as material flows, environmental performance of technology, the linkages between environment and economy, and sustainable use of resources. For the purpose of looking into the potential and particular problems of industrial symbiosis development, we have used a broad analytical scheme focused on three different aspects:

1. An analysis of the *physical preconditions and possibilities* for the development of industrial symbiotic networks, particularly the identification of material and resource flows in the studied case. All industrial symbiotic networks are built on different kinds of energy and material flows. An analysis of the industrial resource flows is therefore necessary for explaining network developments at a particular locality. Complementarity of different flows is the necessary precondition for the existence of a network. This analysis should not, however, be confined to the flows in the actual (realized) network but should also include the potential network: all such flows that occur in the locality.

2. An assessment, on the basis of this analysis, of *the economic and environmental effects* of the realization of different potential exchanges, both for the involved companies and for the local region. The benefits created by the networking activities for the involved companies and for the local region consist of direct and indirect profits. The analysis of direct profits establishes the concrete economic and environmental efficiency gains in terms of economic savings and reduced emissions and resource consumption of realized exchange activities. The analysis of indirect profits is related to the potential of the networking activities to increase the attraction value of the local region for industrial companies. Networking activities increase the mutual dependence among industrial companies. This seems to keep the companies in the locality. But it is also possible that the networking may attract new companies that will be able to profit from the local network and cooperation culture.

3. An analysis of the *central conditions and mechanisms behind the development* of symbiotic networks, for example, the identification of barriers and inertia that influence the development of these activities. The fact that a mass flow analysis often shows a much larger potential symbiotic network than that which actually exists indicates that there are factors other than physical and economic ones that influence the creation of networking activities. These factors relate to technological, institutional, organizational, economic, and mental conditions that in complex combinations form the dynamic behind the industrial symbiotic network activities.

The following presentation of the Kalundborg symbiosis case briefly summarizes the observations of interest in relation to our questions. In this chapter we do not follow in detail the analytical framework just described but focus instead on some of the factors of *special* importance for the development of symbiotic relationships and reflect upon these across the analytical categories previously defined.

11.3   The Kalundborg Symbiosis

The Kalundborg symbiosis is one of the world's best examples of industrial symbiotic networks (see chapter 12 for a discussion of the diffusion of the Kalundborg hype). Since the Kalundborg network has been treated extensively in the international literature (e.g., Garner and Keoleian 1995; Ehrenfeld and Gertler 1997; Erkman 1997; Lowe 1997; Jacobsen 2002; Chertow 2000), we will give only a short update here and will focus particularly on the analysis of the factors behind the development of the network.

Six industrial plants are included in the Kalundborg network based on the reutilization of rest products: are

- Novo Nordisk/Novozymes A/S (production of insulin and enzymes)
- Asnæsværket A/S (production of electricity and heat)
- Statoil raffinaderiet A/S (production of petroleum products)
- Gyproc Nordic East A/S (production of gypsum wallboard)
- Soilrem A/S (microbiological cleaning of polluted soils)
- Asnæs fiskeindustri A/S (culture of trout for consumption)

Currently, among these industries, there are nineteen different exchange activities, based on water exchange (seven projects), energy exchange (six projects), and exchange of solid waste (six projects). An overview of the network is presented in figure 11.1. Besides these "hard" projects, there are also a number of "soft" projects in which the cooperation focuses on sharing storage and laboratory capacity, common contracts with external entrepreneurs, personnel recruiting, jobs for spouses, and similar activities. Characteristic of the development of the Kalundborg network has been its spontaneous evolution combined with its commercial aims. The Asnæs power plant and refinery were started around 1960, but it was not until the 1970s that the first exchange linkages were established. Gas was piped from the refinery to the Gyproc plant from its start in 1972; in 1976 Novo Nordisk began shipping sludge to farmers; in 1979, the Asnæs

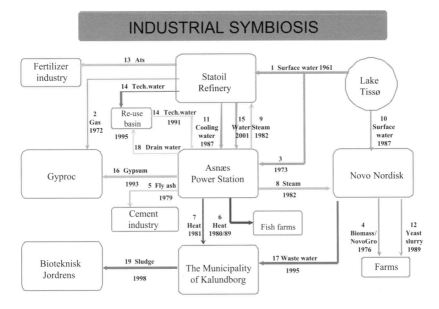

Figure 11.1
The industrial symbiosis in Kalundborg. The years in the figure next to the exchange activities indicate when each activity was initiated.
Source: The Symbiosis Institute in Kalundborg.

power plant began to sell fly ash to cement producers; and in 1981, it started supplying heat to the municipality. Another typical trait of this network is that it is still developing and the exchange relations are changing continuously as an effect of technological change and market relations. The number of exchange relations in the network has doubled since the early 1990s.

11.3.1 Economic and Environmental Savings

The total yearly economic profit of the Kalundborg symbiosis activities for the involved industries adds up to about US$15 million. The total investment has been US$75 million (construction of pipes and various exchange facilities), and the payoff time for the single projects has been approximately five years on average. This economic profit arises primarily as an effect of resource savings. For example, the recirculation of water among companies saves about two million cubic meters of groundwater and one million cubic meters of surface water every year. Via substitution of industrial gypsum from scrubbers for raw gypsum, some two hundred thousand tons of raw gypsum are saved, as well as approximately twenty thousand tons of oil through steam exchange. An example of water exchange can illustrate the efficiency of the industrial symbiotic network. Surface water from Lake Tissø (approximately 1.3 million m^3/year) is used as a coolant in the Statoil refinery. The water is thereafter piped to the Asnæs power plant, where it is used in the turbines and transformed into steam. Then, some of this steam is brought back to the Statoil refinery, where it is used for heating raw oil, and another part of the steam goes to Novo Nordisk, where it is used for the sterilization of raw materials and rest products. Through these interindustrial activities, the surface water from Lake Tissø is used in four processes in three different companies, which increases the material efficiency by a factor of three.

The effects of the symbiosis on the regional economic development have not been looked into systematically. The symbiosis seems to have increased the industrial stability in the region, through making the involved companies more dependent on one another and in a sense more bound to the region. The increased interdependency among the companies seems not to have caused a lock-in situation, since the symbiotic relationships have changed both in terms of quantity and quality over time. However, some companies (particularly the gypsum factory and fish culture) would have difficulties in finding equally good locations in terms of supply of raw materials or energy inputs, but these locational opportunities have probably not, viewed in isolation, had dramatic effects on new investments in the region; no new large industries have been established in the region during the last few decades. This does not mean that industrial symbiotic opportunities in themselves cannot be a locational factor, but it seems to be more fruitful to interpret industrial symbiotic opportunities in a more general frame of traditional locational factors such as local labor markets, infrastructure, centrality, and market access. Opportunities for symbiotic activities must in other words be viewed together with other locational factors, which often have more economic importance and therefore overrule the resource perspective on industrial location and investment attraction. It is, however, interesting to observe that some of the established symbiotic industries increased their investments in the Kalundborg facilities dramatically during the late 1990s. Whether these investments are related to the symbiotic relations and opportunities in the Kalundborg region or to other independent processes is an open question, but the symbiotic network has not been a negative factor in the decision processes.

It should also be mentioned that despite the international attention, the symbiosis and collaboration in the Kalundborg region remains surprisingly unknown in Denmark. Even among people one would expect to be familiar with the symbiosis—power plant engineers, technicians, industrial leaders, planners and researchers—only a minority has even heard about it. This is probably related to the fact that the Kalundborg region is a traditional peripheral area in Denmark, with an undiversified industry and slow-growing regional economy, but very stable both in terms of both population and employment. It is definitely very difficult to market the region as an investment area, which makes promoting regional economic growth and general development a difficult task. The importance of the large companies for industrial jobs, particularly Novo Nordisk and Novozymes, has therefore increased, but as for the country as a whole, an expanding service sector is keeping employment stable in the region. On top of that, the Kalundborg region still remains remarkably unaffected by Denmark's capital region, ninety kilometers away, forms a relatively independent labor market, and—in general—is quite isolated from the knowledge institutions in Copenhagen.

11.3.2 Technological Factors

Technological factors influencing industrial symbiosis consist of concrete technological requirements for realizing the exchange activities. From a technological and material flow point of view, there are some very advantageous traits in Kalundborg, which have also been pointed out by Ehrenfeld and Gertler (1997). In the center of the symbiosis are a couple of large companies located in proximity to one another that have very stable, continuous waste streams and needs for inputs. The companies have also been able to use each others' waste products without much pretreatment. This has made it possible to introduce relatively uncomplicated and inexpensive projects with short payoff times, which was particularly important for establishing the early exchange relationships, concerning flue gas, fly ash, and heat streams. Later exchanges such as sulfur, desulfurized gas, and treated wastewater have had closer linkages to pollution control and would not have been attractive without pollution regulation as part of the explanation. In Kalundborg the very basis for these different exchanges—the input-output match—has in most cases been very good from the very beginning. A wide range of feasibility studies and pilot tests have been undertaken before launching new developments. In connection with both the early primitive projects and later, more sophisticated developments, a number of technological factors have been investigated over and over again. These factors have concerned the quality and the quantity of the possible exchanges, temperature and pressure, flow stability and seasonal variation, time perspectives, implications of exchanges in waste flows for the larger process schemes, monitoring, maintenance, and prospects for regulative changes and technological developments. Supply security has often been emphasized as a particularly critical factor for industrial symbiosis projects. Nevertheless a closer investigation into the initial project phase in many of the exchange relations in Kalundborg indicates that demand security often is subject to much more attention. Temporary lack of demand for a certain by-product in an industrial symbiotic relationship has vital implications for the production process on the producer side. Often no "quick" alternative exists to the recipient of a given by-product. Lack of downstream demand can therefore effect upstream production dramatically, as voluminous by-product streams often have limited storage capacity and alternative recipients are difficult to establish from day to day. Closing down upstream production can be the consequence if demand suddenly fails. On the other hand insufficient supply of a given by-product can mostly be solved easily through alternative sources, backup capacity, or temporary change to traditional inputs from traditional suppliers. Thus, the existence of continuous and large-scale by-product streams from single sources and a secure supply and demand situation are from a technological point of view very critical factors for the realization of industrial symbiosis projects and have been subject to detailed investigations in Kalundborg.

11.3.3   Institutional Conditions

Institutional conditions relate primarily to the legislation and public authorities surrounding industrial activities. Of particular importance is environmental legislation, which often creates need for change, but local planning initiatives can also be quite essential in this context. The development of environmental legislation has definitely been the major driver for the evolution of the Kalundborg symbiosis. Most problems that have been solved through exchange activities have come up as an effect of implementation of national emission legislation or initiatives to save energy and water. The first exchange activities in the 1970s were linked to emission reduction as an effect of the implementation of the national environmental legislation introduced at the beginning of the 1970s. Almost all initiatives in the 1980s were energy-saving, as an effect of the second oil price shock and national energy savings programs, whereas the projects from the late 1980s onward have mostly been concerned with improving waste and water management, which also have been political priorities. Initiatives to use water more efficiently have a background in the relative water shortage in the region, which originates from the (for Danish standards) extreme concentration of large water-consuming processing industries in a region with limited water resources. The flexibility of Danish environmental legislation system, which is based not on fixed technological demands or emission standards, but on a negotiation process between authorities and the companies concerned, has certainly been important for facilitating local problem solving. The exchange activities are an expression of this. This is not the only example in Denmark in which environmental and workplace safety legislation has been able to stimulate technological innovation and development, and this may give support to Porter's (1990) thesis that the institutional setting may play such a role. The community of Kalundborg, which together with the regional government (Vestsjællands amt) is responsible for the implementation and control of the environmental legislation that governs the Kalundborg industries, has generally been well informed and supportive, though not directly involved in negotiations among the companies. An illustrative case from Kalundborg in which regulation has been an important driver for implementation of industrial symbiotic relations is the case of industrial gypsum. The Asnæs power plant was, at the beginning of the 1990s, forced by the regulative authorities to install a large desulfurization plant. Among different technical alternatives for desulfurization, a process producing industrial gypsum as a by-product was chosen. This technical choice was influenced by the fact that the industrial gypsum produced could be used by the colocated plasterboard factory Gyproc as raw material. The contract with Gyproc was signed before the installation of the desulfurization plant, and today more than 75 percent of the plasterboard produced by Gyproc in Kalundborg is based on industrial gypsum from the Asnæs power plant. This case shows how regulation often has been the primary reason for the Kalundborg companies to engage in a symbiotic relationship, but it also indicates that the real implementation of a business-to-business by-product exchange requires more than a regulative framework, because a symbiotic response to a regulation is very often only one possible response among other, nonsymbiotic solutions. In the case of desulfurization and industrial gypsum, several other—and more traditional—possibilities were under consideration in the process of selecting the right solution to the desulphurization problem and choosing the right process and the right reuse of the by-product produced. To select industrial symbiosis in response to a regulation seems to require something more than just a profitable technological match, something that has to do with preestablished social relationships, local communication ties, and perhaps even mutual trust in the problem-solving process.

11.3.4 Organizational Conditions

Organizational conditions, in regard to industrial symbiosis, concern the forms of cooperation involved in the organization of the network but may also involve the internal organization of the companies and relationships between industry and authorities. The public authorities have not been directly involved in setting up the different exchange activities in Kalundborg but have only agreed to them. The ideas have been introduced within the companies, and the setting up of different linkages has involved only the concerned companies. For a long time, the symbiosis consisted only of a number of bilateral exchange activities based on contracts between involved companies, but gradually in the 1980s and 1990s, a more common and integrated platform with the Environmental Club and the Symbiosis Institute evolved. The Environmental Club was a response to the increased concern for the environment in the 1980s and may be viewed as a forerunner to green networks and other collaborative efforts among local authorities, industries and sometimes popular organizations that became widespread in the 1990s. The establishment of the Symbiosis Institute was partly a response to the increasing numbers of visitors who came to Kalundborg as a result of the international attention the area's cooperative exchanges had attracted. The Environmental Club and the Symbiosis Institute, which now is partly responsible for developing the symbiosis, introduced new dimensions to the cooperation, such as the soft cooperation projects, that are more multilateral, as well as linkages between the industrial cooperation and regional sustainable development efforts, involving both successful and competitive industry and a responsible and efficient use of natural resources. It seems obvious that the trust and social connections built up by the earlier experience of mutual dependence among the companies has been quite important for this development. It has also made it possible to handle more complicated exchanges with more diffuse common payoffs. All involved in the Environmental Club, including environmentalists, stress the unusually good relations and common understanding among industrial companies, the local authorities, and the environmental organizations in the region (Uldall 2001). The gradual development of the symbiotic network in

Kalundborg and the partnership between industry and public authorities can be viewed as a process of institutionalization of the industrial symbiosis concept. The process previously described, from individual standing bilateral by-product exchanges to a broader symbiotic network coordinated by the Symbiosis Institute and the Environmental Club, apparently shows a development by which a common understanding of the idea of industrial symbiosis constituted itself among the different actors. The central actors in the different companies gradually formulated a common goal and strategy for development (e.g., as shown in the 2003 objectives for the Symbiosis Institute), and through this process of interorganizational "sense making," a set of common values and beliefs evolved in concert with the symbiotic relations and guided the further development of the different by-product exchange projects. In this process of institutionalization it became evident that the increased social interaction and common beliefs in the network not only laid the groundwork for development of new projects in Kalundborg and other interorganizational arrangements (soft symbiosis projects), but also paved the way for the diffusion of the industrial symbiosis idea to other localities worldwide (see chapter 12). The conceptualization of industrial symbiosis turned the concept into an action-guided framework for the actors and organizations in Kalundborg and for other organizations in other localities.

11.3.5 Economic Conditions

Economic conditions influencing industrial symbiosis are related to the context of economic decisions in the companies, which influence the assessment of economic feasibility of different possibilities. These can be closely related to the branch and market situation or to ownership, but also to the general performance of the companies. Do they have substantial funds for investments? What length of payoff times can be allowed? In Kalundborg, there has been important decision power or a sufficiently short distance among the decision makers to make the essential economic decisions in the companies and linkages to the essential decisions. Several of the local industrial managers have been very committed to the development of the symbiosis. The central plants are the most important within their companies in the country, and all the companies have been relatively successful, expansive, and resourceful. Novo Nordisk and Asnæsværket are Danish, whereas the Statoil refinery was first part of an American, and later on a Norwegian, corporation, and Gyproc has been Swedish-owned but has now been bought by a British company. With the centralization of power within international corporations during recent years, large parts of the decision making have been moved from local plants to head or divisional offices. This development seems also to have influenced the situation in Kalundborg. Today, all the Kalundborg plants are part of growing international companies, and both the ownership and decision-making structures are changing continuously. These changes might have some impact on the companies' willingness to invest in new symbiosis projects with low rates of return and high risks as the local commitment and local trust-based relationships among the facilities are weakened.

The economic gains resulting from the symbiosis in Kalundborg have never been analyzed in depth because of a lack of data and insights into the economic contractual agreements. But as mentioned previously, the total annual savings arising from the symbiosis projects are estimated to be around US$15 million, with an average payback time of five years. Economic viability has been one of the most important drivers behind each of the exchange relations in Kalundborg, but still the economic gains are rather marginal compared to core business gains in the different companies. The Novo Nordisk group has a net turnover of DKK 25 billion, Novozymes a net turnover of more than DKK 5 billion, and Energy E2 (which owns Asnæsværket) a net turnover of DKK 7 billion. Some of the profitable water exchange relations may illustrate the contribution of the industrial symbiosis projects in relation to the overall economic performance of the companies. First, by using surface water instead of groundwater as coolant in the Statoil refinery, a price reduction of 40 percent is obtained, from DKK $10/m^3$ to DKK $6/m^3$. Second, by reusing this surface water (e.g., between the refinery and the power plant) the price again is reduced by 50%, which gives a price on recycled water of DKK $3/m^3$. In relation to cooling water (recycled surface water) these price reductions give an economic benefit of DKK 1.5 million, with around five hundred thousand cubic meters of cooling water recycled each year. In comparison with the net turnover of some of the companies, these gains are marginal on first view. However, additional economic benefits are found in the fact that the different recycling initiatives on water simply created an essential precondition for continuous expansion of production capacity in the water-consuming symbiosis industries. Increased efficiency through recycling laid the groundwork for production expansion (and its related profits) that had never previously been permitted by the authorities because of scarce groundwater resources in the region. Thus, the indirect economic benefits caused by the Kalundborg industrial symbiosis have been substantial and a very central precondition for growth (or at least stability) in the regional economy. As discussed earlier in the chapter, it seems important to emphasize that the benefits of industrial symbiosis should not be seen in isolation, but as a part of a broader economic and environmental reality, which the water-sharing example also illustrates. This is supported by the fact that many of the exchange relations in Kalundborg are driven by a complex combination of economic, environmental, and social factors.

11.3.6 Mental Factors

Mental factors that affect industrial symbiosis relate to barriers connected to reorganization, testing new ideas, and establishing collaboration over branch and company borders, and also to how these barriers can be overcome. To overcome such barriers, communication is central, and it is in this context that the preconditions for, and the establishment and forms of, collaborative networks become important. In Kalundborg, the preconditions were favorable: The industrial executives involved all knew each other, and they also had good contacts in the regulatory authorities. On this basis, the increasing collaboration and its success have gradually strengthened the network of both formal and informal relationships. However, fortunate structural preconditions are not a sufficient explanation for the development; one also needs to emphasize that it seems to have been the result of the right people being in the right place. Developed networks may also inhibit change, and things do not move without the engagement of individuals in key positions who develop and spread ideas and are able to convince opponents, and this has been the case in Kalundborg.

The Kalundborg industrial collaboration has shortened the distance both among the different industrial companies and among industry, the authorities, and different interest organizations in society. The characteristics of the resulting network (it is horizontal, based on trust, reciprocity, and sharing of ideas and information) are all typical for progressive and innovative networks (Putnam 1993). The Environmental Club, and informal relationships among companies and between companies and the community, have gradually strengthened the sense of community in terms both of a common interest in the economic and environmental development in the region and of a level of communication, acceptance, and understanding between industrial and environmental interests that is far beyond what one often finds in other regions (Uldall 2001). The concept of industrial symbiosis, introduced around 1990, has also contributed to strengthening this common platform. This concept not only gave a common name to the many bilateral agreements, but it also contributed to a common understanding and vision and a further challenge to make industry more compatible with nature. As stated earlier, this can be interpreted as an institutionalization process, but it may also be considered as a constructed "storyline" (Hajer 1996) in which elements from different spheres (symbiosis from ecology applied to industry) are combined to create a common understanding. This storyline seems both to have strengthened the common ground internally and also to have been a forceful branding tool, particularly directed to an international audience.

The challenge for further development of the symbiosis in Kalundborg lies in improving the coordination and flexibility in the whole network and creating a system that also can make use of the small flows of waste products of industrial processes. For further development of the symbiosis, especially the symbiosis viewed as a part of a local sustainable development, the ongoing mobilization and interest of industrial actors seem to be the major challenge. Their engagement has of course been a central prerequisite before and during the realization of the industrial exchange activity projects, but industrial symbiosis has also become a part of business as usual. The generation of industrial leaders who were involved in establishing the network has gradually been replaced. Even if the community has been part of the organizational development and through this has developed good contacts with industry, it seems that there is still need for ongoing initiative from industry and local and regional authorities, eventually supported by national initiatives, to develop the symbiosis and local sustainable development plans.

11.4   The Industrial Symbiosis on Avedøre Holme

Avedøre Holme is a peninsula in the southern part of greater Copenhagen. It is the location of the major power plant in the area of the capital, which was opened in the 1970s, and a number of various industrial activities. Thirty industrial companies in this large industrial district were part of a pilot project under the state energy agency (Energistyrelsen) in 1998–2001, with the aim of developing industrial symbiosis among the factories in the area. A consulting firm made an inventory of possible exchanges and developed proposals for projects. These proposals cover biogas, cooling of groundwater, utilization of returns from the district heating network, and reuse of wastewater among the companies. If these projects were implemented, they would result in reductions of seven thousand tons of CO_2 emissions and savings of approximately ten thousand cubic meters of water per year. Which of these project ideas will be realized is not yet clear, but the pilot project has shown, as in many other cases, that there are profitable symbiosis opportunities and therefore a local potential to reduce resource consumption.

An overview of these potential economic and environmental savings in the Avedøre symbiosis project is given in table 11.1. In comparison with the projects in Kalundborg, the potential savings are smaller, but still significant.

The first question that immediately arises in connection with Avedøre concerns why these projects, with possible and profitable exchanges, were not realized spontaneously, prior to the study, as in Kalundborg. Although there are no important differences between Kalundborg and Avedøre in regard to technological or economic preconditions, Kalundborg might nevertheless be quite a different case. It is true that in Avedøre, as in Kalundborg, there are possible and profitable exchange options; and the companies in Avedøre are often even closer to the central offices than the ones in Kalundborg; and they, like those in Kalundborg, are generally economically successful, since Avedøre is a rather modern industrial district. Moreover, the institutional setting in the two locales should be about the same, even if regional authorities in peripheral areas, like Kalundborg, sometimes can be more sensitive to industrial demands than those in more central locations. In Avedøre, however, there seems to be a very low degree of communication and limited or no sense of community in the industrial district, something that has also been observed in other, similar districts. A possible explanation for this lies in the separation between the social sphere and the production sphere, of which an industrial park is an expression, that is, people do not work where they live. An industrial park creates a limited basis for social interaction crossing company borders when the production sphere is geographically separated from the civil society. This explanation is based on the assumption that social interaction is a premise for creating interindustrial relations, which of course needs to be further investigated. From an environmental viewpoint, it might even be a bad idea to locate industry in "special" areas where it is isolated from the rest of society and does not feel responsible for other parts of society. Is the concept of industrial districts not compatible with the sense of community that is central to the concept of a common sustainable future?

In the small town of Kalundborg, the separation between the production sphere and the civil society has not been nearly as rigid as in other industrial districts. People in the different industries have become much

Table 11.1
Economic and environmental savings in the Avedøre symbiosis project

	Reduced CO$_2$ emissions (t/year)	Reduced water use (m^3/year)	Reduced landfill (t/year)	Reduced societal costs (DKK 1,000/year)	Reduced company costs (DKK 1,000/year)	Investments (DKK 1,000/year)
Heat exchanges	2,900	No	No	2,600	482	7,700
Cooling exchanges	—	Yes	No	45–103	108–172	1,000–2,000
Energy exchanges	3,900	No	No	1,700	1,400	5,000
Other resources	3	10,000	100	45	190	55–80
Total	7,100	10,000	100	Approx. 800	Approx. 2,500	Approx. 17,000

Source: COWI (2002).

closer socially, often belonging to the same clubs and social circles, and have much more sense of community than in the industrial district in the big-city region. This has probably been essential for the establishment of communication linkages and the creation of collective identity and a sense of the industrial plants' being a part of the local society.

Since networking activities have not developed spontaneously in Avedøre, there are good reasons to ask how such networking activities best can be promoted. In some similar contexts, it has been emphasized that a facilitator is necessary for encouraging collaboration among industries that do not cooperate spontaneously. This seems, of course, logical, but it is highly questionable whether it can be successfully accomplished via a process in which industry makes data accessible, but otherwise does not take part in the process, which is in the hands of the facilitator. And will networking activities initiated through this type of facilitation process create sufficient commitment to invest in cooperative arrangements of exchange of material flows among the companies, or will the not-invented-here syndrome make them resistant in a possible implementation phase?

In contrast to the Kalundborg symbiosis, that proposed in Avedøre is a top-down project, facilitated by an external consulting firm and financed by the country's energy agency. This external consultant has used a classical approach, first an inventory of the branch pattern, then green accounting to get an overview of resource use, and lastly interviews with companies to get complementary details. On the basis of these data, project proposals have been developed and then presented to the industrial companies. Industry has not been part of the process, with the effect that these companies are often critical, not very motivated, and even resistant to the projects when the implementation phase is reached. It seems, therefore, to be essential to ask how an external actor can approach an industrial district in a way that enables the actor to introduce new ideas and gain companies' acceptance for and engagement with these ideas. It would be interesting to look at different strategies that initiators have employed in efforts to create synergies among industrial companies. How have initiators handed over projects to the companies after working out proposals? Through which channels should the projects be introduced and initiated? Should one ideally start from the top level, from the midlevels, or from the workshop floor? Is this different for different branches or companies? Is it necessary for the initiator to know the company extremely well to be able to identify projects that have the potential to generate environmental and economic profits in the symbiosis concept? It would also be essential to look at the motivation of the facilitator. Is he motivated primarily by the money from public authorities, or has he other wider motivations?

11.5   Conclusions

In all industrial districts—whether more heterogeneous or more homogenous—opportunities may be found for exchange activities that will be profitable from both an economic and environmental viewpoint. Of course, these opportunities vary from location to location and from plant to plant: Each plant generates different waste products and has different possibilities for making profitable use of other plants' waste materials and waste heat. Such potential exchange activities can be realized if there is sufficient understanding and trust among the companies involved. Communication seems to be an essential aspect to enable cooperative projects to meet legislative demands (institutional aspects), to overcome resistance in or among the companies (mental and economic aspects), and to create a basis for more stable collaboration among companies, authorities, and the public (organizational aspects).

The development and the context in Kalundborg are in many ways unique and not directly transferable, but still the example of industrial cooperation that has developed in Kalundborg may provide inspiration for developing and strengthening cooperative networks in other regions. As often is pointed out in the literature (e.g., Garner and Keoleian 1995; Ehrenfeld and Gertler 1997; Lowe 1997), the Kalundborg symbiosis was not planned; it was established through a stepwise development of bilateral arrangements between the companies. It is true that economic rationality has played a more important role than environmental awareness, but economics has not been the only "prime mover." The role of legislation and other administrative and societal demands should not be underestimated. The exchange activities that were implemented in the 1970s and 1980s can, to a considerable extent, be described as creative solutions to problems introduced by regulatory and societal demands to reduce pollution and waste and to save energy and water. The flexibility of the implementation gave the companies the opportunity to search for their own solutions, and together they were able to find solutions that were also economically profitable. The collaborative solutions would not have been possible without good contacts and communication among the companies. The success, as well as the trust and the dependency, that have been created by the established collaboration have been prerequisites for the further development of the symbiosis, which has gradually become more driven by a desire to combine economic savings with environmental awareness.

Industrial districts located in small communities in peripheral areas may have better conditions for establishing industrial symbiotic networks than those located in central urban areas. Such better opportunities for industrial symbiotic networks is connected to the nearness of industry and civil society. This is an interesting observation that should be encouraging for small communities, but it also presents a particular challenge in connection with industrial districts in big city regions. It will be interesting to follow the further developments in Avedøre, but it would also be interesting to compare these with initiatives in similar regions, like the Rotterdam area, that are considered partly successful (e.g., Baas 2000).

Industrial symbiotic networks may generate positive economic and environmental development for the companies involved, and also for the locality where they take place. The case of Kalundborg shows a gradual strengthening of communication and understanding among various local interests, and therefore the platform for making different types of development plans for the sustainable development of the region is much stronger than in many other similar regions. Strong networks and good communication in and among different parts of society are definitely important ingredients and even prerequisites for local sustainable development. Therefore, stimulating eco-industrial networking from the outside must be viewed not as a top-down process, but as a bottom-up development with those who are concerned with playing a major role from the start.

The proposed symbiosis project at Avedøre Holme has been similar to other projects introduced by authorities to set up cooperative efforts among companies and also seems to suffer from similar problems. Symbiosis projects that operate in a vacuum between companies have problems concerned with making the organizations willing to change and try something that has never been tried before and that has not developed within the organization. Of course, this implies that there might be environmental potential in stimulating such cooperation, but perhaps would it be more efficient to realize this via legislation.

The analytic framework of section 11.2 has helped to structure our analysis, but it is problematic that the different aspects in the analysis are often closely intertwined, and this makes it hard to make and keep consistent delimitations. Structures of various kinds are setting the constraints, but when change is contemplated, the particular persons involved and their ability to communicate, develop trustful relationships, and realize initiatives together become central. This is why the social context, the methods, and the persons involved must be much more focused in connection with industrial symbiotic development. Perhaps regional and local environmental cooperation should really start at some place other than on a flowchart showing possible exchange relations?

Note

1. The Symbiosis Institute is a center of knowledge and information established by the six partners of the Kalundborg symbiosis. The major tasks of the institute consist of gathering data concerning industrial symbiosis and communicating the experiences of the symbiosis project; contributing to the creation of new symbiosis projects; acting as a coordinator for studies of industrial symbiosis; cooperating with various international institutions and consultancies; and assisting in courses concerning industrial symbiosis and related subjects.

References

Asheim, B. T. (1996). Industrial districts as "learning regions": A condition for prosperity. *European Planning Studies* 4(4): 379–400.

Ayres, R. (1999). *Zero Emissions Forum*. Tokyo: United Nations University Press.

Baas, L. (2000). Developing an industrial ecosystem in Rotterdam: Learning by ... what? *Journal of Industrial Ecology* 4(2): 4–6.

Boons, F., L. Füssel, and S. Georg. (2000). Deconstructing the Kalundborg Myth. Working paper 2000.10, Institut for organisation og arbejdssociologi, Copenhagen Business School.

Chertow, M. (2000). Industrial Symbiosis: Literature and Taxonomy. *Annual Review of Energy and Environment* 25: 313–337.

Cohen-Rosenthal, E., and T. N. McGaillard. (1998). Eco-Industrial Development: The Case of the United States. Report no. 27, Institute for Prospective Technological Studies, Seville, Spain. Available at http://www.jrc.es/iptsreport/vol27/english/COH1E276.htm.

COWI. (2002). Industriel Symbiose på Avedøre Holme (Industrial Symbiose on Avedøre Holme). Final report.

Desrochers, P. (2002). Eco-Industrial Parks and the Rediscovery of Inter-Firm Recycling Linkages. Working paper, Institute for Policy Studies, Johns Hopkins University. Available at http://www.mises.org/journals/scholar/Eco6a.PDF.

Dunn, B. (1999). Industrial ecology for sustainable communities. *Journal for Environmental Planning and Management* 41(6): 661–672.

Ehrenfeld, J., and N. Gertler. (1997). Industrial ecology in practice. *Journal of Industrial Ecology* 1(1): 67–79.

Erkman, S. (1997). Industrial ecology. *Journal of Cleaner Production* 5(1–2): 1–10.

Garner, A., and G. A. Keoleian. (1995). Industrial Ecology: An Introduction. Working paper, National Pollution Prevention Center for Higher Education, University of Michigan, Ann Arbor. Available at http://www.umich.edu/~nppcpub/resources/compendia/ind.ecol.html.

Hajer, M. A. (1996). *The Politics of Environmental Discourse: A Study of the Acid Rain Controversy in Great Britain and the Netherlands*. Oxford: Clarendon.

Huggins, R. (2000). The success and failure of policy-implanted inter-firm network initiatives: Motivations, processes and structure. *Entrepreneurship and Regional Development* 2: 111–136.

Jacobsen, N. B. (2002). The industrial symbiosis in Kalundborg, Denmark: Industrial networking and cleaner industrial production. In P. Lens, L. Hulshoff Pol, P. Wilderer, and T. Asano (eds.), *Water Recycling and Resource Recovery in Industry*, 664–671. London: International Water Association.

Lowe, E. A. (1997). Creating by-products resource exchanges: Strategies for eco-industrial parks. *Journal of Cleaner Production* 5(1–2): 57–65.

Maskell, P., E. Heikki, I. Hannibalsson, A. Malmberg, and E. Vatne. (1998). *Competitiveness, Localised Learning and Regional Development*. London: Routledge.

Nemerow, N. (1997). *Zero Pollution for Industry: Waste Minimization through Industrial Complexes*. Cambridge, MA: MIT Press.

O'Rourke, D., L. Connelly, and K. P. Koshland. (1996). Industrial ecology: A critical review. *International Journal of Environment and Pollution* 6: 89–112.

Porter, M. (1990). *The Competitive Advantages of Nations*. Cambridge, Mass.: Harvard University Press.

Putnam, R. D. (1993). *Making Democracy Work: Civic Traditions in Modern Italy*. Princeton: Princeton University Press.

Pyke, F., and G. Becattini (eds.). (1992). *Industrial Districts and Inter-firm Co-operation in Italy*. Geneva: International Institute for Labour Studies.

Schwartz, E. (1995). *Industrial Recycling Network: A Model to Integrate Ecological Aspects in a Production Economy*. Leverkusen, Germany: Deutscher Universitäts Verlag.

Storper, M. (1992a). The limits to globalization: Technology districts and international trade. *Economic Geography* 68(1): 60–93.

Storper, M. (1992b). Regional "worlds" of production: Learning and innovation in the technology districts of France, Italy and USA. *Regional Studies* 27(5): 433–455.

Uldall, F. (2001). Natursyn under forandring? (Is nature perception changing?) Master's thesis, Institute of Geography, University of Copenhagen.

12 The Myth of Kalundborg: Social Dilemmas in Stimulating Eco-Industrial Parks

Frank Boons and Marco A. Janssen

12.1 Introduction

In the field of environmental sciences, the concepts of "industrial symbiosis" and "industrial ecology" refer to the idea that the negative ecological impact of economic activities may be reduced more efficiently and effectively if the boundary of the system submitted to environmental management is drawn not around an individual firm, but instead around a group of firms. By looking at a larger system, it is possible to prevent problem shifting: the possibility that efforts to reduce negative ecological impact in one part of the system will create additional impacts in other parts of the system. In a collective approach, firms achieve a competitive advantage through the physical exchange of materials, energy, water, and by-products (Chertow 2000).

The system boundary can be defined in various ways: a sector of industry, firms that are part of the life cycle of a product, or a set of firms situated in a certain geographical area. Drawing the boundary in one of these ways opens up a wider range of technical and social options to reduce negative environmental impact. The social element is important, since one of the main issues is the coordination of activities of the economic actors that are part of the system in deciding about and implementing efforts to reduce environmental impact (Boons and Baas 1997). In this chapter, we focus on the geographical system boundary.

On the basis of the experiences in the Danish town of Kalundborg, the idea of industrial ecosystems, or *eco-industrial parks* (EIPs), has received enormous attention from practitioners as well as scientists during the last ten years (Gertler 1995; Ehrenfeld and Gertler 1997; also chapter 11). Through an interesting process of dissemination, the concept of an EIP has spread rapidly to places such as the Netherlands, Canada, Hong Kong, and the United States (Boons, Füssel, and Georg 2000). As a result, participants in eco-industrial parks have tried to develop their parks toward sustainability, and governments have launched programs to promote such EIP initiatives. This has given rise to a number of conceptual and empirical studies of these initiatives.

There is some controversy about the stimulation of eco-industrial parks. Some scholars argue that there is nothing new about eco-industrial parks and that Kalundborg was nothing more than a rediscovery of old economic principles (Desrochers 2001, 2002). Others state that eco-industrial parks need to be designed at a larger organizational level to derive even more economic and environmental returns than self-organized parks like Kalundborg (Hawken 1993; Baas and Boons 2004). We suspect that both perspectives miss an important point, namely, the need to overcome social dilemmas in order for industries to invest in infrastructure that allows systematic and cost-effective exchange of material flows.

The organization of this chapter is as follows. The history of Kalundborg is briefly reviewed in section 12.2. Next, section 12.3 examines how the Kalundborg experience spread and generated new policy initiatives elsewhere. Section 12.4 argues that Kalundborg is not unique, but a rediscovery of old principles. The results of policy initiatives are discussed in section 12.5, in which it is shown that it is difficult to design and stimulate eco-industrial parks that work in practice. In section 12.6 the social dilemmas relevant to most eco-industrial parks are studied. The chapter concludes with a discussion on which types of policy might stimulate eco-industrial parks.

12.2 Diffusion of the Kalundborg EIP Concept

The long-term development of the Kalundborg eco-industrial park has been described in detail in chapter 11. Currently it functions as a model for advocates of eco-industrial parks. It is not easy to track the dissemination of the Kalundborg concept. Boons, Füssel, and Georg (2000) reconstructed the process of dissemination by tracking Internet sources. The story of Kalundborg can be seen as a first example of the diffusion of the idea of EIPs.

The term "industrial ecosystems" was coined in a paper presented at the 1977 annual meeting of the German Geological Association by an American geochemist (Erkman 1997). The idea resurfaced more than a decade later in an article in *Scientific American*, written by two employees from General Motors (Frosch and Gallopoulos 1989). A few years afterward, this article was summarized in the language of business by a consultant from Arthur D. Little (Tibbs 1991). According to Erkman, this document was instrumental in disseminating the idea of industrial ecology within business circles.

Fueled among other things by the discussions following the Brundtland Commission report on sustainable development (Brundtland 1987), the Frosch and Gallopoulos article, according to Erkman (1997), "sparked off strong interest.... The article manifestly played a catalytic role, as if it had crystallized a latent intuition in many people, especially in circles associated with industrial production, who were increasingly looking for new strategies to adopt with regard to the environment" (5). Apparently, industrial ecology became a new metaconcept that seemed to hold the promise of embracing existing techniques and practices as well as developing new and more effective ones designed to decrease the environmental impact of production and consumption activities. In short, industrial ecology has acted as an energizing and mobilizing concept.

In the early 1990s, a group of U.S.-based scientists and businesspeople formed the Vishnu group, which saw as its role the dissemination of the concept of industrial ecology. After a visit to Denmark, one of its members had a student write a doctoral thesis on Kalundborg. This study (Gertler 1995), which was also published on the Internet, appears to have been crucial in spreading the Kalundborg story. It was also instrumental in corroborating the idea of industrial ecology and showing that it could, indeed, be more than wishful thinking and conceptual desk research. In addition, the study showed the evolutionary character of the Kalundborg symbiosis, stressing the complex interplay of technical and social forces in its origin and development: economic efficiency as a motivation for the various actors taking part in the symbiosis; the environmental regulatory regime's role as facilitator for innovative solutions; and, finally, the locally embedded network of agents for sharing ideas, information, and solutions to common problems. The Gertler version of the Kalundborg symbiosis thus contained lessons and visions as to the whys and hows of industrial ecology.

Separate from the Vishnu group, the idea of EIP was picked up by Cornell University in the United States, which had a long-standing tradition in the social aspects of community development. The idea of industrial ecology, and more specifically EIP, fit well with this tradition. These developments took on national policy relevance in the United States in

1995, when the President's Council on Sustainable Development devoted itself to the theme of EIP. Here, the different strands in U.S. activities around this idea were brought together.

More recently, the concept of industrial ecology (IE, of which EIP is an important element) has manifested itself in a journal, the *Journal of Industrial Ecology*, and an international society, the International Society for Industrial Ecology.

Some interesting issues have emerged relating to the way in which the EIP and IE concepts are disseminated:

• The IE and EIP concepts are defined very differently by their adopters. Some use IE as a label to denote the technical linkage of production processes and the use of wastes in production processes, whereas others use it to refer mainly to processes of cooperation, development, and management of geographically bounded areas or communities. Both concepts thus serve as a *boundary object* (Adolffson 2001): an idea or object that acts as a bridge between two different social groups.
• Certain actors use concepts in a strategic way. By linking existing activities to a concept, they can profit from the popularity of the concept. This seems to be the case with the researchers from Cornell University.
• By themselves, concepts are fluid. They can be made visible, and therefore more suitable for rapid and effective diffusion, by writing them down. The Gertler study is an example of this; it is widely cited and seems to have been the basis for many other descriptions of Kalundborg, even when not cited as such.

12.3   Putting the Kalundborg Story into Practice in the Netherlands

The previous section provided insight into the ways in which "Kalundborg" has been transmitted. At the receiving end of the communication line, there are practitioners in different countries who have picked up the idea and have based their actions on it. By way of illustration, this section describes how this has taken place, and is currently taking place, in the Netherlands.

12.3.1   Fertile Soil

For an idea like EIPs to find a successful destination, there must be some sort of connection possible with existing activities. We have found two such activities in the Netherlands: (1) revitalization of industrial parks and (2) improving energy efficiency through cooperation.

12.3.1.1 Revitalization At the beginning of the 1990s, efforts were undertaken to "revitalize" industrial parks in the Netherlands. In the 1960s, these parks had been established by municipalities to move industrial activities away from the centers of towns and concentrate them in parks located at their borders. In the years that followed, these parks evolved as certain companies left and others came, and the parks were enlarged to make room for growing industrial activities. At the end of the 1980s, a substantial proportion of these parks had developed into fragmented areas with no coordination of activities, often dangerous traffic situations, and deteriorating buildings and infrastructure. The revitalization effort, directed at renovating, consisted of a national subsidy program to enable municipalities to invest in the industrial parks. In a number of cases, this led to the establishment of coordination mechanisms among companies, such as foundations for collective representation and bureaus for "park management." Apart from their function of coordinating activities of the firms located in the industrial park, these mechanisms aimed at "speaking with one voice" to municipal officials.

12.3.1.2 Improving Energy Efficiency In the same period, the Ministry of Economic Affairs was looking for options to increase the energy efficiency of Dutch industry. These efforts originated in the early 1970s, when Dutch companies suffered as a result of a boycott of oil from the Oil Producing and Exporting Countries (OPEC) undertaken because of the pro-Israel stance of the Dutch government. This boycott made companies aware of their high level of energy use and initiated efforts to undertake energy saving. Throughout the 1980s such energy-saving and energy awareness activities were further developed. One of the options for further efficiency improvements was to consider groups of companies instead of individual firms, notably, through the exchange of process heat. However, a test case initiated by the ministry was largely unsuccessful, mainly because it was difficult to organize cooperation among firms.

12.3.2 Win-Win Solutions: Looking for Ideas
In 1997, the Ministry of Economic Affairs issued, together with the Ministry of Environmental Affairs (EZ) and the Ministry of Transport and Water Management, a policy note, *Environment and Economy (Nota Milieu en Economie)* (EZ 1998). This policy paper had as its central aim to stimulate activities that combine ecological and economic benefits (win-win changes). In preparing this policy paper, public servants from EZ visited many firms to meet people involved in activities that could be

incorporated into the study. One of the activities they encountered was brought to their attention by the province of Noord-Brabant. Within this province there was a small group of public servants organized into what they themselves called the Project Innovation Team (PIT). Their aim was to initiate innovative activities in which provincial authorities play a nontraditional (nonlegislative) role, often leading to a public-private partnership. One of the group's projects concerned an industrial park, known as Rietvelden/de Vutter (RiVu). The environmental coordinator of the dominant company in this park, a plant of Heineken breweries, had been inspired by Kalundborg and tried to develop and implement similar ideas about cooperation in RiVu. It found a willing partner in the coordinator of the PIT. Together they visited Kalundborg and wrote a report on it. They used the label *duurzaam bedrijventerrein* (eco-industrial parks), the name under which such initiatives are now known in the Netherlands.

This initiative was taken on board by the EZ people, who started looking for, and found, similar projects in other parts of the Netherlands. A common characteristic of these projects was that firms located in the same geographical area were often cooperating to reduce their environmental impact and at the same time reduced the costs of their activities.

The writing of the policy note was coordinated by a professor in environmental management at Erasmus University, Rotterdam. Upon hearing about the enthusiasm of the public servants for eco-industrial parks, he linked it to the activities he was involved in with his research institute, which had been working on a similar initiative in the Europoort/Botlek area of the Rotterdam harbor area, started some years before (Baas 1998). From the 1990s on, companies in the Europoort/Botlek area had joined forces to develop environmental management systems and had been able to obtain governmental funding in support. As the end of this funding was approaching, environmental officials were looking for new financial sources to continue the development of these environmental management systems. As it happened, the national government had just issued a stimulation program for the improvement of the environmental performance of product chains. This stimulated the companies to search for options to develop cooperative efforts. The researchers, one of whom had contacts with Kalundborg, linked the ideas of Rotterdam companies to the Kalundborg example. In developing the Rotterdam initiative, called INES (Industrial Ecosystem), the directors of the Kalundborg energy plant and the Dutch company Novo Nordisk, another Kalundborg participant, shared their experiences. These visitors were given a tour of the Europoort/Botlek area and stated that it would be possible to develop initiatives for the area similar to those in Kalundborg.

EZ's policy note was based on a number of showcases (*boegbeelden*) of the win-win philosophy. In selecting these, the Ministry of Economic Affairs consulted the national association of Dutch industries (VNO/NCW). Although they were initially worried that additional legislation might ensue, in general the theme of eco-industrial parks met with great enthusiasm of the VNO/NCW people. In the version of the policy note that was eventually sent to the Dutch parliament, eco-industrial parks were one of the showcases, with RiVu and the Rotterdam harbor area as the two main examples.

12.3.3 Implementing Ideas

After the policy paper was issued, EZ decided that it would be good to install a steering group that would monitor the diffusion and implementation of the eco-industrial park concept. This group had a diverse membership: apart from representatives of three ministries (EZ, the Ministry of Traffic and Transport, and the Ministry of the Environment and Spatial Planning), there were representatives of industry, local and regional authorities, and the two main initiatives showcased in the paper, that is, RiVu and INES. This group developed several projects to stimulate the diffusion of the EIP concept, the most important of which was a program aimed at providing financial resources for initiatives taken by local communities. As EZ provided most of the money, the program had a focus on energy efficiency. At the same time, it focused on the process of cooperation. In preparing its activities, the steering group had asked a large consulting firm, KPMG, to provide it with input. This resulted in an overview of examples, as well as a framework for how to go about making an industrial park more sustainable (KPMG 1998). Part of the consultants' input was the observation that the cooperative process was the main bottleneck in developing eco-industrial parks; not the development of new technical possibilities, but their acceptance and organization, was the main barrier to success.

The Dutch government's stimulation program for improvement of the environmental performance of product chains has been in place since 1999, and the number of projects that have been submitted has surprised all the members of the steering group. It was decided to apply the principle "let a thousand flowers flourish," that is, give a chance to any initiative that is taken by either local authorities or local groups of companies.

Table 12.1
Type of park in supported projects

New parks	Existing parks	Other (virtual, unknown)
26	34	2

Table 12.2
Phase of supported projects (projects usually span more than one phase)

Initiation	Orientation	Design	Decision making	Implementation
22	34	28	15	5

Selection among the proposals submitted was based on the following criteria:

1. Projects selected should have clearly developed and measurable goals for the project period.
2. There was a slight bias toward projects that included the implementation of energy-saving schemes. This criterion was a result of the fact that as noted previously, the organization responsible for implementing the subsidy scheme, EZ, had energy saving as an important focus.
3. The set of projects selected should cover different types of initiatives; both existing industrial parks and parks to be developed should be included; and the initiatives should cover different phases of development.

The first round of the EIP program, coordinated by NOVEM (2001), resulted in the cofinancing (usually 50 percent of calculated costs) of sixty-two projects. Tables 12.1 and 12.2 provide some insight into the type of projects that were granted support. In particular, they show the coverage of different types of initiatives.

Table 12.2 shows that most projects selected for support were in the early phases of development of EIPs: The development covered the initial or orientation phase for fifty-six out of the sixty-two projects supported. In these phases, the establishment of organized park management, as a basis for further activities, was the main goal. Later phases covered in supported projects, such as the design and implementation of technical linkages among companies or decision making about sharing utilities, often were activities that had already been initiated and subsequently were put in project form for the purpose of obtaining public support through a subsidy. The projects supported show a wide diversity in terms of who was the initiating actor: sometimes this was one firm, seeking to establish a linkage with one or more other firms; in some cases it was a local governmental authority; and in other cases it was a group of entrepreneurs.

12.4   Is Kalundborg Unique?

Because the Kalundborg example has diffused and inspired policy elsewhere, the question emerges whether Kalundborg was really unique. If so, what were the reasons? If not, what are other examples? Desrochers (2001) argues that EIPs are not new and in fact are nothing more than a rediscovery of interfirm recycling linkages. In the nineteenth century, waste recovery and exchange among independent firms was widely practiced in the Western world. Perhaps changes in labor costs, environmental regulation, and globalization during the past several decades have made it less attractive to exchange materials at a local scale.

After the discovery of Kalundborg, scholars started to look for other examples and found these all over the world (Desrochers 2002): within the Austrian province of Stryria, the Ruhr region of Germany, and the Jyväskylä region of Finland, and in the petrochemical complexes of Los Angeles, Houston, and Sarnia (Canada).

According to Desrochers (2001), cost-benefit considerations cause firms to look for the most cost-effective way to deal with waste. This often leads to reuse of waste by the same firm or by other firms. Entrepreneurial firms are creative in finding new ways of reusing and recycling their waste in monetarily beneficial ways. Complex interfirm recycling linkages are not easily designed top-down by public agencies. Some scholars argue that a design of interfirm recycling linkages leads to better symbiotic relationships compared with self-organized linkages. Design of linkages can benefit from locating and specifying industries and factories according to a grant scheme (Hawken 1993). However, bureaucrats and public planners have only limited access to the information that is required to organize profitable synergetic relationships among firms. An important problem, according to Desrochers, is that environmental policies that define the conditions for waste treatment restrict firms from being innovative with regard to reuse and recycling. Environmental regulation can therefore in fact act as a barrier to the emergence of eco-industrial parks.

Increased complexity of production processes and environmental regulations of inputs and outputs ultimately requires investments in interfirm linkages. When firms need to make significant investments in equipment to treat waste to an extent sufficient for the waste treatment output to be able to serve as an input to other firms, the economic or profit incentives for establishing interfirm linkages are low. The requirement of significant investment costs can be seen as a social dilemma that prevents firms from exchanging material flows. Before we discuss this social dilemma, the impact of the Kalundborg example on policymakers will be considered.

12.5   Lessons from a Dutch Case: The INES Eco-Industrial Park

One of the industrial parks in the Netherlands with a fairly long history is situated in the Rotterdam harbor area and is home to firms from the processing industry. This park has been studied by Baas (1998, 2000).

At the end of the 1980s, the Dutch government and Dutch industry agreed on a voluntary scheme to implement environmental management systems in industrial firms until 1995. Companies in the Europoort/ Botlek area, located in the Rotterdam harbor, decided to work together for this purpose. Coordination was made available by the regional industry association, Europoort/Botlek Interests (EBB). Consultants as well as researchers were involved in the project, which came to be known as INES (Baas 1998).

The network that subsequently was created served as the basis for identifying options to diminish the environmental effects of the companies involved. This occurred through looking at possible linkages of production processes and the sharing of utilities. The network was formed from contacts among environmental managers of the various companies. They took Kalundborg as an explicit model for their effort. Representatives of the Kalundborg region were invited to Rotterdam to provide information about its success story.

The INES project had the following goals (Baas 1998, 191):

• to stimulate cleaner production approaches within individual companies
• to perform network analyses of activities, material, and energy streams and of options to reuse materials, by-products and energy
• to develop a knowledge infrastructure to support the development of an eco-industrial system

The project, which formally started in 1994, was expected to last until 1997. It was divided into three phases:

1. Communication of goals of the projects to, and building support for the project among companies.
2. Conducting prefeasibility studies.
3. Designing selected projects for implementation.

An important point on which all companies participating in INES agreed was that any activity that would be implemented had to be at least cost neutral. In other words, companies did not want to invest in activities without being certain that return on investment would be acceptable and that no additional operational costs would result. The fact that EBB was able to obtain subsidies for implementing projects provided substantial support, because the pressure of project costs was alleviated.

The network analysis gave insight into technically feasible activities. From this set, three activities were selected: compressed air utility sharing, wastewater treatment, and reduction of biosludge. Baas (2000) reports that it took five years to implement one of these activities, the sharing of compressed air. Although there was an initial level of trust among the companies in the region, it took time to build support for the specific activity. Moreover, the information initially collected about the technical aspects of materials flows from the companies proved to be inaccurate. In addition, the supplier of compressed air initially involved in the project revised its priorities and withdrew from participation. Ultimately, another supplier of compressed air took advantage of the opportunity and installed a system that has been operational since the beginning of 2000.

Currently, a second INES project is being implemented. This followed a period of over two years in which there was no involvement from EBB, the coordinating actor. But eventually the INES philosophy was taken up again, this time with the explicit goal of developing a more strategic approach to building an eco-industrial system. In order to move beyond the technical-operational approach that dominated the first project, the second project has chosen to start a dialogue on strategic issues, such as "the future of fossil fuels," an important topic, as a significant number of the companies involved are related, directly or indirectly, to the oil industry. This dialogue involves actors' operating at strategic levels within government and industry. Getting actors interested in this dialogue, and linking a more strategic discussion to concrete activities, remains a major challenge for the industrial park.

12.5.1   Coordination and Cooperation

The previous examples show that cooperative activities can emerge among firms, possibly stimulated by governmental policy. Examples of successful cooperation are, however, scarce. The development of EIPs is often hampered by problems of coordination and cooperation, which manifest themselves in different forms:

1. Although located in geographical proximity, firms in an industrial park often have no close relationship with one another. There are local industrial clubs that can serve as a starting point for coordination, but they are insufficient in terms of commitment, membership, and level of trust to function as a tight coordination mechanism. Developing commitment, a shared vision, and the requisite level of trust is a difficult and time-consuming process. Taking ideas like Kalundborg and transplanting them into a totally different context therefore often will not work. Gertler's (1995) analysis of the social network that underlies the technical linkages among firms in Kalundborg shows how long it has taken for these linkages to evolve.

2. Firms in an industrial park are often production plants of large businesses. Decisions concerning their activities are made in some headquarters far away. This implies that even if site managers are willing to consider linkages with other firms in the park, they still need to convince company managers who are not part of the local social group.

3. A problem discussed more by scientists than practitioners relates to the adequate system boundary for eco-industrial parks. In general terms, the question is, what system can be best optimized? A cleaner production approach focuses on optimizing the production system of the individual firm from an environmental perspective. Looking for an optimum within the group of firms in an EIP may cause individual firms to stop considering the prevention of waste within themselves, as they are looking collectively for ways to use waste of one firm as an input for another firm in the park. In addition, firms are part of larger product chains beyond a local EIP. This may mean that various stakeholders, such as suppliers, consumers, and nongovernmental organizations, put pressure on firms to contribute to reducing the environmental impact of the product chain as a whole. Together, the local EIP and wider product chain context may lead to contradictory demands on environmental actions and strategies undertaken by particular firms.

4. Over the past ten years, many firms have changed dramatically, focusing on core activities and outsourcing of noncore activities. On the one

hand, this has made them more experienced in developing partnerships, because they often need to control the activities they have outsourced by way of a mechanism that allows more influence than a market. On the other hand, it makes them more sensitive to the fact that each additional link, be it technical or organizational, makes them more dependent on other firms. The idea that organizations have a fundamental need to reduce dependency has been developed theoretically (Pfeffer and Salancik 1978) and is also recognized in practice.

The INES case shows how the coordination of activities of firms situated in an eco-industrial park can be hampered by the fact that these firms are part of other systems (a multinational corporation, a product chain) and thus are subject to multiple and often conflicting incentives to adapt their activities and their environmental effects. Developing an eco-industrial park requires that firms, individually and as a group, develop capabilities to deal with these conflicting demands. In that sense, the main lesson from the Kalundborg example is that such a development requires a long period of time.

12.6   A Collective Action Theory of Eco-Industrial Parks

The contrast between the emergence of cooperation in Kalundborg and cases in which it did not happen presents a fundamental puzzle in the study of social organization. Conventional economic theory assumes that economic actors make decisions in their own interest. However, cooperative behavior is observed that does not fit with the concept of the selfish individual. Mancur Olson (1965) states that "rational, self-interested individuals will not act to achieve their common or group interests" (2). The reason for this claim is that when interests are shared, rational actors should prefer to free-ride, that is, to let others pay the cost of goods that will also benefit the free-riders. If we nevertheless do see groups acting to further their joint interests, this can be explained in terms of private incentives relating to rewarding contributors or punishing noncontributors.

Since the 1980s, empirical evidence that individuals are able to develop cooperative solutions has been abundant. Many examples can be given in which people have organized themselves to achieve much higher outcomes than is predicted by the conventional theory (Ostrom 1990). Laboratory experiments show that communication is a crucial factor for deriving cooperative behavior (Ostrom, Gardner, and Walker 1994).

Furthermore, the ability of participants to determine their own monitoring and sanctioning system is critical to sustaining such behavior (Ostrom, Gardner, and Walker 1994). The reasons that these factors are important are not precisely known, but the hypothesis is that they relate to the development of mutual trust during interactions among resource users.

Transferring these insights to industrial symbiosis, we see a number of individual firms that can derive a better performance by cooperation. Here we do not just mean buying and selling of waste, but adjusting production processes in such a way that neighboring firms can be connected. According to Olson's argument, firms want to invest in adjustments in their production process only when they directly derive a financial benefit or when governmental regulation prescribes such adjustments. Firms, which have adjustment investments with long payback times, experience uncertainty about the actions of neighboring firms. Will they be able to deliver the anticipated waste flows according to agreed-upon or required quality standards? What if the other firms find new cost-effective ways to reduce waste? What if the neighboring firms go bankrupt or move to another location? Structural and costly adjustments in the production process to enable cooperation with other firms are therefore very risky. The question comes down to under what circumstances firms are willing to make such investments.

The success of Kalundborg and other self-organized EIPs might indicate that collections of firms are able to overcome the dilemma just sketched. Table 11.1 concerning the Avedøre symbiosis shows high investment levels that are paid back only after a number of years. The insight that the network of social interactions among companies has been crucial in the development of social capital can be explained by the arguments of Elinor Ostrom (2000). The existence of norms in a group that place group interests above those of individuals gives individuals the confidence to invest in collective activities, knowing that others will do so too. Reciprocity and trust are important social norms that can be developed in a group (Ostrom 2000). Another important norm is to agree on sanctions for those who break the rules. Finally, social norms can be developed during repeated interactions but can decay easily as a result of cheating. In the case of an industrial park, this could take the form of not delivering or not accepting waste flows once other firms have made costly investments.

It has been rumored that the success of Kalundborg actually relates to the frequent gathering of managers in a local pub. The local pub might

		Firm B	
		Exchange	No exchange
Firm A	Exchange	$(b_a - c_a,\ b_b - c_b)$	$(b'_a - c_a,\ -c_b)$
	No exchange	$(-c_a,\ b'_b - c_b)$	$(0,0)$

Figure 12.1
Formalizing industrial symbiosis as a collective action problem.

well have acted as one of the places where agents repeatedly interact. Another element is that the managers of most of the firms came from the local community. This may have meant a high level of initial mutual trust.

We will now describe the problem of creating industrial symbiosis as a formal model (figure 12.1). If a firm i invests in providing output as an input for another firm, it will cost an amount c_i. If this other firm j derives the inputs from firm i, it will benefit by an amount b_j. The resulting problem can be described in a payoff table. A firm can decide to invest in exchanging waste or not. If $b_i > c_i$, then the benefit for the park will be that both firms exchange waste, but since neither knows what the other firm will do (e.g., uncertainty in future activities of the other firm, how long it will exist), the best action from the perspective of individual rationality is not to invest. The values of b and c might be influenced by policies like tax on waste or subsidies for adjustment of processes. But more is needed than only changing the payoff matrix. Suppose both firms make an investment, and one of the firms experiences a better opportunity, $b' > b$, to recycle its waste. Then, for example, firm A is faced by $b'_a - c_a > b_a - c_a$, and firm B will not receive the benefits of the interfirm linkages, but only their costs, $-c_b < b_b - c_b$. Given current knowledge, cooperation is beneficial for both, but uncertainty faced by the other firm leads to a dilemma and may ultimately prevent the decision to invest.

Insights from the literature on collective action and on the evolution of cooperation provide some guidelines to overcome mutual defection of the players (Axelrod 1984). The players need to repeatedly interact in build up mutual trust relationships. This mutual trust contributes to the willingness of both players to cooperate. If the players do not interact frequently, an attempt to cooperate may fail, since they do not trust each other enough to enter into cooperative action. Creating interaction in an industrial park might coincide with already-ongoing collective activities like security or insurance, energy supply, and infrastructure investments. If mutual trust increases, new projects might be initiated that have longer payback times. The fact that in the nineteenth century, the recovery of waste was more common than it has come to be since then might be due to more local economic interactions and lower mobility of firms. Nowadays, many firms are players in a global market and interact less with physically nearby neighbors.

One of the main benefits of self-organization of collective action is the strong commitment of local actors. Especially when actors are able to define their own monitoring and sanctioning regimes, long-term cooperative solutions can follow.

The story of Kalundborg is mainly a story of self-governance. In order to mimic the success of Kalundborg, one needs to create the conditions for self-governance. It is not just a matter of technological feasibility. How can policy stimulate self-governance? One possibility is to focus on the connectedness of the firms in an industrial park. We can distinguish among the following levels of connectedness:

• Economic actors have exchange relationships, and thus there is coordination among their activities, which is based on autonomous decisions by dependent units. This is a normal situation in most industrial parks. At the outset, there may have been some planning, but over the years, companies have come and gone, and existing ones have changed their direction, have grown, or have declined. The majority of industrial parks in the Netherlands are of this type (Lambert and Boons 2002).
• Companies located in an industrial park strive toward some collective goals, which introduces the idea of group coordination. The individual elements are still autonomous, but they can examine whether joint activities lead to win-win outcomes. Participation in these activities is determined on a project-to-project basis.
• Companies in an industrial park function as the constituents of an ecosystem as it is conceived by advocates of industrial ecology. This makes it possible to implement options that may be detrimental to some or most companies but nevertheless contribute to the functioning and stability of the ecosystem as a whole.

These levels of connectedness can be related to the notion of collective action in the following way. At the first level, there is nothing indicating collective action. Actors interact, but at a low level of intensity, and there is no social structure in terms of trust, and there are no mechanisms for monitoring and sanctioning that could be helpful in bringing collective action about.

The second level can be seen as a pre-collective-action phase. Actors develop joint activities that are based on the direct payoff that cooperation brings (hence the term "win-win"). These joint activities do not constitute collective action. Nevertheless, they can be successful only if some coordination mechanisms are developed. Thus, they lead to the emergence of a basic level of trust, as well as mechanisms that can be used as sanctioning and monitoring instruments. Such a pre-collective-action phase corresponds to the historical analysis of organizations that produce public goods (such as labor unions) (Hechter 1981); they seem to have started out by providing private goods and to have developed later into collective action organizations. The first period in the INES case is an example of this level of connectedness, since only investments with very short payback times were made.

The third level of connectedness constitutes genuine collective action. On the basis of coordination mechanisms and trust among parties, it is possible to develop actions that have differential costs and benefits for the actors. Given this social structure, the actors are willing to accept individual costs in the present because they expect collective benefits in the future, since they have confidence—mutual trust—that the other partners will be committed to the agreed-upon cooperation. The current state of affairs in the INES case seems to be an attempt to attain this level.

These levels of connectedness are implicitly recognized in the more general literature on "network development" (Chisholm 1998) and "systems development" (Checkland 1981). This literature aims to develop tools for actors in communities, or more generally, social systems, to move from the modest level of coordination of autonomous activities toward connectedness that serves longer-term strategic goals for the network as a whole. Such a process involves a recurring cycle of action (cooperative projects among network members) and reflection (evaluation of projects and reformulation of network goals, structure, and processes) (Checkland 1981, 163). One of the central insights of this literature is that the process of network development is successful only if the actors in the network themselves learn how to improve their connectedness.

12.7 Conclusion

Kalundborg has functioned as a successful and influential illustration of an EIP within the field of industrial ecology. The Kalundborg case is used by policymakers, scholars, and consultants as an exemplary project of how to redesign industrial parks. However, the success of planned or designed EIPs has been limited so far. The conditions to create an EIP are context dependent. In fact, the Kalundborg development occurred in response to specific local social circumstances that stimulated the building of mutual trust among industries and created an environment for cooperative action.

To stimulate long-term cooperation in industrial parks, local firms need to build up a mutual trust relationship, and the technical advantages of cooperation need to be significant. Current policies are to a great extent aimed at improving the technical conditions and not the social conditions or the composition of firms in a park. The time horizon for decision making is usually so short that long-term goals cannot be expected to be met.

A top-down design of eco-industrial parks by bureaucrats and public planners is unlikely to be the most successful avenue to follow. As research on collective action problems shows, such top-down arrangements are often not effective in creating sustained cooperative arrangements. Furthermore, bureaucrats and designers may not have the knowledge required to see entrepreneurial opportunities to reduce waste flows. More can be expected from incentives for self-organized interactions, based on stimulating repeated interactions and lowering the investment costs by reducing environmental regulation or providing subsidies for investments for interfirm linkages. The current focus of EIPs is on top-down design of technical solutions, which has had only limited success. Long-term solutions are expected to emerge once the social dilemmas involved in creating EIPs are overcome.

References

Adolffson, P. (2001). Ways of Measuring and Interpreting the Urban Environment: The Creation and Use of Representation of the Air and Water in the City Organizing. Paper presented at the European Group for Organizational Studies conference, Lyon, France, July.

Axelrod, R. (1984). *The Evolution of Cooperation*. New York: Basic Books.

Baas, L. W. (1998). Cleaner production and industrial ecosystems: A Dutch experience. *Journal of Cleaner Production* 6(3–4): 189–198.

Baas, L. W. (2000). Developing industrial ecosystem in Rotterdam: Learning by ... what? *Journal of Industrial Ecology* 4(2): 4–6.

Baas, L. W., and F. A. Boons. (2004). An industrial ecology project in practice: Exploring the boundaries of decision-making levels in regional industrial systems. *Journal of Cleaner Production* (special issue: Applications of Industrial Ecology). In press.

Boons, F. A., and L. W. Baas. (1997). Industrial ecology: The problem of coordination. *Journal of Cleaner Production* 5: 79–86.

Boons, F. A., L. Füssel, and S. Georg. (2000). Inside Bruno's Travelling Agency: Deconstructing the Kalundborg Myth. Paper presented at the 16th European Group for Organizational Studies colloquium, Helsinki, July.

Brundtland, G. (ed.) (1987). *Our Common Future: The World Commission on Environment and Development*. Oxford: Oxford University Press.

Checkland, P. (1981). *Systems Thinking, Systems Practice*. Chichester, England: Wiley.

Chertow, M. R. (2000). Industrial symbiosis: Literature and taxonomy. *Annual Review of Energy and the Environment* 25: 313–337.

Chisholm, R. F. (1998). *Developing Network Organizations: Learning from Practice and Theory*. Reading, MA: Addison-Wesley.

Desrochers, P. (2001). Eco-industrial parks: The case for private planning. *Independent Review* 5(3): 345–371.

Desrochers, P. (2002). Regional development and inter-industry recycling linkages: Some historical perspectives. *Entrepreneurship and Regional Development* 14: 49–65.

Ehrenfeld, J., and N. Gertler. (1997). Industrial ecology in practice: The evolution of interdependence at Kalundborg. *Journal of Industrial Ecology* 1(1): 67–80.

Erkman, S. (1997). Industrial ecology: A historical view. *Journal of Cleaner Production* 5(1–2): 1–11.

EZ (Economische Zaken). (1998). *Nota Milieu en Economie*. The Hague: Straatsdrukkerij/Uitgeverij.

Frosch, R., and N. Gallopoulos. (1989). Strategies for manufacturing. *Scientific American* (September): 144–152.

Gertler, N. (1995). Industrial Ecosystems: Developing Sustainable Industrial Structures. Master's thesis, Massachusetts Institute of Technology, Cambridge, MA.

Hawken, P. (1993). *The Ecology of Commerce*. New York: Harper Business.

Hechter, M. (1981). *Principles of Group Solidarity*. Berkeley and Los Angeles: University of California Press.

KPMG (Klynveld Peat Marwick Goerdeler). (1998). *Duurzame bedrijventerreinen. Handreiking voor het management van bedrijven en overheid* (Sustainable industrial parks. A helping hand for management of industry and government). The Hague: Economische Zaken.

Lambert, A. J. D., and F. A. Boons. (2002). Eco-industrial parks: Stimulating sustainable development in mixed industrial parks. *Technovation* 22: 471–484.

NOVEM (Nederlandse organisatie voor energie en milieu). (2001). *Werken aan duurzame bedrijventerreinen* (Working on sustainable industrial parks). The Hague: Economische Zaken.

Olson, M. (1965). *The Logic of Collective Action: Public Goods and the Theory of Groups*. Cambridge, MA: Harvard University Press.

Ostrom, E. (1990). *Governing the Commons: The Evolution of Institutions for Collective Action*. New York: Cambridge University Press.

Ostrom, E. (2000). Collective action and the evolution of social norms. *Journal of Economic Perspectives* 14(3): 137–158.

Ostrom, E., R. Gardner, and J. Walker. (1994). *Rules, Games and Common-Pool Resources*. Ann Arbor: University of Michigan Press.

Pfeffer, J., and G. R. Salancik. (1978). *The External Control of Organizations*. New York: Harper and Row.

Tibbs, H. (1991). *Industrial Ecology: A New Environmental Agenda for Industry*. Brochure. Boston: Arthur D. Little.

VI Conclusions

13 Policy Implications: Toward a Materials Policy?

Book title

Jeroen C. J. M. van den Bergh, Harmen Verbruggen, and Marco A. Janssen

Q53

(OECD)

Q58

13.1 Introduction

Economic activities transform natural resources into materials, with the purpose of producing goods and services that contribute to economic welfare. The structural by-products of this transformation are the generation of waste and pollution, as well as the disturbance of the natural environment. Although many countries have in recent decades formulated environmental policies to cope with these by-products, coherent and all-embracing policies specifically oriented toward the sustainability of materials use have not yet been developed. Materials can nevertheless be regarded as the core element of a policy approach that aims to stimulate sustainable development. Such a policy and traditional environmental policy are obviously intertwined but nevertheless can be distinguished from one another on the basis of specific goals having to do with the time horizon and the type of environmental problems addressed. For example, noise externalities are usually regarded as an environmental problem, but not necessarily as a serious threat to a sustainable development. In this final chapter of the book, we will discuss the specific role that material and waste policies can play in the context of sustainable development.

The main purpose of this study was to integrate economics and industrial ecology. In chapter 1, it was argued that the main literature on industrial ecology, represented by Socolow et al. (1994), Graedel and Allenby (2003), Ayres and Ayres (2002), and the *Journal of Industrial Ecology*, is very weak on the economic dimension, suggesting that industrial ecology is all about planning and design. In particular, the literature largely neglects economic angles and considerations, economic methods of analysis and their applications, and related policy dimensions. The aim of this book was to arrive at an improvement of policy advice by adding economic elements to theory, methods, and applications in industrial ecology. An added value of the book lies in the inclusion of multiple studies with similar objectives but employing different methods or models. The synthesis in the current chapter tries to create another added value: namely, by bringing together the elements (waste treatment, recycling, and dematerialization), the levels (eco-industrial park, region, country, world), the instruments, and the methods. However, evidently, integration of all these in a single, formal model is beyond our (and anyone's) ambition and capability.

13.2   Integrative Methods for Policy Analysis: A Summary

This section summarizes the methods for integrating economics and industrial ecology that have been proposed, discussed, and applied in the previous chapters. It has been shown that by adding an economic context to industrial ecology in the form of costs, benefits, investments, market distortions, international trade, and so forth, policy realism is enhanced, because the direct and indirect effects of policies focusing on material throughput depend very much on economic mechanisms like substitution in production, markets, international trade, and economic growth. Different methods focus on specific combinations of these elements.

Statistical-historical modeling (chapter 3) at the country level makes it possible to confront a number of aggregate exergy and material use indicators with the traditional aggregate income (GDP) indicator. This method is powerful, as it shows that the type of aggregate indicator matters for the policy conclusions drawn. A main problem of applying this method is that in order to arrive at a time series that is sufficiently long to allow for rigorous statistical regression analysis, comparable (i.e., mutually consistent) data need to be collected or reconstructed for a long period of time. An important finding of the study in chapter 3 is that the U.S. economy turns out not to be "dematerializing" to a degree that is relevant for environmental policy goals. The study implies that in order to be effective, policy should focus not so much on directly reducing the total mass of materials consumed, but on reducing the need for consumables, notably intermediate products. When time series are shorter, modeling or descriptive analysis can still render insightful conclusions, as illustrated by chapter 5, which employs econometric estimates, and chapter 10, which offers descriptive indicators of foreign trade and recycling. Again, the aggregation and classification schemes adopted are crucial, notably, the distinction between secondary and primary materials and between developed and developing countries.

Two types of complex systems modeling were illustrated in chapters 5 and 6. Chapter 5 combined engineering, capital vintage, and econometric analysis, resulting in a dynamic computer model that was applied to industrial energy use in the United States and focused on the iron and steel, pulp and paper, and ethylene industries. The approach is a type of integrated industrial systems analysis, which steps away from standard economic equilibrium analysis, in which markets clear through prices (see this chapter's discussion of chapters 7 and 9). Instead, this approach takes for granted that industrial systems are constantly changing and are therefore in a persistent state of disequilibrium. Optimal investment is frustrated by market imperfections, uncertainty, and bounded rationality—myopia in combination with time lags of investment consequences. An important finding is that each policy instrument triggers particular kinds of responses, depending on the industry studied. Examples are shifts in production among the various segments of the industry concerned, changes in the fuel mix, alterations in the use of intermediate products, transitions toward renewable energy sources, reductions in total energy use, and reductions in carbon emissions. Therefore, a mix of policy instruments is the most effective way to realize policy goals. Yet to allow for an appropriate analysis, the models applied must include enough industry-specific features to allow linkage of policy instruments.

Chapter 6 offered two dynamic models: namely, the Australian Stocks and Flows Framework and the OzEcco embodied energy flows model. The first is a large stock flow model of the Australian economy that accounts for important physical transactions in mass units in the Australian economy. The second operationalizes the notion of embodied energy, integrates the driving forces of population, lifestyle, organization, and technology, and translates these into environmental impacts. Together, they can be seen as a systems dynamics representation of Australia's metabolism that tries to operationalize the idea that the physical economy should conform to the physical laws of thermodynamics and mass balance. The nature of these models for Australia makes it possible to study integrated policies—through scenario analysis—that affect material flows through the economy, including energy policy, climate policy, and even land regulations. Such models show that incorporating details of physical realities and dynamics, like capital vintages, leads to slow reactions of the economic system to policy incentives. This resembles insights of chapters 3 and 5, notably, that dematerialization occurs in only a few sectors, because of slow technological adaptations that lag behind volume increases.

Two types of modeling presented in the book made use of input-output (I/O) data and techniques, an old and proven approach to realizing the integration of economic and environmental information, as well as to finding a compromise between bottom-up information and top-down economic modeling. A technique that has been around for some time but has not been widely applied, namely, dynamic input-output (DIO) modeling, was illustrated in chapter 8. It enables an analysis of changes between sectors, as well as between regions, resulting from technological changes. All these are reflected in changed I/O coefficients. A more recent technique, namely, structural decomposition analysis (SDA), was applied in chapter 4. Both techniques make it possible to step away from the rigid framework of constant coefficients that characterizes static I/O analysis. SDA enables the decomposition of changes in certain material use indicators over a given period of time into a range of effects, including I/O-related structural or sector-shifting effects. This requires that comparable I/O tables be available for at least two points in time. The two approaches, SDA and DIO, are complementary, in that SDA results can be translated into dynamic specifications of the DIO framework. This is something that certainly deserves more attention in future research. Whereas SDA offers relevant information for ex post policy evaluation, forecasting using SDA information and DIO is useful in assessing potential overall or macro effects (aggregate income, demand, and sector output) of certain material, waste, or recycling policies. The SDA approach presented here is innovative in two respects. First, it analyzes hybrid-unit I/O tables, which include a mix of physical and monetary data. Second, it uses the SDA results in forecasting scenario analysis. Finally, perhaps the most important innovation is that DIO and SDA solve the traditional problem of including technological change in I/O models in a sophisticated way.

Two types of integration of industrial ecology with equilibrium analysis, for a long time the most popular technique among mainstream economists, have been illustrated here. Partial equilibrium analysis was discussed and applied in chapter 7, and general equilibrium analysis was illustrated in chapter 9. Chapter 7 discussed STREAM, a partial equilibrium model for material flows in Europe, with emphasis on the Netherlands. The model provides a consistent framework for analyzing material use scenarios and for related environmental policy analysis of dematerialization, recycling, input substitution, market and cost prices, and the international allocation of production capacity. The model structure makes it possible to deal with very specific instruments, such as taxation of primary materials, performance standards for energy and emissions, and deposit money for scrap. Chapter 9 presented a general equilibrium model of the waste market. Such a technique is suitable for studying market distortions, in this case focused on flat-fee pricing. A stylistic application to the Netherlands demonstrated that introducing a unit-based price for waste collection will stimulate both the prevention and the recycling of waste and can improve welfare, even if implementation costs and enforcement costs are taken into account. A technique related to equilibrium modeling, referred to here as an international-material-product-chain model, which focuses on static optimization, was proposed and applied in chapter 10. That chapter offered policy analysis at the level of international trade in materials. General equilibrium tools are strong in addressing the question of economic efficiency. However, they are top-down oriented and often simplistic in the dynamics and physical detail that might be required for concrete policy advice.

Institutional analysis at the level of the eco-park was the focus of chapters 11 and 12. These chapters were nontechnical, in contrast to the other chapters, because of the type of analysis they present, which emphasizes institutional, organizational, stakeholder, and evolutionary aspects. The chapters stress the opportunities to develop eco-industrial parks, the mechanisms through which these opportunities are offered, and the lessons that can be learned from Kalundborg, in terms of both the economic limitations of the Kalundborg symbiosis and the critical economic factors that contributed to its success. Chapter 12, in addition, elaborated on the idea that the eco-park approach is a special case of a collective action problem and that governments should refrain from planning and tight regulation and instead foster self-organization processes by means of assistance in network building and possibly subsidies.

In conclusion, all chapters together show the value of following a pluralistic methodological approach to the integration of economics and industrial ecology. Although the first economic models that included physical dimensions appeared in the 1960s and 1970s, there is still no unified methodology for such models. This is not a serious problem, as different approaches allow different questions to be tackled. Only by using a variety of approaches can one understand the various economic aspects of industrial ecology, and the different levels and scales of policy, physical and economic processes. This in turn enables an assessment of the potential of, and barriers to, important transitions that will reduce environmental problems caused by the use of materials. We certainly do not claim to have covered all aspects of the up-and-coming economic branch of industrial ecology, but we do think we have covered all useful methodological approaches. In terms of policy relevance, the studies presented in this book's chapters reveal the disadvantages or ineffectiveness of many policies, which result from the lack of incentives to achieve a reduction in material use or from the presence of perverse incentives that stimulate displacement and illegal dumping. Alternatively, physical realities can make it difficult, if not impossible, to replace the historically inherited physical infrastructure of the economy in the short term.

13.3 Sustainability and Material Flows

Against the background of the insights obtained by merging economics with industrial ecology in various case studies, we will, in this and the following sections, explore possible policy implications with regard to material flows. Obviously, the current use of many materials is not sustainable. The depletion of nonrenewable resources perhaps comes first to mind. But the degradation and depletion of renewable resources are possibly more serious threats to sustainable development, because over-exploitation of renewable resources reduces the ability of future generations to derive welfare from those resources. In addition, the recovery and processing of nonrenewable resources, as well as the exploitation of renewable resources, cause degradation of the environment. This involves the loss of ecosystems and biodiversity and the pollution of air, soils, and waters. A possible consequence is the reduction of ecosystem services for future generations. In other words, the use of primary material resources indirectly leads to sustainability problems.

This underpins the relevance of policy aimed at the sustainable use of natural resources and materials. A concrete concept in this context is "dematerialization" (see chapter 2). One suggestion for operationalizing this includes the notion "Factor Four," interpreted as doubling wealth while simultaneously halving resource use (von Weizsäcker, Lovins, and Lovins 1997). This concept indicates, for example, that in an economy subject to annual growth at a rate equal to 3.5 percent, which means a doubling of income over a period of twenty years, the use of materials needs to be reduced at a rate equal to 6.7 percent per year, implying a quarter of the original resource use per unit of income after twenty years. Assuming that wealth is proportional to income, wealth will have doubled, whereas total resource use will be 50 percent (income × materials use per unit of income = 2×0.25) of the total resource use at the beginning of the period. Chapter 3 presents a very original historical quantitative-empirical study of dematerialization. The chapter concludes that—counter to expectation—in most sectors of the United States, there has been no dematerialization during the last hundred years.

The goal of dematerialization is based on the philosophy that less throughput in the economy leads to less depletion and overexploitation of natural resources, as well as less pressure on the environment. From a traditional economic perspective, this general goal cannot be regarded as sufficiently well defined, as it would most likely be an inefficient way to realize higher goals, such as maximum social welfare. Traditional environmental economics considers the allocation of natural resources over time and future generations to lead to maximum welfare when all external effects related to the use of materials are adequately reflected in prices and constraints faced by economic agents. Such an optimal policy would consist of a combination of the Hotelling rule for resource scarcity and the Pigovian tax (e.g., Lusky 1975) to internalize external environmental costs. A set of constraints needs to be imposed to prevent the depletion of essential resources, because the Hotelling rule guarantees an optimal use of the resource in time, but not a sustainable availability of the resource. However, whereas the policy outlined above may sound ideal from an economic-theoretical perspective, it is very unlikely to be implemented. A major problem with such a policy is that making the necessary trade-offs—private and social benefits and cost calculations—requires constructing and solving an extremely complex empirical model (Kandelaars 1998). Future external effects are particularly uncertain. In fact, it is not even straightforward to evaluate environmental effects solely in physical terms by using life cycle analysis. This indeed implies a number of considerations:

- Which resources and materials need to be used economically? And how should these be selected? For example, it is difficult, or even impossible, to judge whether substitution of a material that generates considerable pollution—like a heavy metal, aluminum (energy use), or a synthetic material (PVC)—for a scarce material (like tropical timber) is beneficial in terms of the net environmental consequences.
- The number of economically useful reuse and recycling options to deal with material waste has increased over time. Some applications of recycled materials already face a shortage at the national scale in some countries, which has given rise to a sharp rise in international trade in secondary materials over the last decade (see chapter 10).

• Dematerialization means prevention of waste, which is preferable to material reuse and recycling. Closing material cycles seems less complex than realizing significant dematerialization. Reuse and recycling reduce the need for new (virgin) materials but do not necessarily lead to a reduction of the volume of material flows through economic systems. One consequence is that, as opposed to dematerialization, reuse and recycling continue to put pressure on the environment through freight transport. Another is that recycling activities, such as secondary metal production, are often associated with high levels of energy use and various types of pollution.

• Lengthening the life of products can give rise to using a larger amount of, as well as more advanced, materials. In addition, it may conflict with easy decomposability of products for the purpose of recycling.

Even when there is a clear policy perspective on the sustainability objectives and environmental externalities of particular material and product flows, the question remains where policy should attack, and through which instruments it should be implemented.

13.4   Sustainability Policy for Materials

Sustainable recovery or exploitation of a natural resource is a form of resource stock management and hence will have to be regulated from the supply side. Depending on the type of resource and the prevailing property and use rights, resource suppliers need to agree on recovery and exploitation. For example, in the case of tin, with an estimated resource availability of about forty years, suppliers on the global market—notably Bolivia, Brazil, China, Indonesia, Malaysia, Thailand, and Peru—need to formulate a joint market strategy. This should take into account the essential demand for tin for economic transformation processes, the substitution possibilities that are available, and the range of possible technological developments. An international resource agreement seems to be the most appropriate way to arrange this joint market strategy. Regulation of the supply will lead to price increases and consequently to the availability of tin for more people in the future.[1]

This type of regulation (aimed at sustainability) of the supply side is currently already operational for some resources, notably forests and endangered species (through the Convention on International Trade in Endangered Species of Wild Fauna and Flora, or CITES), and is being developed for others, like water resources. However, important progress is still needed, as for many resources no regulation exists at all. Important barriers are presented by the absence or obscurity of property and use rights, governmental failures, and a lack of international coordination. Even existing regulation of resource supply is not always satisfactory, as in the case of fish stocks. Although urgency is evident to all participants, the allocation of total allowable catch and related compensation generally leads to severe political conflicts and consequently to ineffective resource management strategies. These effects are strengthened by the scientific uncertainty about the level of sustainable harvesting, which has often caused politicians to define total catch levels that are too high to sustain the harvested resource. In addition, the inadequacy of sanctioning instruments has allowed the continuation of free-riding.

The effectiveness and efficiency of sustainable resource management is at risk when regulation adopts a demand-side perspective, for example, by levying taxes on tin, timber, and fish. The resulting reductions in profits can ultimately lead to price increases, which in turn will stimulate higher levels of exploitation. This is an example of the rebound effect. A similar effect can occur in the case of recovery and reuse of materials. Here, in addition, demand regulation can be undone by the supply of secondary materials competing with the supply of primary materials.

13.5 Waste Policy

Although materials policies have not seen much progress, waste policies have been widely implemented in OECD countries. In many countries, a waste hierarchy has dominated the formulation of waste policy. For example, the "U.K. Waste Hierarchy Policy" is characterized by a preference for waste prevention over waste reduction, reuse, recycling, recovery, and landfill, in that order (Phillips et al. 2002). The Dutch waste policy follows a similar but slightly different ranking: prevention has the highest priority, followed by recycling, and then combustion and dumping. The U.S. Environmental Protection Agency (EPA) uses a similar reduce-reuse-recycle hierarchy.

These hierarchies are aimed particularly at preventing negative external effects related to combustion and dumping. In countries with scarcity of land, combustion is often preferred over dumping. Irrespective of national differences, the objective is that use of materials be reduced as much as possible so as to reduce end waste. Cascading is employed to close remaining waste cycles. Waste policy has been characterized by a national, or even regional (provincial/state), context, aimed at efficiency in physical-technical terms. The concept of industrial symbiosis has become popularized as a result of the well-known example of Kalundborg, an industrial park in Denmark. However, few studies have examined why Kalundborg happened, and how to apply this industrial symbiosis to other industrial parks. Instead, it has often been assumed that bottom-up calculations were enough to identify potential benefits, ignoring transaction costs (chapter 12). Subsequently, applications to other industrial parks have not been very successful (chapters 11 and 12). In general, economic efficiency has received far less attention than efficiency in physical-technical terms. Indeed, to date, insights from cost-benefit analysis of different waste treatment options have not had much impact (chapter 9). Instead, idealistic or altruistic considerations seem to have dominated (Ackerman 1997).

In recent years, four developments have led to a reconsideration of waste policy formulation. First, governments and utility companies in the waste sector are abandoning their traditional position as a regulator and capacity planner. Second, the market for waste is subject to "internationalization." For example, the national borders within the European Union no longer hold for waste treatment and useful applications of waste. Moreover, regulation of international transport of waste by the Basel Convention and the European Union is inadequate to avoid undesirable international shifting of environmental problems. Third, waste policy is being considered in a broader context, linked to environmental issues other than those that are material related. In particular, energy policy goals may cause a reconsideration of the hierarchy in waste policy. Waste combustion for electricity generation ("thermo-recycling") changes the priority usually given to prevention. Fourth, enlarging cascades through repeated use of products can increase particular types of environmental pressure. The sustainability of a reused product, including its production and packaging, and the processing of the waste it generates, is at stake here.

The economic perspective on waste policy is not simple. Many studies so far have adopted a partial approach that does not render definite insights. Both theoretical and empirical studies have been able to show that the generation of waste can be very sensitive to user fees, if combined with programs that enlarge public awareness of the waste problem. Most studies disapprove of a flat-fee pricing system in which the tariff is independent of the amount of waste supplied. But economic studies provide different results in terms of what an optimal policy looks like. The main choice is between upstream and downstream taxes. The first can take the form of a deposit refund system or a "waste tax" on the consumption good to internalize the waste treatment costs in the price of the product. The second can be implemented as a unit-based pricing system, in which the fee can depend on the actual amount of waste generated or on proximate indicators, such as the number of persons in a household. The disadvantages of a downstream tax is that enforcement costs are high and that one often ends up with illicit dumping, burning, or other unintended forms of disposal—dumping waste in the neighbor's bin or in the disposal at work. Such behavior is, from an economic welfare perspective, unattractive, as it involves high social costs. Some studies have therefore gone as far as to argue in favor of subsidizing legal waste disposal. Empirical studies have shown that significant levels of illegal disposal are not a hypothetical consequence of price-based waste policies: up to 30 percent of any reduction in waste generation resulting from the implementation of such policies may be caused by an increase in illegal disposal.

Chapter 9 was based on a general equilibrium approach, which is particularly useful for analyzing price-based instruments and welfare impacts that include environmental externalities. The broad perspective thus adopted allowed the chapter to show that even with costs of illegal disposal being taken into account, a downstream tax can be more efficient to tackle the waste problem than an upstream tax. A downstream tax means that private households feel a very strong and direct incentive to prevent and recycle waste, whereas an upstream tax provides no real incentive to increase recycling and only a very weak incentive to reduce the generation of waste. This weak incentive is the result of the fact that the price incentive is so small in comparison with consumer product prices that it has no significant impact on consumption expenditure patterns. The conclusion of chapter 9 is then very clear: The introduction of unit-based pricing is an inevitable component of any economically defensible policy aimed at the reduction of waste generation. As always, this is not the end of the story. Further research needs to be undertaken to examine what is precisely optimal from an empirical perspective, based on a weighting of environmental gains from waste reduction and consumption-related welfare losses resulting from regulation. This weighting may differ among regions and countries, depending on consumer preferences and external costs of waste.

13.6   The International Dimension of Waste and Recycling Policy

International flows of waste, which arise from the treatment of waste in a country other than the one in which it was generated, are controversial. Within the European Union a country is not allowed to question the environmental standards of countries to which it exports its waste. As a result, some countries create a barrier against trade in waste, thus reducing economic opportunities for efficiently treating waste.

From the perspective of international trade theory, it is self-evident that trade in waste can lead to cost reductions and welfare maximization. Currently, a significant amount of waste is being traded and recycled internationally, notably between the North and the South, but also among countries in the South. In recent decades, international trade in most secondary materials has increased faster than the production of such materials. The total trade volume of secondary aluminum, lead, copper, zinc, and paper increased from 2.5 billion kilograms to 21.5 billion kilograms during the period 1970–1997 (chapter 10). International reuse is dominated by iron scrap and steel scrap, the trade of which increased during the same period from 20 billion kilograms to 37 billion kilograms. These developments have mainly been driven by significant differences in recycling costs and benefits between poor and rich countries. Moreover, there is an oversupply of secondary materials in rich countries and a shortage of high-quality secondary materials in poor countries. As a consequence, domestic prices for waste are relatively low or even negative in rich countries and relatively high in poor countries. Large differences in the price of labor are responsible for these cost and price differences. This holds, of course, especially for recycling activities that are relatively labor-intensive. An example is the manual disassembly of computers. The reduction of transaction and transport costs has further contributed to the "globalization" of trade in secondary materials. Cost factors are important here, because these trade flows primarily concern materials with a relatively low value. In order to manage these trends, the regulation of international trade in secondary materials needs to be improved. In particular, there is a serious need to regulate working conditions in recycling activities involving dangerous wastes. Such regulation may be implemented through an adequate revision of the Basel Convention.

13.7   Indicators for Dematerialization

Before discussing the desirability and possibility of an integrated "sustainability and environmental policy" for material resources and waste, it is useful to examine whether there are sufficient and reliable data for designing and supporting such a policy.

An important problem to be solved is that it is not always immediately clear how dematerialization focused on specific physical indicators contributes to the availability of critical resources in the future or to a reduction of environmental pressure in the present. In particular, it is not evident which dematerialization indicator(s) need to be formulated and used. It is clear that an aggregated indicator in kilograms, such as the total material requirement (TMR) proposed by the Wuppertal Institute, makes little sense from economic, environmental, or welfare perspectives. This indicator has been developed around the notion of materials inputs per service unit (MIPS) (von Weizsäcker, Lovins, and Lovins 1997) and is based on the debatable assumption that one can simply add all kinds of different materials (measured in kilograms) used during the life cycle of a product or service to formulate an aggregate environmental indicator (also in kilograms). Virtually all (environmental and resource) economists regard the TMR indicator as conveying information that is completely irrelevant for solid environmental policymaking, because so many materials with entirely distinct environmental effects per kilogram of material used are lumped together in one indicator without applying a careful weighting procedure. Economists would advise, for example, that materials be weighted in accordance with the amount of externalities they generate per kilogram (other weighting approaches have been proposed by environmental scientists). This is a good example of a situation in which adding an economic dimension to industrial ecology changes the judgment about a method, with possibly serious implications for derived policy suggestions. Chapter 3 presents as alternative indicators (1) mass per capita and exergy per capita; (2) mass and exergy per unit of GDP; (3) embodied exergy per unit of mass; and (4) groupings of material flows (fossil fuels, metals, agricultural products, construction materials, and chemicals).

It seems much wiser to employ either a set of various, homogeneous indicators or to aggregate different materials using a well-motivated weighting scheme. From the perspective of sustainability, such a scheme could be based on the relative scarcity of different materials. From an economic welfare perspective, weighting could be based on marginal external costs assessed for each type of material. Through the economic (monetary) valuation of external effects, dematerialization can be directly linked to a reduction of particular environmental impacts (Ackerman 1997). Nevertheless, it should be noted that economic valuation as well as the optimization of external effects is based on a number of microeconomic assumptions, such as rational agents and perfect information, that mostly do not hold in reality. Therefore, a set of purely physical indicators is preferred. Similarly, these same considerations suggest that price instruments will be less efficient and effective than is often stated in the standard economic literature on environmental policy (Baumol and Oates 1988). In other words, optimal policy is an illusion and at best a theoretical benchmark. Physical targets for dematerialization policy are then a possible substitute. In economic terminology, this is an example of the inevitable "second-best policy."

A dematerialization indicator can be formulated as a fraction, the denominator of which reflects an economic category (GNP, sectoral production levels, product, consumer). Dematerialization at a high level of aggregation is partly an autonomous process caused by an increase in the share of services in economic production at higher levels of income. However, measurement of the change in services is difficult for a variety of reasons (Verbruggen 2000):

- New services often show a fluctuating price pattern.
- The Baumol effect—higher prices of services as a result of greater demand—will cause an increase in the share of services in the GNP without a real increase in service production (Baumol 1967).
- Because of international competition, labor-intensive industries like textiles and clothing, shoes, and shipbuilding are shifting from OECD countries to non-OECD countries. That is, even without a change in consumption patterns, OECD countries are dematerializing their supply side of the economy.
- Services like design, image, and quality will make up an increasing part of the price of material products. Since material products fall outside the category "services," their increasing service component is subject to measurement problems.
- Production processes incorporate an increasing service component, as a result of these processes' becoming more information intensive. Such services do not receive adequate attention in traditional sector measurement categories.
- The only empirical fact that has been well supported so far is that on average, consumers spend a higher share of their income on services when their income increases. In other words, the income elasticity of the demand for services is larger than one.

To distinguish services from materials, we need, in addition to better categories of production sectors and product groups, the systematic collection and analysis of physical data, preferably in close connection with national accounts (see chapter 4). Physical data provide more information about structural changes than monetary data, because they reflect the physical technological structure of the economy. Of course, the availability of physical data will allow a very interesting comparison to be made with associated monetary data or, more concretely, the tracing of indicators over time that provide information on monetary value per kilogram of material used.

GNP has, by definition, no physical or material dimension, as it is an indicator of value added. This relates to the fact that economic value, reflected in the prices on the market, ultimately depends on the services delivered by products. It is therefore somewhat arbitrary which physical aspects one assigns to a specific service, that is, only the material contained in the physical product that directly generates the service, or also the waste and emissions caused in its production chain (backward)? And how should one account for materials in durable products versus materials in other products (food, detergents, etc.) and packing materials? Chapter 3 (see note 7) shows that during the production of one kilogram of computer chips, almost a symbol of dematerialization, indirectly more than two hundred kilograms of materials are used.

Against this background, it is evident that what a dematerialization indicator precisely measures or reflects remains questionable, especially when it is an aggregated indicator of material use. It would already be a significant step toward a material policy if we were to determine to what degree economic sectors contribute to the use of different materials and how their contributions change over time. A detailed analysis along these lines requires the use of physical (or hybrid) input-output tables for a number of years. Subsequently, structural decomposition analysis can detect direct and indirect material use of production processes and their changes over time (see chapter 4).

13.8   Toward an Integrated Policy

Is it desirable and feasible to develop a sustainability policy for materials and waste? Might it not be better and more efficient, given the uncertainties about available resources, technological developments, substitution possibilities, and ultimate environmental impacts of materials, to regulate only the environmental effects of materials over the whole production cycle? This is already an ambitious goal, especially if linked to the coordination of national policies at an international scale. When environmental policy is clear and effective, market mechanisms will cause necessary adaptations, technological developments, and new applications.

If that should take place, there would be no need for a general demateri-alization policy. In addition, international agreements are needed to sup-port suppliers of critical natural resources. This support should then be aimed at assisting the suppliers to develop sustainable management of exploited resources.

However, a number of objections can be raised against this line of thought. The environmental effects of the recovery, exploitation, and processing of materials lead, per unit of material, to more environmental damage and pollution than during the consumption and waste phases of the production cycle. This is caused mainly by the increasing complexity of economic processes, characterized by a long trajectory of intermediate products and components. Furthermore, the production of primary materials often leads to significantly higher environmental pressure than that of secondary materials. Aiming at a general dematerialization will thus contribute to reducing the environmental consequences of material use. In the long run, one might aim for 100 percent recycling of non-renewable resources, referred to as "waste mining" (Ayres and Ayres 1996). This will, however, require an increasing amount of energy input.

Economists traditionally are not in favor of physical goals, of which dematerialization is an example. One should note, however, that physical goals do not imply physical regulation and are consistent with price-based and other policy instruments. Furthermore, the idea of optimizing exter-nalities, through maximization of social welfare including external costs, is based on a number of microeconomic assumptions, some of which may be too restrictive and unrealistic to provide an accurate and robust de-scription of human behavior. This holds especially for bounded rational-ity and imperfect information and uncertainty about investments (in dematerialization). As a result, price instruments will be less efficient and less effective than often suggested and are really what economist refer to as imperfect, second-best instruments. In other words, "optimal policy" is an illusion and at best a theoretical benchmark. Moreover, effectiveness is at stake when technology, trade, and consumption patterns are histori-cally "locked in." Price incentives are then insufficient to realize social objectives.

These considerations suggest that a general dematerialization policy among national governments, as well as at the level of international governance—European Union, United Nations, international agreements —is meaningful and an almost inevitable element of a second-best policy that aims at effectiveness through unlocking and stimulating large-scale transitions. A public dematerialization goal can then translate into dematerialization objectives and strategies at the level of the private sector. Firms have many opportunities for realizing dematerialization but are often kept away from these opportunities by profit concerns and various types of public regulation. In view of this, and given the social benefits and the nonobvious nature of a dematerialization approach, governments certainly should take the lead in setting in motion a dematerialization process.

Dematerialization and waste policy support each other in the long run, even if in the short run they are often conflicting. A dematerialization policy might be implemented that covers the whole product chain and is focused on reducing lock-in situations caused by established technological trajectories, organizational relations, and institutions. The aim is to facilitate technological innovation and the transition toward a material-poor economy. Such a policy can make use of the physical requirements of production processes and products, even when, from an economic point of view, this is second-best. Other strategies for dematerialization are the stimulation of material cascading through cooperation in the product chain; the introduction of price corrections in waste policy; and the enlargement of the international allocation of waste management, recycling, and trade in secondary materials, subject to appropriate international regulation through international agreements.

Such a dematerialization policy requires physical data in combination with relevant economic information, to create a basis for the development of dematerialization indicators. The contributions in this book have provided a starting point for the creation of tools and methodology to promote a better understanding of the long-run relations among economic structure, international trade, and material flows.

Acknowledgement

We are grateful to Stefan Anderberg for helpful comments.

Note

1. The notion of international commodity-related environment agreements is related (Kox 1991; Linneman and Kox 1995). Such agreements are aimed at stimulating countries involved in the export of resources or simple commodities to implement production methods that cause less environmental pressure. Price support or reduced competition is the means for accomplishing this. So far, however, the regulations of the WTO conflict with these agreements, as the WTO regulations do not allow differential treatment of identical commodities that are produced in different ways.

References

Ackerman, F. (1997). *Why Do We Recycle? Markets, Values, and Public Policy.* Washington, DC: Island Press.

Ayres, R. U., and L. W. Ayres. (1996). *Industrial Ecology: Towards Closing the Materials Cycle.* Cheltenham, England: Elgar.

Ayres, R. U., and L. W. Ayres (eds.). (2002). *A Handbook of Industrial Ecology.* Cheltenham, England: Elgar.

Baumol, W. (1967). Macroeconomics of unbalanced growth. *American Economic Review* 57: 415–426.

Baumol, W. J., and W. E. Oates. (1988). *The Theory of Environmental Policy.* 2nd ed. Cambridge: Cambridge University Press.

Graedel, T. E., and B. R. Allenby. (2003). *Industrial Ecology.* 2nd ed. (1st ed., 1995). Upper Saddle River, NJ: Pearson Education, Prentice Hall.

Kandelaars, P. P. A. A. H. (1998). Economic Analysis of Material-Product Chains: Models and Applications. Ph.D. diss., Thesis Publishers/Tinbergen Instituut, Amsterdam (also published by Kluwer Academic Publishers, Dordrecht, the Netherlands).

Kox, H. L. (1991). Integration of environmental externalities in international commodity agreements. *World-Development* 19(8): 933–943.

Linneman, H., and H. L. M. Kox. (1995). International Commodity-Related Environmental Agreements as an Instrument for Sustainable Development. Summary report, Economics Department, Vrije Universiteit, Amsterdam.

Lusky, R. (1975). Optimal taxation policies for conservation and recycling. *Journal of Economic Theory* 11: 315–328.

Phillips, P. S., K. Holley, M. P. Bates, and N. P. Freestone. (2002). Corby waste not: An appraisal of the UK's largest holistic waste minimisation project. *Resources, Conservation and Recycling* 36: 1–31.

Socolow, R., C. Andrews, F. Berkhout, and V. Thomas (eds.). (1994). *Industrial Ecology and Global Change.* Cambridge: Cambridge University Press.

Verbruggen, H. (2000). Voorwaarden voor een duurzame kenniseconomie (Conditions for a sustainable knowledge economy). *Naar een Duurzame Kenniseconomie* (Toward a sustainable knowledge economy). Advisory Council for Research on Spatial Planning, Nature and the Environment (RMNO), Report no. 145, Rijswijk, the Netherlands.

von Weizsäcker, E., A. B. Lovins, and L. H. Lovins. (1997). *Factor Four: Doubling Wealth–Halving Resource Use.* Report to the Club of Rome. London: Earthscan.

Contributors

Anthony Amato
Environmental Policy Program
School of Public Affairs
University of Maryland
2100 Van Munching Hall
College Park, MD 20742, USA
aamato@wam.umd.edu

Stefan Anderberg
Institute of Geography
University of Copenhagen
O.Voldgade 10
DK-1351 Copenhagen
Denmark
sa@geogr.ku.dk

Leslie Ayres
Centre for the Management of
Environmental Resources
INSEAD
Boulevard de Constance, 77305
Fontainebleau, France
Leslie.Ayres@insead.edu

Robert Ayres
Centre for the Management of
Environmental Resources
INSEAD
Boulevard de Constance, 77305
Fontainebleau, France
Robert.Ayres@insead.edu

Heleen Bartelings
APE BV
Lange Voorhout 94
2514 EJ Den Haag
The Netherlands
h.bartelings@ape.nl

Frank Boons
Department of Environmental Studies
Erasmus University
P.O. Box 1738
3000 DR Rotterdam
The Netherlands
boons@fsw.eur.nl

Brynhildur Davidsdottir
Department of Geography and Center
for Energy and Environmental Studies
Boston University
675 Commonwealth Avenue
Boston, MA 02215, USA
bdavids@bu.edu

Rob Dellink
Environmental Economics and
Natural Resources Group
Wageningen University
P.O. Box 8130
6700 EW Wageningen
The Netherlands
rob.dellink@wur.nl

Barney Foran
CSIRO Sustainable Ecosystems
GPO Box 284
Canberra, ACT 2601
Australia
Barney.Foran@csiro.au

Rutger Hoekstra
Statistics Netherlands
P.O. Box 4000
2270 JM Voorburg
The Netherlands
RHKA@cbs.nl

Annemarth Idenburg
Office for Environmental Assessment
National Institute for Public Health
and the Environment
P.O. Box 1
3720 BA Bilthoven
The Netherlands
Annemarth.Idenburg@rivm.nl

Noel Brings Jacobsen
Department of Geography and
International Development Studies
Roskilde University
P.O. Box 260
DK-4000 Roskilde
Denmark
nbj@ruc.dk

Marco Janssen
Center for the Study of Institutions,
Population, and Environmental
Change
Indiana University
408 North Indiana Avenue
Bloomington, IN 47408-3799, USA
maajanss@indiana.edu

Hein Mannaerts
Netherlands Bureau for Economic
Policy Analysis (CPB)
Postbus 80510
2508 GM Den Haag
The Netherlands
H.J.B.M.Mannaerts@cpb.nl

Franzi Poldy
CSIRO Sustainable Ecosystems
GPO Box 284
Canberra, ACT 2601
Australia
Franzi.Poldy@csiro.au

Matthias Ruth
Environmental Policy Program
School of Public Affairs
University of Maryland
3139 Van Munching Hall
College Park, MD 20742, USA
mruth1@umd.edu

Pieter van Beukering
Institute for Environmental Studies
Free University
De Boelelaan 1087
1081 HV Amsterdam
The Netherlands
Pieter.van.Beukering@ivm.vu.nl

Jeroen van den Bergh
Department of Spatial Economics
Faculty of Economics and Business
Administration and
Institute for Environmental Studies
Free University
De Boelelaan 1105
1081 HV Amsterdam
The Netherlands
jbergh@feweb.vu.nl

Ekko van Ierland
Environmental Economics and
Natural Resources Group
Wageningen University
P.O. Box 8130
6700 EW Wageningen
The Netherlands
Ekko.vanIerland@alg.shhk.wau.nl

Harmen Verbruggen
Institute for Environmental Studies
Free University
De Boelelaan 1087
1081 HV Amsterdam
The Netherlands
harmen.verbruggen@ivm.vu.nl

Benjamin Warr
Centre for the Management of
Environmental Resources
INSEAD
Boulevard de Constance, 77305
Fontainebleau, France
benjamin.warr@insead.edu

Harry Wilting
Office for Environmental Assessment
National Institute for Public Health
and the Environment
P.O. Box 1
3720 BA Bilthoven
The Netherlands
Harry.Wilting@rivm.nl

Index